MW01486751

Development and Decolonization in Latin America

Written in an accessible language, this book is a fully updated and revised edition of *Latin American Development*, a text that provides a comprehensive introduction to Latin American development in the twenty-first century and is anchored in decolonial theory and other critical approaches.

This new edition has been revised and updated in a way that takes into account recent changes in political leadership, the retreat of the Pink Tide, the Colombian peace accords, new forms of political and territorial mobilization, the intensification of extractivism, murders of environmental defenders, major disasters, and the new contours of feminist and anti-patriarchal struggles. It features new chapters on decolonial theory, Latin America in the world, disastrous development, Afrodescendant struggles, and the Latin American city. The book emphasizes political, economic, social, cultural, and environmental dimensions of development and considers key challenges facing the region and the diverse ways in which its people are responding, as well as providing analysis of the ways in which such challenges and responses can be theorized. It explores the region's historical trajectories, the implementation and rejection of the neoliberal model, and the role played by diverse social movements.

It is an indispensable resource for students and university lecturers and professors in development studies, Latin American studies, geography, anthropology, sociology, political science, economics, and cultural studies. In addition, it provides an invaluable introduction to the region for journalists and development practitioners.

Julie Cupples is Professor of Human Geography and Cultural Studies at the University of Edinburgh, UK. She has been working in Latin America since 1990 and has done research in Nicaragua, Guatemala, Costa Rica, Mexico, and Colombia, including the San Andrés Archipelago. Her publications have dealt with a range of themes including revolution and conflict, Indigenous and Afrodescendant media practices, gender and sexuality, elections, and disasters and environmental risk.

Routledge Perspectives on Development
Series Editor: Professor Tony Binns
University of Otago

Since it was established in 2000, the same year as the Millennium Development Goals were set by the United Nations, the Routledge Perspectives on Development series has become the pre-eminent international textbook series on key development issues. Written by leading authors in their fields, the books have been popular with academics and students working in disciplines such as anthropology, economics, geography, international relations, politics and sociology. The series has also proved to be of particular interest to those working in interdisciplinary fields, such as area studies (African, Asian and Latin American studies), development studies, environmental studies, peace and conflict studies, rural and urban studies, travel and tourism.

If you would like to submit a book proposal for the series, please contact the Series Editor, Tony Binns, on: jab@geography.otago.ac.nz

Education and Development
Simon McGrath

Postcolonialism, Decoloniality and Development, 2nd edition
Cheryl McEwan

South-South Development
Peter Kragelund

Gender and Development, 3rd edition
Janet Momsen

Aid and Development
John Overton and Warwick E. Murray

Theories and Practices of Development, 3rd edition
Katie Willis

Non-Governmental Organizations and Development, 2nd edition
David Lewis, Nazneen Kanji and Nuno S. Themudo

Development and Decolonization in Latin America, 2nd edition
Julie Cupples

For more information about this series, please visit: www.routledge.com/series/SE0684

Development and Decolonization in Latin America

Second edition

Julie Cupples

Routledge
Taylor & Francis Group

LONDON AND NEW YORK

Cover image: © Lucy Brown (loca4motion) / Alamy Stock Photo

Second edition published 2022
by Routledge
4 Park Square, Milton Park, Abingdon, Oxon, OX14 4RN

and by Routledge
605 Third Avenue, New York, NY 10158

Routledge is an imprint of the Taylor & Francis Group, an informa business

© 2022 Julie Cupples

First edition published by Routledge 2013

British Library Cataloguing-in-Publication Data
A catalogue record for this book is available from the British Library

Library of Congress Cataloging-in-Publication Data
A catalog record has been requested for this book

ISBN: 978-0-367-62543-6 (hbk)
ISBN: 978-0-367-62708-9 (pbk)
ISBN: 978-1-003-11045-3 (ebk)

DOI: 10.4324/9781003110453

Typeset in Bembo
by Newgen Publishing UK

For Ángel Gahona and all those who risk their lives making media

Contents

Figures

Tables

Boxes

Acknowledgements

In the past decade, since I wrote the first edition of the book, much has changed in Latin America and an updated edition is therefore necessary. Hugo Chávez and Fidel Castro were still alive, as were Berta Cáceres and Marielle Franco. The Colombian peace accords had not yet been signed and Venezuela had not yet descended into political and economic chaos. Ríos Montt had not yet been convicted of genocide in a Guatemalan court. Nicaragua was dealing with growing authoritarianism, but still did not have political prisoners and Ángel Gahona and Álvaro Conrado had not yet been assassinated. Hurricane Maria had not yet brought its devastating impact to the island of Puerto Rico and Fuego had not yet buried the Guatemalan community of San Miguel Los Lotes. As these events have unfolded, sophisticated decolonial analyses have proliferated that have challenged and shaped my thinking and my approaches.

Back then, I was still living in Aotearoa New Zealand, where I spent a big chunk of my adult life and raised my kids and from where I did much of my early Latin American and especially Nicaraguan research. I wrote the first edition in the aftermath of the 2010–2011 Christchurch earthquakes that shattered our lives and sent me back to the UK after the University of Canterbury enacted its own kind of disaster capitalism. The move to the University of Edinburgh in 2013 gave me many more opportunities to work in and on Latin America, through my teaching and research, and also through a range of institution-building activities that I have been involved in. The second edition was however also written during a disaster, in this case the global coronavirus pandemic of 2020–2021, so like 2011, it has also involved much improvisation in order to deal with inaccessible libraries and other multiple disruptions.

My involvement in Latin America started with the Nicaraguan revolution in the 1980s. Working in and with Nicaragua has been the most rewarding political education that is as much emotional, spiritual, and erotic as it is intellectual. I remain indebted to so many Nicaraguan friends who have fought for a better world through revolutionary means and have continued to do so, long after the so-called leaders betrayed their revolutionary ideals. My current research stretches beyond Nicaragua, especially into Colombia and the San Andrés archipelago, Costa Rica, and Guatemala. In the past years, I have also had the opportunity to engage with interesting initiatives in Chile, Ecuador, and Mexico. I'm particularly grateful to my long-term collaborators; academics, activists, civil society leaders, and mediamakers, that I've been able to work with over many years, some across multiple projects and in more than one place. There are too many people to mention but my deepest gratitude goes to the following people for their practical, emotional, and intellectual support and for the contribution they have made to the activism, research and ideas that underpin this book.

In Nicaragua, I'd like to thank above all Irving Larios and Dixie Lee, as well as María Luisa Acosta, Amy Bank, Juana Bilbano, Deborah Bush, Shaun Bush, Madeleine Caracas, Janett Castillo, Avelino Cox, Lottie Cunningham, Raúl Davies, Tamara Dávila, Neyda Dixon, Saúl Fuñez, Daisy George West, George Henríquez-Cayasso, Johnny Hodgson, Evelyn Flores, Eva Hodgson, Alta Hooker, Guillermo Incer Medina, Hazel Lau, Paul Leiva, Larry Montenegro-Baena, Dolene Miller, Sofía Montenegro, Nora Newball, Sergio Ramírez, Auxiliadora Romero, Cleveland Webster, and Yuri Zapata. And Patrick Welsh who now lives in Glasgow but remains thoroughly *pinolero*. Special thanks too to three friends and collaborators who are no longer with us, namely Ángel Gahona, Sadie Rivas, and Sergio Sáenz. In San Andrés and Providence, I'd like to thank Jimmy Archbold, Ignacio Barrera Kelly, Sergio Bent, Emiliana Bernard, Tanisha Brown, Corinne Duffis, Dean Mashiin Hyman, Penndale Humphries, Derrick Martínez Ward, Jairo Rodríguez Davies, and Edgar J. Stevens. In mainland Colombia, thanks go to Liliana Angulo, Derby Arboleda Quiñonez, Jhojan Cano, Mauricio Mitchell, Jhon Narváez, Carlos Perea, Ramón Perea, Sedney Suárez, and Aurora Vergara-Figueroa. Thanks too to David Sierra, a *paisa* in Edinburgh. In Costa Rica, I am grateful to Rina Cáceres, Kendall Cayasso-Dixon, Mike Joseph, and Carlos Sandoval García. In Guatemala, thanks go to Erika Ajuchán Ruíz, Carmen Soledad Azurdia, Roberto Carlos Castellanos, Elena Chichival, Gustavo Chigna, Carla Chun Quinillo, Ale Colom, Rüdiger Escobar Wolf, Leo Espinosa, Dolors Ferrers, Eufemia García, Otilia García, Sandra Patricia Gómez Zacarías, Juan Manuel Hernández Puac, Sofía Letona, Urbano Lorenzo, Aracely Martínez, Abraham Paz, Mischa Prince, Álvaro Revenga, and Diego Vásquez Monterroso. In Chile, thanks to Lindsay Carte, Sebastian Dube, Sandra Fernández, Ernesto López Morales, Manuel Prieto, and María Esperanza Rock Núñez, and especially to Marcela Palomino-Schalscha and Cristian Leaman-Constanzo, Chileans in Aotearoa New Zealand, for adventures and collaborations in both locations. In Ecuador, thanks to Patricia Carrera, Estefania Pereiros, and Norman Wray. In Mexico, I'd like to thank

Antonio Gallardo, Mario López González Garza, and Emma Morales. I'd also like to thank Roberto Zurbano in Cuba, Cecilia Moreno Rojas in Panama, and Zulma Valencia de Suazo in Honduras.

At the University of Edinburgh, I have many dear friends and collaborators with whom I've also the good fortune to work with in Chile, Colombia, Ecuador, Guatemala, Mexico, Nicaragua, San Andrés and Providence, as well as South Africa. I'd like to extend extra special thanks to Eliza Calder, Charlotte Gleghorn, Jan Penrose, Raquel Ribeiro, and Tom Slater for their love, friendship, collaboration, political commitment, intellectual insight, and good humour in the field, and for making my working life in the neoliberal university so much better. I'm also grateful to Josep Almudéver Chanzà, Amelia Bain, Andrew Bell, Melisa Miranda Correa, Soledad García-Ferrari, Hamish Kallin, Alistair Langmuir, Lisa Mackenzie, Dalinda Pérez, Margherita Scazza, Neil Stuart, and James Smith. And thanks to my former Head of School Simon Kelley for his support of critical scholarship, his genuine interest in my work, and for still believing that the contemporary university does not have to be an unhappy place to work.

Thanks also go to the many friends and academics from other universities and organizations for research, writing and workshop collaborations, support for funding bids, invitations to speak, and their intellectual contributions. I am especially grateful to Ramón Grosfoguel, whose influence on my thinking and this book will be more than apparent. Thanks too to Paul Adams, Teresa Armijos, Andreza Aruska de Souza Santos, Luciano Baracco, David Beel, Tony Binns, Marney Brosnan, Ian Bruff, Anne Carruthers, Ryan Centner, Valentín Clavé-Mercier, Dunya Fehimovic, Dolores Figueroa, Hilary Francis, Miguel González, Eduardo Gudynas, Juliet Hooker, Rosaleen Howard, Mo Hume, André Jansson, Emma Kelly, Sara Kindon, Sara Koopman, Patricia Lagos Redondo, Simon Lambert, Nick Lewis, Chris Lukinbeal, Kate Maclean, Ashiya Mendheria, Nicholas Mirzoeff, Diana Ojeda, Eline van Ommen, Tobías Palma Stade, Lisa Parks, Eric Pawson, Dick Peet, Viviana Ramírez, Dennis Rodgers, James Sidaway, Lauren Sinreich, Ileana Selejan, Geoff Stahl, Marcin Stanek, Amanda Thomas, Matthew Watson, and Francesca Zunino. Thanks to Tame Iti of the Tūhoe Nation from whom I've learned so much about settler colonialism and decolonial activism.

I would also like to thank all of the University of Edinburgh students who have taken my course Development and Decolonization in Latin America over the past five years with whom I have developed the material for this book. It is such a privilege to work with students who recognize the importance of non-Eurocentric ways of knowing and are outraged at the injustice inflicted on people in the name of development.

I must also express my gratitude to all of the bodies that have funded my research and made frequent travel to Latin America possible. These include AHRC, British Council/Newton Fund, ESRC, Marsden Fund of the Royal Society of New Zealand, and NERC.

Thanks also to the team at Routledge, especially Faye Leerink, Andrew Mould, and Egle Zigaite, for their encouragement, support, and patience.

Thanks too to Thomas Klein for frequent motivational check-ins, fun selfies, and uplifting emojis.

Finally, I would like to thank my wonderful family, who have done much of this with me and tolerated my frequent absences when they couldn't. Thanks to my husband, collaborator, and frequent co-author, Kevin Glynn, and my (now very much grown up) children, Tash and Ruben, for memorable times in León caliente and many other locations. Thanks too to Lola and Callum, and Lloumi, Leila, and Quincy.

Abbreviations

ACDEGAM	Asociación Campesina de Ganaderos y Agricultores del Magdalena Medio/Association of Middle Magdalena Ranchers and Farmers
ACLU	American Civil Liberties Union (US)
AILLA	Archive of the Indigenous Languages of Latin America (US)
ALBA	Alianza Bolivariana para los Pueblos de Nuestra América/Bolivarian Alliance for the Peoples of Our America
AMLO	Andrés Manuel López Obrador (President of Mexico)
ANTRA	Associacão Nacional de Travestis e Transexuais/National Association of Transvestites and Transsexuals (Brazil)
AUC	Autodefensas Unidas de Colombia/United Self-Defences of Colombia
B-19	Barrio 19/19th Street Gang (El Salvador/US)
BID	Banco Interamericano de Desarrollo (in English IDB)
BRI	Belt and Road Initiative (China)
BRICS	Brazil-Russia-India-China-South Africa
BRT	Bus Rapid Transit
C169	ILO Convention 169
CABO	Central American Black Organization (in Spanish ONECA)
CAFTA	Central American Free Trade Agreement
CEB	Comunidades Eclesiástica de Base/Comunidades Eclesiais de Base/Christian Base Communities

CEPAL	Comisión Económica para Latinoamérica y el Caribe (in English ECLAC)
CERJ	Consejo Étnico *Runejel Junam* – *Runejel Junam* Council of Ethnic Communities (Guatemala)
CGWIC	Chinese Great Wall Industries Corporation
CIA	Central Intelligence Agency (US)
CICIG	International Commission against Impunity in Guatemala/Comisión Internacional contra la Impunidad en Guatemala
CIDH	Comisión Interamericana para los Derechos Humanos (In English IACHR)
CONAIE	Confederación de Nacionalidades Indígenas de Ecuador/Confederation of Indigenous Nationalities of Ecuador
CONAQ	Coordenação Nacional de Articulação das Comunidades Negras Rurais Quilombolas/ National Commission of Articulation of Black Rural Quilombola Communities (Brazil)
CONAVIGUA	Comité Nacional de Viudas de Guatemala – National Committee of Widows of Guatemala
CONRED	Coordinadora Nacional para la Reducción de Desastres/National Coordination for Disaster Reduction (Guatemala)
COPINH	Consejo Cívico de Organizaciones Populares e Indígenas de Honduras/Council of Popular and Indigenous Organizations of Honduras
CPR	Comunidades de Pueblos en Resistencia/ Communities of Peoples in Resistance (Guatemala)
CUC	Comité de Unidad Campesina/Committee for Peasant Unity (Guatemala)
CUD	Coordinadora Unica de Damnificados/Overall Coordinating Committee of Disaster Victims (Mexico)
ECLAC	Economic Commission for Latin America and the Caribbean (in Spanish CEPAL)
ENSO	El Niño Southern Oscillation
DEA	Drug Enforcement Administration (US)
DESA	Desarrollos Energéticos Sociedad Anónima (Honduras)

DREAM	Development Relief and Education for Alien Minors (US)
DT	Dependency theory
ELN	Ejército de Liberación Nacional/National Liberation Army (Colombia)
ENSO	El Niño Southern Oscillation
EPL	Ejército de Liberación Popular/Popular Liberation Army (Colombia)
EUG	Entrepreneurial Urban Governance
EZLN	Ejército Zapatista de Liberación Nacional/Zapatista Army of National Liberation (Mexico)
FARC	Fuerzas Armadas Revolucionarias de Colombia/ Revolutionary Armed Forces of Colombia (Colombia)
FBI	Federal Bureau of Investigation (US)
FDI	Foreign Direct Investment
FMLN	Frente Farabundo Martí para la Liberación Nacional/ Farabundi Martí Front for National Liberation (El Salvador)
FPIC	Free, Prior and Informed Consent (part of ILO C169)
FSLN	Frente Sandinista para la Liberación Nacional/ Sandinista Front for National Liberation (Nicaragua)
FTA	Free Trade Agreement
FTAA	Free Trade Area of the Americas
FTZ	Free Trade Zone
GAM	Grupo de Apoyo Mutuo/Group of Mutual Support (Guatemala)
GM	genetically modified
HDI	Human Development Index
HIPC	Heavily Indebted Poor Countries Initiative
IACHR	Inter-American Commission for Human Rights (In Spanish CIDH)
ICE	Immigration and Customs Enforcement (US)
IDB	Inter-American Development Bank (in Spanish BID)
IDP	Internally Displaced People
ILO	International Labour Organization
IMF	International Monetary Fund
INADI	Instituto Nacional contra la Discriminación, la Xenofobia y el Racismo/National Institute Against Discrimination, Xenophobia, and Racism (Argentina)

ISDS	Investor–State Dispute Settlement
ISHR	International Service for Human Rights
ISI	Import Substitution Industrialization
KCA	Kappes, Cassidy and Associates
LGBTQ+	Lesbian, gay, bisexual, transgender, queer and other identities
M-19	Movimiento 19 de Abril/19th of April Movement (Colombia)
MAQL	Movimiento Armado Quintín Lame/ Quintín Lame Armed Movement (Colombia)
MAS	Movimiento al Socialismo/Movement to Socialism (Bolivia)
MAS	Muerte a Secuestradores/Death to Kidnappers (Colombia)
MCD	Modernity/Coloniality/Decoloniality
MNR	Movimiento Nacionalista Revolucionario/ Revolutionary Nationalist Movement (Bolivia)
MS-13	Mara Salvatrucha (El Salvador/US)
MST	Movimento dos Trabalhadores Rurais Sem Terra/ Landless Workers Movement (Brazil)
NAFTA	North American Free Trade Agreement
NGO	Non-governmental organization
NIDL	New International Division of Labour
NTAE	Non-traditional agricultural export
OAS	Organization of American States (in Spanish OEA)
OEA	Organización de Estados Americanos (in English OAS)
OFRANEH	Organización Fraternal Negra Hondureña/Fraternal Black Organization of Honduras
ONECA	Organización Negra Centroamericana (in English CABO)
OPEC	Organization of Petroleum Exporting Countries
OPIP	Organización de Pueblos Indígenas de Pastaza/ Organization of Indigenous Peoples of Pastaza (Ecuador)
PCN	Proceso de Comunidades Negras/Process of Black Communities (Colombia)
PNBV	Plan Nacional para el Buen Vivir/National Plan for Buen Vivir (Ecuador)

PRI	Partido Revolucionario Institucional/Institutional Revolutionary Party (Mexico)
PROMESA	Puerto Rico Oversight, Management, and Economic Stability Act
PT	Pink Tide
PU	Planetary urbanization
RAE	Real Academia Española/Royal Spanish Academy
SDGs	Sustainable Development Goals
SICA	Sistema de Integración Centroamericana/Central American Integration System
TINA	There is no alternative (i.e to neoliberalism)
TIPNIS	Territorio Indígena y Parque Nacional Isiboro Sécure/Isiboro-Sécure National Park and Indigenous Territory (Bolivia)
UCA	Universidad Centroamericana/Central American University (El Salvador)
UCP	United Constitutional Patriots (US)
UDEFEGUA	Unidad de Protección a Defensoras y Defensores de Derechos Humanos de Guatemala/Protection Unit for Human Rights Defenders in Guatemala
UFCO	United Fruit Company
UN	United Nations
UNASUR	Unión de Naciones Suramericanas/União de Nações Sul-Americanas/Union of South American Nations
UNDP	United Nations Development Programme
UNIA	United Negro Improvement Association
URNG	Unidad Revolucionaria Nacional Guatemalteca/ Guatemalan National Revolutionary Unity (Guatemala)
USAID	United States Agency for International Development
WB	World Bank
WGIP	Working Group on Indigenous Populations
WHO	World Health Organization
WSF	World Social Forum
WTO	World Trade Organization
ZB	Zone of Being
ZNB	Zone of Non-Being

Glossary

1492 the year of the conquest of what is now Latin America by European powers.

Abya Yala América/the Americas. A precolonial and decolonial reference for the Latin American continent from the Gunadule language that means "mature earth" and contests the colonial term "New World" (see also Anáhuac and Tawantinsuyu).

abyssal thinking according to Boaventura de Sousa Santos who draws on Frantz Fanon, Eurocentrism is characterized by abyssal thinking that divides up the world hierarchically into a Zone of Being and a Zone of Non-Being.

Afrodescendant a Black person or person of African descent.

aguayo a carrying cloth used by Aymara and Quechua people in the Andes.

alabao a Colombian wake and funeral song sung to help the deceased on their way to eternity that has been reworked in the Pacific to contest political violence.

alterity in the work of Enrique Dussel, alterity is the space and lived experience from which the oppressed Other fight for liberation from a colonizing system (alterity is the underside of the totality).

América the whole American continent including Canada, the United States, Central and South America, and the islands of Caribbean. While it is an everyday concept in Latin America, it also challenges US hegemony and the tendency in English to use America to refer only to the United States.

Anáhuac a precolonial and decolonial term for Mesoamerica (see also Abya Yala and Tawantinsuyu).

ayllu a network of families with a common ancestor and a shared territory and the basis of government in Quechua and Aymara communities in the Andes.

baroque ethos an eccentric form of modernity that enables marginal and subversive knowledges and politics to assert themselves (de Sousa Santos). It emerges in part because of the weakness of the colonizing powers (Spain and Portugal).

barrio a popular, low-income, or irregular neighbourhood.

blanqueamiento whitening in order to improve the race (in Portuguese branqueamento).

body–politics and geopolitics of knowledge an approach that sees knowledge as embodied and produced in a particular context (unlike Eurocentric knowledge that pretends it is disembodied and universal).

border thinking a concept introduced by Gloria Anzaldúa and developed by Walter Mignolo that emphasizes the knowledge that results from the struggle against coloniality. Border thinking involves making use of and engaging with dominant and Eurocentric knowledges but mixing them with alternative cosmovisions and epistemes from native or Black traditions.

branqueamento whitening in order to improve the race (in Spanish blanqueamiento).

Buen Vivir an Aymara and Quechua concept that signifies a decolonial non–capitalist approach to the world. It refers to good life in a broad sense and rejects models focused on economic growth (also known as suma qamaña, sumak kawsay).

campesino a small or subsistence farmer.

Candomblé a diasporic syncretic religion that combines West African and Catholic elements and that was developed in Brazil as a result of the slave trade.

caracoles self-governing autonomous Zapatista territories.

caudillo a strong man leader that tolerates no dissent and who operates through clientelistic means.

caudillismo the phenomenon of caudillo leaders.

chicana/chicano a US American of Mexican descent.

chinampa an Aztec form of agriculture in which crops are grown on gardens floating in a lake.

ch'ixi an Andean concept that emphasizes the juxtaposition of different elements and how different cultures with different temporalities co-exist.

cholo a person of Indigenous descent in the Andean region who has migrated to the city, used to be derogatory but is now being resignified.

clientelism the exchange of votes for favours.

codex, códice a native American cultural text of significance. Many of them were destroyed by the colonizers.

colonialism the formal administration of one part of the world by another after conquest and occupation. It usually involves genocide and the theft of land and resources.

coloniality   colonial attitudes, practices, and policies that exist with or without the presence of a formal colonial administration. Racism is the basis of coloniality.

colonial matrix of power   a concept developed by Aníbal Quijano that emphasizes the oppressive and hierarchical outcomes that coloniality exerts through race, gender, sexuality, spirituality, and authority.

comedor popular   community-run soup kitchen.

conquistador   conqueror.

corporatism   the top-down state control and manipulation of interest groups.

cosmopolitanism   cosmopolitanism is a form of counter-hegemonic globalization that can unite oppressed peoples in different parts of the world (de Sousa Santos).

coyote   a person who helps undocumented migrants to cross the Mexican border into the US for a fee.

Creole   sometimes a European born in Latin America (especially in Spanish as *criollo*). Usually, and especially in the Caribbean, a person of mixed Black and European descent. Black Creoles in Nicaragua, Costa Rica and elsewhere often speak Creole English as well as or instead of "standard" English and Spanish.

criollo   a person of Spanish descent born in Latin America (see also Creole).

disappearance, to disappear   in the context of the dirty wars and human rights abuses of the 1970s and 80s, disappearance refers to the forced deprivation of freedom without due process by state or paramilitary forces. Latin American human rights activists began to use the verb "to disappear" as a transitive verb to capture its state-led nature.

ejido   communal land holding in Mexico based on usufruct rights.

encomienda   a colonial labour system in which colonial settlers were granted land and the labour of a group of colonized Indigenous groups.

epistemicide   the killing or destruction of knowledge.

exteriority   a concept that Enrique Dussel takes from Emmanuel Levinas to focus on the spaces that have not been fully colonized by Eurocentrism. These spaces contain forms of thinking of the oppressed Other that reside outside of or partially outside of the dominant system and that are a source of liberation and epistemic justice.

Eurocentrism   the problematic idea that knowledge produced in Europe and/or by European thinkers is universal, objective, and superior. Eurocentric thought often endorses colonialism or glosses over its effects and atrocities, it reproduces white supremacy, and fails to recognize the existence and significance of other kinds of knowledges in the world. Eurocentrism is dominant in Latin American universities as well as in European ones.

favela   a Brazilian shantytown or irregular settlement.

femicide   the murder of women.

feminicide/feminicidio the murder of women because they are women in a context of gendered power structures.

filibustero someone who engages in an unauthorized military expedition in a foreign country to seize power or foment insurrection.

Fue el estado It was the state.

indianismo an anti-colonial ideology that is focused on Indigenous liberation and resists assimiliation in the form of indigenismo.

indigenismo a nationalist ideology and political movement that attempts to recognize but often to also assimilate Indigenous culture within the nation-state. Indigenismo is often compatible with the dominant sense of mestizaje and can work to silence African and Afrodescendant heritages.

interculturality Communication and exchange between different cultures and knowledge system in a way that promotes dialogue and understanding and mutual respect.

Latin America A European geographical and geopolitical construction for the continent formerly referred to as Abya Yala, Anahuác, or Tawantinsuyu. Now appropriated by many people who live in the continent.

kuagro an African political and cultural system in place in the palenque that ensures its political and cultural survival.

ladino another word for *mestizo*, used in Guatemala.

latifundio a very large piece of privately owned land or estate (see minifundio).

locus of enunciation the place from which one speaks. This could be from a disembodied Eurocentric position that is not self-reflexive, or from a decolonial one that takes the geopolitics and body-politics of knowledge into account.

machi a Mapuche spiritual leader.

machismo an exaggerated form of masculinity that expresses itself in sexist behaviour such as domestic violence and the desire to have many sexual partners.

maquiladora an export processing factory.

marianismo a traditional gender role for women that draws on the ideal of the Virgin Mary to promote submissiveness and women's moral superiority to men; the female counterpart of *machismo*.

mestiza consciousness a term developed by Gloria Anzaldúa that captures the ways in which colonialism and patriarchy have impacted on Chichana women and the forms of resistance that emerge from this lived experience.

mestizaje racial mixing, especially between European and native or Indigenous peoples. It is a process that has been central to nation-building in most Latin American countries. Mestizaje as an ideology tends to involve a hierarchical colonial performativity and is usually placed above both Indigeneity and Blackness. It works to exclude Indigenous knowledge and the dominance of mestizaje means that Indigenous and Black Latin Americans do not enjoy the same privileges and forms of mobility enjoyed by mestizo Latin Americans. For Gloria Anzaldúa, mestizaje is a site of critical feminist agency.

mestizo   a mixed-race person, but especially between European and Indigenous (but see mestizaje).

metrocable   a cable car form of public transport.

minga   a Quechua word for collective work that benefits the whole community.

minifundio   a very small plot of land used for subsistence agriculture (see latifundio).

mulato   a term used in the colonial era to designate someone of Black and Indigenous mixed race.

muxe   a third gender in Zapotec culture.

neoliberalism   an economic model based on a belief in free markets.

neplanta   a Nahuatl word used by Gloria Anzaldúa meaning in-between space.

ni perdón, ni olvido   We neither forgive nor forget.

ni una menos   Not one (woman) less.

normalista   a student that attends rural schools known as *Normales* created after the Mexican revolution.

Nos están matando   They are killing us.

Nos quitaron tanto que nos quitaron hasta el miedo   They took so much from us, they even took away our fear.

Nuestra América   a term coined by Cuban nationalist writer José Martí that posits a counter-hegemonic America (or América) full of eman-cipatory potential to the hegemonic imperialistic European America that oppresses and dispossesses. Nuestra América has emerged from the violence of conquest and the mixing of Indigenous, African, and European blood. Embracing Nuestra América means drawing on local knowledges not imported European ones.

olla común   community-run soup kitchen.

pachakuti   an Andean non-linear concept of time that means the overturning of space-time in a way that involves either catastrophe or positive renewal. The word is composed of *pacha* meaning earth and *kuti* meaning to turn back or turn over. Long periods of time are punctured by *pachakuti*, events that are totally transformative. The conquest was a *pachakuti*.

Pachamama   a revered Andean fertility goddess sometimes referred to as World Mother or Time Mother.

palenque   a community formed by escaped slaves (Colombia).

palenqueros   the inhabitants of a palenque.

pollera   a traditional layered skirt worn by Indigenous peoples in Bolivia and Peru.

plurinationality   the recognition as in the constitutions of Ecuador and Bolivia that there is more than one nationality within the state.

pluriversalism   the recognition that there are multiple worlds and multiple ways of knowing. Pluriversalism means recognizing these other worlds and not acting as if there was just one world. A pluriverse is a world in which many worlds fit.

populism   a political culture and style forged by charismatic political leaders of different political persuasions who are able to appeal and connect with people's everyday lives.

quilombo   a community formed by escaped slaves (Brazil).

race   a hierarchical concept invented by the colonizers in the 15th and 16th centuries that designates colonized peoples as inferior and less than human and white Europeans as superior.

racism   a cultural practice rooted in violence and discrimination that reproduces racial difference.

Raizal   a native Afrodescendant inhabitant from the San Andrés archipelago. Raizal people speak Creole English, instead of or as well as Spanish.

Requerimiento   a colonial statement read to native peoples that established Spain's right to conquer them and take their lands.

Santería   a diasporic syncretic religion that combines West African and Catholic elements and that was developed in Cuba as a result of the slave trade.

societal fascism   a process that permanently excludes or rejects people from the social contract.

sociology of absences   a concept developed by Boaventura de Sousa Santos to understand the ways in which Eurocentrism marginalizes and silences non-Eurocentric ways of knowing and being. It results in certain things not appearing to exist because they have been successfully suppressed by Eurocentric knowledge formations. It is confronted by the ecology of knowledges.

Suma Qamaña   See Buen Vivir.

Sumak Kawsay   See Buen Vivir.

Tawantinsuyu   the Quechua word for the Incan Empire; tawa = four, suyu = region. A decolonial way of referring to Latin America (see also Abya Yala and Anahuác).

teleférico   a cable car form of public transport.

territorio/territory   the production of space through and for political means especially by bottom-up or subaltern forces. More expansive than the conventional term as it is used in English, territorial struggles are often about cultural and legal recognition, the collective fight for autonomy, the appropriation of space in search of a dignified life, the defence of life, and the disruption of Eurocentric forms of assimilationist or extractivist practices.

testimonio   a narrative genre in which subordinated people talk or write in their own words in order to reveal a collective experience of oppression that has been silenced or erased.

tico   a colloquial word for Costa Rican.

transmodernity   A concept developed by Enrique Dussel based on an embrace of a form of modernity that is at the service of humanity and the needs of the oppressed Other harmed by Eurocentric modernity. Transmodernity is opposed to the modernity that has brought violence and oppression.

Vivos se los llevaron, vivos los queremos   They took them alive, we want them back alive.

CHAPTER 1

Made in conquest

The making of contemporary Latin America

THE FAILURE OF DEVELOPMENT

Political protest, rebellion, and cultures of solidarity have been a central feature of life in the region we now call Latin America since the conquest in
1492. In 2019, just a few months before lives the world over were disrupted
by the global coronavirus pandemic, Latin America exploded. Across the continent, people turned out onto the streets in their thousands to protest at
persistent inequality, socio-economic exclusion, authoritarianism, corruption,
extractivism, political violence, racism, and misogyny. While each protest has
its own set of historical and geographic specificities, collectively they can be
understood as the result of the miserable failure of the development project
and at the incomplete status of decolonization. The 2019 struggles revealed
the "protagonism created by the collective action of civil society" (Cotto
Morales, 2020: 130) and drew attention to structural power in its myriad
forms. In February, Haitians mobilized en masse against corruption and austerity measures. In July, hundreds of thousands of Puerto Ricans took to
the streets to demand the resignation of the governor Ricardo Rosselló. In
October, Indigenous people in Ecuador forced the government of President
Lenín Moreno to back down on a set of austerity measures that would have
hit the most vulnerable hardest. A week later in Chile, a rise in the price
of a Santiago metro ticket escalated into full-blown protests against three
decades of inequality and neoliberal policies. These protests, which left 20
dead and thousands injured, included a feminist performance about gender-
based violence that went viral (see Chapter 6). Furthermore, they resulted
in a referendum in which people voted for a new constitution. In Bolivia,
after years of Indigenous empowerment and social progress along with the
contradictory promotion of extractivist activities, in November polarizing
left-wing president Evo Morales was forced to resign and leave the country
after his opponents accused him of electoral fraud. His removal from power in

DOI: 10.4324/9781003110453-1

a divided country led to the temporary installation of a Christian fundamentalist government, focused on undermining racial equality initiatives. Morales' party, the MAS, returned to power in fresh elections held at the end of 2020. Also in November, Colombians took to the streets in large numbers to protest at the economic policies of President Iván Duque and at the failure of the peace process, initiated after peace accords were signed in 2016 to bring an end to a four-decade long violent civil war. The murder of 18-year-old Dilan Cruz shot by the police enraged the protestors (see Figure 1.1). At around the same time, Peruvians protested at the removal of President Martín Vizcarra after he attempted to shut down a congress widely viewed as corrupt. There was ongoing dissatisfaction too in Brazil, Venezuela, and Nicaragua, where authoritarian leaders had implemented a range of anti-democratic measures in order to stay in power, including policies that work against the interests of the popular sectors or against environmental protections. In 2018, a popular anti-government rebellion in Nicaragua led in large part by students was quashed by the government of Daniel Ortega and left hundreds dead and many political prisoners. In Brazil, right-wing populist Jair Bolsonaro declared war on the environment, Indigenous peoples, and LGBTQ+ populations – although not so much on the Covid-19 pandemic – and began to undo decades of social progress achieved under the previous Workers Party governments of Luiz Inácio Lula da Silva (Lula) and Dilma Rousseff. In the process, huge swathes of the Amazon were left to burn. Venezuela has been mired in economic and political chaos since the death of Hugo Chávez in 2013. His successor, President Nicolás Maduro has remained in power despite widespread food shortages, electricity blackouts, a serious refugee crisis, and the opposition leader Juan Guaidó unilaterally declaring himself interim president.

Many Latin Americans are fighting for peace, but political violence remains widespread. In Colombia, Guatemala, Honduras, and Mexico, environmental defenders who oppose destructive hydroelectric dams or mining projects are often criminalized or assassinated with impunity. The murders of Berta Cáceres in Honduras and Marielle Franco in Brazil, killed because they challenged these colonial dynamics, continue to reverberate around the continent (see Chapters 4 and 8). The Mexican "war on drugs" has claimed thousands of lives, as cartel members, local politicians, and corrupt police officers work together.

While many Latin Americans were protesting, others were on the move, trying to escape violence and searching for better economic opportunities. Starting in late 2018, migrant caravans composed of hundreds of would-be migrants from El Salvador, Guatemala, Honduras, and Nicaragua formed and began to move together towards the Guatemala–Mexico border and the Mexico–US border (see Figure 1.2). These migrations led to the further militarization of the Mexico–US border under the anti-immigration Trump presidency. In the final week of his presidency and just before the inauguration

Figure 1.1 Homage to student Dilan Cruz, killed in the protests in Bogotá in 2019

Source: futbolero, CC BY-SA 4.0, via Wikimedia Commons

Figure 1.2 Central American migrants charge their phones in the Ciudad Deportiva
Magdalena Mixhuca temporary camp in Mexico City in 2018

Source: ProtoplasmaKid, CC BY-SA 4.0, via Wikimedia Commons

of Joe Biden, around 3,000 migrants left Honduras and El Salvador for the
United States.

THE INVENTION OF DEVELOPMENT

While protest and migratory movements underscore the degree to which
substantial development challenges remain in Latin America, there is also no
doubt that in recent decades Latin America has been a key source of inspiration on how to do politics and development differently. The events outlined
above follow decades of dramatic political change; peace accords, transitions
to democracy, embrace and rejection of neoliberal free market economics, and
high-profile forms of Indigenous mobilizations that have called into question
many of the premises of the Latin American nation-state and the development
models it has embraced. The social movements fighting for change are influential and inspirational.

Yet it wasn't so long ago that Latin America was seen, especially by expert
outsiders, as a key part of the so-called third world and as a region in need
of development assistance and expertise from the first world that had already
developed and knew therefore how to do it. Arturo Escobar (1995) shows

that it was in the period after the Second War World that the idea of development as we currently understand it was brought into being. At this time, the United States, less affected by the Second World War, was enjoying economic ascendancy, a number of nations in Asia and the Pacific were gaining their independence, and new kinds of global relationships were being forged. It was also a time of global optimism about the prospects for economic and technological progress and the United States was anxious to gain access to so-called emerging markets. According to Escobar (1995), the inaugural address of US President Harry Truman in 1949 can be seen as a landmark moment that heralded the contemporary development age. Truman asserted:

> More than half the people of the world are living in conditions approaching misery. Their food is inadequate, they are victims of disease. Their economic life is primitive and stagnant. Their poverty is a handicap and a threat both to them and to more prosperous areas. For the first time in history, humanity possesses the knowledge and the skill to relieve the suffering of these people. (…) I believe that we should make available to peace-loving peoples the benefits of our store of technical knowledge in order to help them realize their aspirations for a better life. (…) Greater production is the key to prosperity and peace. And the key to greater production is a wider and more vigorous application of modern scientific and technological knowledge.
>
> (US President Harry Truman, 20 January 1949,
> cited in Escobar 1995: 3)

So Truman's approach amounted to defining half the world in negative terms, as suffering, poor, and underdeveloped and the other half, the western world, as in command of the knowledges and technologies to address this suffering. This mode of thinking took hold in Europe and the United States and had serious consequences for Africa, Asia, Latin America, and the Caribbean. The idea that with western help, knowledge, and technologies third world countries would modernize and "catch up" with the industrialized countries became pervasive and gave rise to the development industry we know today, based on multilateral and government aid agencies, non-governmental organizations (NGOs), and experts (economists, agronomists, demographers, geographers, sociologists, and so on) based in universities or think tanks. Development rapidly became underpinned by a set of power-bearing discourses that had tangible material consequences. These discourses tended to construct the third world in negative terms as a site of lack, disease, conflict, and political instability. It is important to recognize that discourses bring reality into being and should not be understood as something distinct from reality or from material conditions of life. All discourses are, however, unstable and hegemonic discourses that serve the powerful are always challenged by counterhegemonic discourses that serve the interests of subaltern social groups.

The invention of development at this time was also driven by strong geo-political motivations. This was the start of the Cold War and concerns were developing within the US administration about the spread of communism and the appeal of communist ideas to disenfranchised and impoverished groups of people in the third world. One of the most influential contributions to modernization thinking was Walt Rostow's *The Stages of Economic Growth: A Non-Communist Manifesto* published in 1960. The geopolitical underpinnings of modernization theory are clearly depicted in the book's subtitle. Rostow (1916–2003) advised on national security during the presidencies of John F. Kennedy and Lyndon B. Johnson and was a committed free-market cap-italist and anti-communist. Capturing the hearts and minds (and economies) of the third world and integrating these economies into a global capitalist system was viewed as an important strategy, which would prevent Latin American countries turning to socialist or communist thought and ideals. Rostow's influential thesis posited that the countries of the third world were much like the pre-industrial communities of Europe and therefore they had to experience the conditions of economic transformation and social change that European countries had already experienced during the industrial revolution. Indeed, for Rostow, modernization comprised five distinct stages; traditional society, preconditions for take-off, take-off, drive to maturity, and high mass consumption. Development was thus seen as a linear process towards the level and type of development found in the West. The big difference between developed and developing countries was that the developing countries had to be *helped* on their path to development by the industrialized nations, while the first world had developed by its own means. Becoming developed meant adopting first world practices, ideas, knowledges, and investment.

Modernization thinking assumed that Indigenous and non-European cul-tural traditions would be abandoned as they developed, and people adopted modern values and practices. In many ways, it was a ludicrous proposition, as all nations are trying to "develop" in the sense that they work to promote eco-nomic production and try to secure better health care, education, transport, and standards of living. There are plenty of people in "developed" countries such as the UK or the US who live in poverty, rely on foodbanks, or don't have decent housing. In other words, almost 80 years after the inauguration of the development era, the first world is still far from having achieved prosperity and well-being for all.

Furthermore, the modernity/tradition binary on which the concept of development rested is also deeply flawed as all countries are engaged in the negotiation of cultural traditions and the implications of cultural change. All traditions are, as early 20th century Peruvian socialist, José Carlos Mariátegui (1970[1927]:161), in his revolutionary reclaiming of the concept of trad-ition from the traditionalists and in his anti-colonial appeal to Peruvians to peruanize themselves, wrote "alive and mobile." But as a discourse that had been centuries in the making, development was powerful and compelling, and it took hold, in the targets of development as well as those working to implement it in both policy and practice. In the minds of the general public, it became a common sense. But modernization thinking quickly came under challenge from Latin American dependency theorists inspired by socialist and anti-imperialist thought and informed by their own economic conditions (see Chapter 3).

By the 2000s Latin American variants of participatory democracy, constitutionalism, activism, and the claiming of rights for Nature as well as for people were disrupting the colonial and Eurocentric models embraced by local elites and setting examples to the rest of the world. It is now clear that scholars, activists, and policymakers in the Global North have more to learn from Latin America than Latin America has to learn from them (Cupples et al., 2019a). As John Holloway writes of Bolivia in the foreword to a book by Raúl Zibechi (2010: xv), Bolivia was once seen as "a backward, underdeveloped country which could hope, if it was lucky, to attain the development of a country like Germany one day in the future" whereas now Bolivia is a source of political hope and people in Europe are inspired by what they have seen the people in Cochabamba and El Alto achieve.

Development then is more contested than ever and there have been calls by postdevelopment and anti-colonial theorists to jettison the whole concept as a neocolonial process that does more harm than good (Esteva, 1985; Escobar, 1995; Rahmena, 1997). Nonetheless, development remains persistent and top-down development initiatives including the UN-led Millennium Development Goals (2005–2015) and the Sustainable Development Goals (SDGs) (2015–2030) continue to have important discursive and therefore material effects. SDG thinking has become hegemonic in many sites of development theory and practice and while they might prove to be a mechanism for positive change and will certainly do some good, it is crucial that as a top-down development initiative they are subject to critical scrutiny (see Figure 1.1 and Box 1.1).

BOX 1.1 THE SUSTAINABLE DEVELOPMENT GOALS

The 2015 Sustainable Development Goals, also known as Agenda 2030, are made up of 17 mainstream development goals (see Figure 1.1) and 169 specific targets that are designed to underpin the work of states and other public and private development actors. They are posited by the UN as "a blueprint to achieve a better and more sustainable future for all by 2030." They constitute an attempt to combine development goals around poverty and inequality with ambitions to tackle climate change and protect the environment. All the governments of Latin America have ratified the goals and it is possible that they will encourage governments in the region to implement some poverty reduction and environmental protection measures in order to meet some of the targets. The goals have, however, been widely criticized and there is no doubt that they do little to disrupt the developmentalist and Eurocentric status quo. The Goals themselves are also contradictory and compete with one another. For example, Goal 8 continues to emphasize the pursuit of economic growth rather than wealth distribution which is likely to compromise the ability to tackle climate change or protect biodiversity. Jason Hickel (2019) has shown that the sustainable use of natural resources and the necessary reduction of carbon emissions required to keep us below a level of global warming of no more than 2°C are incompatible with the annual global GDP growth rate of 3% that Goal 8 seeks to achieve. The Goals also continue to maintain the conceptual separation between the environment and people, so are less compatible with the territorial movements in Latin America led by Indigenous and Afrodescendant groups which are working to defend environments and livelihoods simultaneously (Hope, 2020). Indeed, Indigenous and Afrodescendant groups and their worldviews are not referenced in the SDGs in spite of their role in rearticulating the dominant meanings of development. The Goals also universalize development goals and fail to take account of regional and cultural differences. While they emphasize environmental protection far more than the Millennium Development Goals which preceded them did, they are also very much data-driven and statistics can easily be methodologically manipulated by states and other actors. There is also often little coherence across the region in terms of the data collected and whether these are accurate or adequately compiled. While some of the targets are of course highly desirable and living standards would rise were they to be achieved, Agenda 2030 has largely failed to identify the means and tools by which such targets will be reached and so it is possible that they "will remain wishful thinking, at best implemented in an ad hoc and piecemeal fashion, contradicting the stated intention that it should be a universal, comprehensive and indivisible agenda" (ECLAC, 2016: 9–10, cited in Willis, 2019: 124). Given that extractivist and neo-colonial development models continue to be endorsed and supported by many governments in the region and the environmental defenders who oppose such activities, continue to be criminalized and assassinated, it is likely that the SDGs are merely strengthening capitalist and neoliberal development models "at the expense of the decolonial territorial agendas that are proving key to challenging frontiers of resource extraction" (Hope, 2020: 2). There is no evidence to date that ratifying the SDGs has curtailed the ambitions of Latin American states to continue to promote unsustainable development. Like other top-down development agendas, the SDGs are, however, a mechanism around which social movements can mobilize in pursuit of their own ambitions (see Figure 1.3).

Sources: Willis, 2018; Hickel, 2019; Hope, 2020

Figure 1.3 The Sustainable Development Goals

Source: United Nations

Development is not, however, a monolith. Its meanings are not fixed; it takes variegated forms in different places and is often embraced and accommodated by bottom-up forces as well as resisted and rejected. Postdevelopment thought has as a result attracted substantial analysis and critique (for discussion see Nederveen Pieterse, 2000; Ziai, 2019). Furthermore, it is clear that postdevelopment does not just always mean outright rejection of the concept of the development but can focus on the ways in which development is reworked and rearticulated by actors on the ground in geographically specific and sometimes empowering ways. Even if development has produced some negative and insidious outcomes in countries that Eurocentric geopolitical imaginaries have designated as third world, the subjects or "agents of development can never be finally captured as the fixed subjects of this or that discursive formation, since sociodiscursive dynamics can no more be finalized than the passage of time or history can be frozen" (Cupples et al., 2007: 788). From a Gramscian perspective, we can say that material conditions of everyday life that many Latin Americans face – hunger, inadequate access to health care, underemployment, flimsy housing, an inability to complete school, disasters, urban insecurity, and failed harvests – act to disrupt the hegemonic social formation or common sense, replacing it with a counterhegemonic "good sense." In other words, people do have a sense of what they're up against and do try to create a better and more dignified life for themselves. They might then *re-articulate* the dominant discourses of development, to come up with something more empowering and inclusive. Moreover, this struggle can only take place within the existing ideological terrain (Cupples et al., 2007). It would therefore be pointless to turn up in a low-income rural or urban community in Latin America and inform people who are mobilized to improve their standard of living that development was a big con and they need to overthrow it, because this group of people will already be in the process of taking ownership of development, rearticulating the concept in a way which *makes sense to them*.

This is a particularly useful approach for both studying and practising development in Latin America. As experienced development practitioners know, "just as one problem is attenuated, others arise; as soon as one group of people finds a measure of satisfaction, others begin to express grievances" (Cupples and Glynn, 2013: 1012). Whether the tenets of development and postdevelopment are accepted or not, it is valuable to recognize that evoking development is a tricky business because it always means asserting some kind of power relationship, either about how the world is or how the world should be. What constitutes good development for one group or individual might be rejected as inappropriate by another. Consequently, development does not have an end point and is continuously being reworked.

Another valuable approach to this question is that posed by Joel Wainwright (2008) who draws on the work of Indian postcolonial feminist Gayatri Spivak and asserts with a double negative that "we cannot not desire development." Wainwright shows how criticizing, deconstructing, and interrogating development is necessary but doing away with development in spite of its flaws is just not possible in the current conjuncture. In other words, development might have emerged in the context of colonizing and modernizing processes, it might be a neocolonial technique of power in many ways which has indeed had negative, disempowering, and insidious outcomes for many, but we still have a moral responsibility to respond to conditions of suffering in the world.

WHAT IS LATIN AMERICA?

America/América as a geocultural entity was created in the 15th and 16th centuries as a result of the conquest (Mignolo, 2000). In the millennia prior to European conquest and settlement, Latin America was home to many large, complex, and diverse civilizations, including the Zapotec, Mixtec, Olmec, Toltec, Aztec, and Mayan cultures of Mexico and Mesoamerica, the Valdivia culture of Ecuador, the Chavín, Moche, and Inca cultures of Peru, and the San Agustín and Muisca cultures of Colombia. There were nomadic hunter–gatherer tribes as well as peoples who practised agriculture and cultivated maize, manioc, squash, potato, chillies, and cotton. Many Indigenous civilizations had advanced knowledge systems. The Mayans, for example, had a fully developed writing and calendar system, which linked to agricultural, cosmological, astronomical, religious, and mathematical knowledges (see Chapter 8). The Aztecs were skilled warriors and had sophisticated architecture and modes of urban planning. The Incas had well-developed engineering skills in transport systems and irrigation for agriculture.

Some of the Indigenous civilizations which existed prior to the Spanish conquest collapsed for reasons that are quite poorly understood, although climate change, drought, deforestation, famine, political rebellion, and war are likely contributing factors. Others, such as the Aztec empire of Mexico and the Inca empire of Peru, flourished and expanded and became resilient cultures and mighty empires. It is important to acknowledge the heterogeneity of the Indigenous population and to recognize that pre-Columbian Latin America comprises complex histories which defy easy categorization.

Prior to conquest, the region was referred to by a number of different names including Abya Yala, Tawantinsuyu, and Anáhuac. After conquest, these names were subject to erasure as the Spanish colonizers began to refer to the region as the Indias Occidentales or West Indies, a term that "denied [the inhabitants] of the possibility to be where they were" (Mignolo, 2011: 88). The term Latin America was not widely used until the 19th century and its use emerged as a means to distinguish the region from Anglo America and to express criticism of US imperialist actions, including the taking of a large chunk of Mexican territory, and racist attitudes expressed by white US Americans towards people in the region (McGuinness, 2003). It is important to bear in mind that America/América does not refer only to the United States, but also to Bolivia, Chile, Guatemala, and Paraguay, even though many people seem to understand America as a synonym for the United States. Uruguayan theorist Eduardo Galeano (1971:16), who referred to Latin America as a region of open veins, protested how US hegemony meant that "[a]long the way, we even lost the right to call ourselves Americans" (my translation).

Walter Mignolo (2005) urges us helpfully to think of Latin America as an invention and an idea, rather than just as a geographical location, although even as a geographical location it does not have a fixed definition. Indeed, one cannot definitively assert which countries belong to Latin America and which ones are excluded, despite the attempt to do so in Table 1.1. Indeed, the countries included in or excluded from Latin America as a region depend on territorial, linguistic, or political definitions as well as on histories of colonialism. While the Spanish and Portuguese-speaking countries of the Central and South American mainland are almost always included, these territories also contain English-speaking (Belize, Guyana, the Caribbean coasts of Nicaragua, and Costa Rica), Dutch-speaking (Suriname), and French-speaking (French Guiana) territories, which are sometimes excluded. The island nations of the Caribbean also pose definitional complexities. While almost all of them could be included or excluded, I have included Cuba, the Dominican Republic, Haiti, and Puerto Rico, because of their identification with Latin America by many of their citizens as well as the multifaceted influence of these nations on Latin American politics and development. The exclusion of other Caribbean islands should not be an attempt to deny the important and long-term

Table 1.1 Selected development indicators

	Population millions (CIA World Factbook 2021)	HDI ranking (UNDP 2019)	Life expectancy (UNDP 2019)	Infant mortality rate per 1000 births (CIA World Factbook 2021)	Mean years of schooling (UNDP 2019)	GNI per capita in $US (UNDP 2019)	External debt $US billions (CIA World Factbook 2019)	Percentage of population with access to internet CIA World Factbook 2018/2019	Percentage of population below the poverty line (CIA World Factbook 2018/2019)	Gender Inequality Index (GII) and ranking (UNDP 2019)	CO_2 emissions per capita (World Bank 2016)[1]	Intentional homicides per 100,000 people (UN/World Bank)[2]
Argentina	46	46 (VH)	76.7	9.55	10.9	47,495	278.5	74	35.5	0.328(75)	4.6	5
Belize	0.4	110 (H)	74.6	11.4	9.9	6,382	1.135	47	41	0.415(97)	1.5	38
Bolivia	12	107 (H)	71.5	39.27	9.0	8,554	12.81	43.8	37.2	0.417(98)	2.0	6
Brazil	213	84 (H)	75.9	18.37	8.0	14,263	681.336	67.47	4.2	0.408(95)	2.2	27
Chile	18	43 (VH)	80.2	6.68	10.6	23,261	193.298	82	8.6	0.212(49)	4.7	4
Colombia	50	83 (H)	77.3	12.88	8.5	14,257	135.644	62.26	35.7	0.428(101)	2.0	25
Costa Rica	5	62 (VH)	80.3	8.59	8.7	18,486	29.589	74.09	21	0.288(62)	1.6	11
Cuba	11	70 (H)	78.8	4.19	11.8	8,621	30.06	57.15	NA	0.304(67)	2.5	5
Dominican Republic	10.5	88 (H)	74.1	21.68	8.1	17,591	23.094	74.82	21	0.455(107)	2.4	10
Ecuador	17	86 (H)	77.0	18.55	8.9	11,044	50.667	57.27	25	0.384(86)	2.5	6
El Salvador	6.5	124 (M)	73.3	12.38	6.9	8,359	17.24	33.82	22.8	0.383(85)	1.1	52
Guatemala	17	127 (M)	74.3	26.81	6.6	8,494	22.92	65	59.3	0.479(119)	1.1	23
Guyana	0.8	122 (M)	69.9	22.68	8.5	9,445	1.69	37.33	35	0.462(115)	3.1	14
Haiti	11	170 (L)	64.0	41.29	5.6	1,709	2.762	32.47	58.5	0.636(152)	0.3	7
Honduras	9	132 (M)	75.3	15.39	6.6	5,308	9.137	31.7	48.3	0.423(100)	1.1	39
Mexico	130	74 (H)	75.0	11.64	8.8	19,160	456.713	65.77	41.9	0.322(71)	3.9	29
Nicaragua	6	128 (M)	74.5	19.57	6.9	5,284	11.674	27.86	24.9	0.428(101)	0.9	7
Panama	4	57 (VH)	78.5	11.25	10.2	29,558	101.393	57.87		0.407(94)	2.7	9

Paraguay	7	103 (H)	74.3	23.83	8.5	12,2249	16.662	64.99	0.446 (107)	1.1	7
Peru	35	79 (H)	76.7	19.37	9.7	12,252	81.333	52.54	0.396 (87)	1.9	8
Puerto Rico*	3	...	81.5[3]	6.16	...	...	72	70.6[4]	...	...	21
Suriname	0.6	97 (H)	71.7	26.6	9.3	14,324	1.7	48.95	0.436 (105)	3.1	5
Uruguay	3.4	55 (VH)	77.9	8.48	8.9	20,064	43.705	68.28	0.288 (62)	2.0	12
Venezuela	29	113 (H)	72.1	22.23	10.3	7.045	100.3	72	0.479 (119)	5.5	37

* As Puerto Rico is not an independent nation-state, some data are not available

The Human Development Index (HDI) measures development by combining life expectancy, educational achievement and income into a composite index. The scores enable the UNDP to rank countries from 1-47 (very high human development VH), 48-94 (high human development H), 95-141 (medium human development M) and 142-187 (low human development L).

GNI per capita is total national income in a year divided by the population. It does not reveal how equally that income is actually distributed nor how it is spent.

The Gender Inequality Index (GII) measures and ranks countries according to inequality in achievements between women and men according to three dimensions: reproductive health, empowerment, and the labour market. The GII ranges between 0 and 1. Higher GII values indicate higher inequalities between women and men.

1 For comparison UK is 5.8 and US is 15.5

2 From UN Office on Drugs and Crime's International Homicide Statistics Database via data.worldbank.org. Data range from 2016 to 2018

3 Data for Puerto Rican life expectancy from CIA Factbook

4 These data are from the website of the Financial Oversight and Management Board for Puerto Rico. Puerto Rico's debt had reached $72 billion by 2016 https://oversightboard.pr.gov/debt/ (see Chapter 9)

economic and cultural relationships between the Hispanophone, Anglophone, and Francophone Caribbean. While many Caribbean islands are independent nation-states, some English-speaking Caribbean islands such as Corn Island (Nicaragua) and San Andrés and Providence (Colombia) belong to Spanish-speaking Latin American nation-states. The inhabitants of the Caribbean, both islands and mainland coastal regions, often inhabit diverse modes of regional identification, which for many would certainly include Latin America as well as other regional identifications (the Creole Nation, the Black Atlantic) and some have stronger ties to Britain or France than they do to Spain or Portugal. French Guiana and Puerto Rico do not (yet) have independent status. French Guiana is an overseas region of France, and while it belongs to the European Union and uses the Euro as currency, it is economically integrated in South America. Puerto Rico is a Spanish-speaking unincorporated territory of the United States and has a strong political movement pushing for independence. Puerto Ricans are also US citizens and nearly six million Puerto Ricans live in the United States, many more than live on the island.

DIVERSE GEOGRAPHIES

Latin America covers a large geographic area from the US–Mexican border in the north to the southern shores of South America and as noted above includes a number of Caribbean islands. It is composed of at least 20 nation-states, a land mass of 21 million square kilometres, and a population of almost 600 million people. Some 50 million of these identify as Indigenous and 150 million as Black or Afrodescendant. It is frequently broken up into sub-regions, such as North America, Central America, South America, and the Caribbean (see Figure 1.4).

As Argentinian theorist, Néstor García Canclini (2002) writes from his adoptive home in Mexico City, talking of Latin America as a whole is not easy. As he points out, in any attempt to compare Argentina and Mexico, the divergences often overshadow the similarities. It seems like a cliché to say that Latin America is a continent of contrasts. But there is no doubt that concep-tualizing such a large and diverse region as a single entity is a difficult task. The continent as a whole confounds any straightforward generalizations about the region's development, given the stark and intimate ways that affluence and poverty and modernity and tradition are juxtaposed. A few minutes spent poring over the development indicators in Table 1.1 reveal as many differences and divergences between Latin American countries as similarities. There are also significant economic and cultural differences within countries, regions, and neighbourhoods. All Latin American cities as well as its rural areas are characterized by dramatic socio-cultural heterogeneity.

Most Latin Americans speak Spanish, but Portuguese is also spoken in Brazil, the largest country in Latin America. Creole English is spoken in Belize,

Figure 1.4 Map of Latin America

Source: Marney Brosnan

parts of Nicaragua, Costa Rica, and Panama, and on the Colombian islands of San Andrés and Providence. There are also many Indigenous languages spoken, including Quechua and Aymara in the Andean region and more than 20 different Mayan languages in Guatemala. Mexico alone has more than 200 different languages.

Latin America is far more linguistically diverse than Europe. While Europe has only two language families and one isolate language (a language that is not related to other languages), Latin America has more than 50 language families and more than 70 isolates. While hundreds of different languages are still spoken and some Indigenous languages have more than a million speakers, others are seriously endangered.

As Latin America is composed of a large land mass, it has a diverse range of climates and environments. Much of the region is tropical but it also has some mid-latitude territories and a number of outstanding biophysical features, which include the world's highest settlement, airport, volcano, railroad, and highway, the world's second largest river, as well as the largest tropical rainforest in the world (González, 2011). It is believed that the Amazon rainforest produces around 40% of the world's oxygen supply (Wiarda and Kline, 2011). The range of biodiversity present in Latin America is staggering and includes crocodiles, jaguars, tapirs, spider monkeys, toucans, quetzals, turtles, boas, anacondas, hummingbirds, flamingos, pacas, llamas, and alpacas. But like many of the region's languages, some of these birds and mammals are endangered.

In Latin America, one can find snowy mountains, dry deserts, lush highlands, and fertile coastal lowlands. Alfonso González (2011) has identified 13 different natural regions in Latin America, all with their own unique characteristics and economic and agricultural activities. They include the Amazon basin, the Orinoco basin, the Antilles archipelago, the coastal lowlands of Mexico and Central America, a highland cordillera stretching from Mexico to Bolivia, the Brazilian highlands, the pampas of Northwest Argentina, and the Atacama desert region of Peru. These diverse landscapes have facilitated a range of economic and agricultural activities. Latin America has produced and produces maize, beans, rice, potatoes, rubber, coffee, sugar, cacao, wheat, bananas, citrus fruit, mangoes, flowers; it has mined gold, silver, copper, and tin; herders and ranchers have reared llamas, alpacas, sheep, goats, and cattle, while fishing communities fish for shrimp, lobster, and red snapper. In addition, Latin America's natural landscapes attract large numbers of tourists and visitors who travel to the region to experience its forests, mountains, and wildlife. Much of the continent lies on an earthquake fault line or in a hurricane belt and the combination of those biophysical characteristics with weak socio-economic development and preparedness means Latin America has also been home to some of the world's most devastating disasters. Much of Latin America is also extremely vulnerable to climate change and frequently experiences droughts and floods.

In most Latin American cities, you'll find skyscrapers, luxury shopping malls, and immaculately dressed business people along with informal urban exclusion, poverty, and social unrest. In the countryside, you'll find vast export-oriented estates concentrated in the hands of a wealthy landowner or large agribusiness, alongside subsistence farmers who grow crops for domestic consumption and often feed whole families and communities from small plots and landless campesinos who are forced by their landlessness to work as exploited plantation labour. Inequality, the difference between rich and poor, is as much a development issue as poverty. Attending to Latin American development means paying attention to wealth and affluence, a point which many development practitioners with their understandable focus on poverty tend to overlook. In all large Latin American cities there are attempts to manage inequality through forms of urban segregation, which include high security walls, gated communities, and armed guards at banks and restaurants frequented by the wealthy. Despite these attempts, the rich and poor intermingle constantly. At the traffic lights in Managua, the capital of Nicaragua, poor malnourished children clean the windscreens of Toyota landcruisers belonging to the city's wealthy residents in exchange for a few coins. The rich and poor are also brought into intimate proximity through the provision of labour services, as the poor inhabit the homes of rich on a daily basis to clean their homes, do their laundry, care for their children, and tend their gardens.

What Latin America, both as a geographical location and as an invention, shares is the history of colonialism, genocide, slavery, commodity trade, uneven land tenure, environmental destruction, US imperialism, postwar development, and neoliberal structural adjustment. Different parts of the region are shaped by and have responded to these historical and geographical processes in different ways, but they are all shaped by them in some way.

One of the key messages of this book is that colonialism, and its successor coloniality, was brutal and devastating but it was also challenged and resisted, both in high profile rebellions and in the spaces of everyday life. The conquest didn't completely wipe out native cultures but rather involved complex and creative forms of fusion and hybridization. Since the colonial era, Latin Americans have tactically and selectively blended Indigenous and African cultures with European ones, a process which continues in the present. Despite the undeniable loss of cultural memory and tradition afflicted on Indigenous and African heritages, these cultures continue to assert themselves in the development project, through media and popular culture, religious practices, political struggles, and everyday life. They continue to hybridize European-derived and imposed cultural forms and thus complicate what we mean by development. Even Catholicism, imposed on the native people by the conquistadors, was hybridized and therefore controlled (Rowe and Shelling, 1991) through blending with pre-Columbian and African religions. Contemporary devotion to Candomblé, Santería, and Santa Muerte (see Chapters 8 and 10) is testament to the cultural resilience of non-European religious practices and their survival over time and against the odds. In recent decades, through the liberation theology movement, Latin America has also used Catholicism to reflect on and challenge political injustice. When reflecting on hybridization, it is also important not to lose sight of the fact that Latin Americans, like Europeans, are also very receptive to US cultural influences and consume US cultural products such as music, Hollywood movies, fast food, and Coca-Cola with regularity. US cultural influences, as well as economic and political ones, for better or worse, are also part of the mix. Of course, the US continues to undergo a constant process of "Latinization" (see Chapter 5).

MODERNITY/COLONIALITY/DECOLONIALITY

This book is underpinned by a decolonial approach to development and it draws on insights from the Modernity/Coloniality/Decoloniality (MCD) paradigm as well as on other critical approaches to development studies informed by feminist, postcolonial, postdevelopment, and political economy approaches. It seeks to draw on and contribute to broader moves to decolonize the university (see Box 1.2). The MCD emerged from Latin American cultural studies and has centred the voices and perspectives of Indigenous and Afrodescendant intellectuals and activists and explores the ways in which Latin Americans engage in modes of border thinking as they draw on more than one episteme.

This book will help you to draw on this body of theory in your own work and provide you with a basis from which to develop your own more specialized academic interests. It will also provide you with a broad understanding of a large and diverse continent and the past and present struggles enacted in order to create a better and more dignified life. Engaging with critical theoretical scholarship of this kind can be profoundly intellectually stimulating. It might encourage you to rethink the training and education you have received thus far and help you to approach it and other dimensions of your life (the things you read, the media you consume, the work you do, the local and global politics that affect you and others) with a critical lens.

BOX 1.2 DECOLONIZING THE UNIVERSITY

Universities across the world in both the Global South and the Global North are dominated by Eurocentric thought. Course reading lists are often dominated by the same white Euro-American scholars, who are also more likely to be men, and many students go through their entire degrees without reading scholarship authored by Black and Indigenous intellectuals. Students are often led to believe that they are exposed to diverse perspectives and are encouraged to think critically but often these perspectives all belong to the same Eurocentric world view. Furthermore, Eurocentric thought, especially in the physical and natural sciences, but also sometimes in the humanities and social sciences, is often presented as objective and universal, even though all knowledges are localized and embodied. There is, as Robert Aman (2018: 175) writes, "nothing so ethnocentric, so particularist, as the claim of universality." This Eurocentric dominance is serious, especially for Indigenous students or students of colour who don't find their histories, cultures or heritage adequately represented in their degree programmes or institutions and "find themselves in spaces calibrated to maintain white supremacy"

(Ahmet, 2020: 678). Black students in Britain with family connections to the British empire or Afro-Latinx students in the United States are denied access to the knowledges that might help them to make sense of their own histories and of the forms of racial discrimination they have experienced growing up or on entering higher education. As Gurminder Bhambra et al. (2018: 817) write:

> We may all agree that we have a scholarly duty to introduce students to the broadest possible intellectual landscape, and yet, the exclusions reproduced epistemically in university teaching undermine the integrity of scholarship and the possibilities of academic work engaging a wider range of voices.

But these forms of epistemic dominance do a disservice to all students. All students stand to benefit from the democratization of knowledge that decolonized courses offer (Andrews, 2016). These exclusions constitute forms of epistemic violence and they are linked to the physical violence that results in the premature death of racialized populations, in the Mediterranean Sea and in Grenfell Tower in London as well as in the mines, sweatshops, and plantations of the Global South, and the uneven concern expressed towards these deaths. It is becoming apparent that Eurocentric knowledges alone are insufficient to tackle the challenges that afflict our world such as climate change, often because of subject specialization where those measuring and studying glacial retreat or declining forest cover in the context of climate change are not also analyzing the colonial-capitalist drivers of these changes. Their research projects are often driven by technoscientific concerns rather than by a politics of climate justice. This book is inspired by and seeks to contribute to global movements to decolonize the university. Across the world, through movements such as Rhodes Must Fall (Cape Town and Oxford), Why is my Curriculum White? and Why isn't my Professor Black? students and teachers have been drawing attention to the forms of epistemic exclusion that are reproduced in the westernized university. These movements have emphasized how the neoliberal logics that underpin the contemporary university – based on marketization and student indebtedness and where so-called diversity initiatives in the neoliberal university tend to gloss over racism and sexism rather than work to remove it (see Ahmed, 2012)– also work against the necessary decolonization of the university. It is important to recognize that decolonizing a curriculum is not just about adding more female, Black, or Indigenous authors to a reading list. While this is an important move and makes the reading list more diverse, it doesn't necessary contribute to decolonization. Decolonizing a curriculum means teaching from an-Other locus of enunciation and centring non-Eurocentric ways of knowing. Late Afro-Brazilian activist and intellectual, Beatriz Nascimento (2021a

[1988]: 306, see Chapter 8) describes the repulsion she developed towards Eurocentric scholarship as follows:

> The origin of this repulsion lies in a personal negation of the Western rationalist thought that for so long has been part of my intellectual training. After years of research, and as an expression of 20 years of activism, I have come to a radical rejection of everything that could seem European or erudite. This has accompanied a desire for a rupture with strictly scientific thought. This puts me in an ambivalent position: even as this thought fascinates me (I have been socialised in it, I cannot escape it), I reject it as being premised on colonisation.

It is important to note that not all Latin Americans, and not even all Black and Indigenous Latin Americans, are decolonial thinkers. Most of them are not. The Chicago Boys were Chilean academics based in Santiago who imported a Eurocentric free market theory into Chile after training in the United States (see Chapter 3). By the same token, not all white Europeans are colonial thinkers. For example, Boaventura de Sousa Santos is a Portuguese man who has made major contributions to decolonial thought. Eurocentric bodies of knowledge can be partially decolonized through modification. Marxism is a Eurocentric approach that has been decolonized by non-white intellectuals in many parts of the world, as we see in the work of Angela Davis, Claudia Jones, Cedric Robinson, CLR James, Walter Rodney, and many others. Decolonizing a curriculum does not mean the exclusion of all European and Eurocentric thinkers. Decolonial thought can also be used in combination with Eurocentric thought. I find Foucault and Gramsci very useful, as did Edward Said and Stuart Hall. But a decolonial approach means treating Foucault and Gramsci horizontally with Fausto Reinaga, Sylvia Rivera Cusicanqui, or Linda Tuhiwai Smith. And it also means trying to avoid colonial narratives in our classes and in our writing. For example, it is important to avoid the erasure of decolonial struggle and resistance and its role in making the world we live in today. For example, the British government did not abolish slavery: slavery came to an end because of the resistance and rebellion of the enslaved. But decolonizing does mean confronting head-on the epistemological and political consequences of Eurocentrism and as well as the modern-colonial-capitalist context in which the contemporary university operates and in which we study and work. It means confronting white privilege, class privilege, and male privilege that shape how we teach and how we do research and the role of many of our disciplines, including geography and anthropology, in the making of empire and conquest. It means avoiding "intellectual colonialism" that involves citing a white northern author while ignoring Latin American intellectual traditions that are better suited to

Latin American realities (Gudynas, 2015a). While we seek to engage with decolonizing initiatives, we must not forget as Eve Tuck and Wayne Yang (2012) have emphasized that decolonization is about the restoration of sovereignty and the return of stolen lands, *it is not a metaphor*. They write that "[w]hen metaphor invades decolonization, it kills the very possibility of decolonization; it recenters whiteness, it resettles theory, it extends innocence to the settler, it entertains a settler future" (p.3). Decolonizing work is long-term, and it is disruptive and unsettling.

Sources: Ahmed, 2012; Ahmet, 2020; Bhambra et al., 2018; Cupples and Grosfoguel, 2018; Gudynas, 2015a; Tuck and Yang, 2012

A key starting point is to focus on the close relationships between material conditions and the question of epistemology, namely, how we come to know what we know and how knowledge is produced. This focus provides you with an opportunity to interrogate your own knowledges, the things you already know about Latin America, about gender, environmental politics, or Indigenous peoples, and to also help you to reflect on what you bring to your studies in terms of education, biography and life experiences, and political motivations and learn from these (see Box 1.3). Your teachers, tutors, and classmates will have different starting points, intellectual concerns, and goals and you can learn from them too. I imagine that many of you are here because you are concerned about human suffering, human rights abuses, inequalities, environmental destruction, migration, and the hardening of borders, or racism, sexism, and homophobia. You might be trying to understand whether we need to dismantle capitalism or make it work better, the relationships between planetary destruction, racism, and capitalism, and the role of the state and the market in providing public services. In other words, you want to use your education to intervene positively in the world and so the politics of your actions cannot be evaded. You have to take sides and abandon Eurocentric pretensions of universality, neutrality, and objectivity. Decolonial work is explicitly anti-colonial, anti-racist, anti-capitalist, queer, and feminist in its orientation and it seeks to stand with, work with and learn from the subordinated and oppressed. It doesn't take a "balance sheet" approach to the question of empire or colonialism, rather it is explicitly and deliberately subversive and aimed at liberation, emancipation, and epistemic justice. It is also focused on putting knowledge to work in the ending of oppression and planetary destruction. It is therefore explicitly on the side of communities in struggle across the continent – those defending territory against environmentally destructive practices, those fighting against misogyny and femicide, those working to create economies of solidarity, those who care for the land and grow subsistence crops, and those who educate to give the next generation a decent change in life. Our focus should be on the places below the abyssal line that separates the Zone of Being from the Zone of Non-Being (de Sousa Santos, 2007) in order to affirm "the collective humanity of the dehumanized in decolonization struggles" (Maldonado-Torres, 2017: 124, see Chapter 11).

BOX 1.3 THE ROLE OF BIOGRAPHY

If you are not Latin American and are based outside of the continent, it is worth exploring your own motivations for studying Latin American development, as personal biographies and trajectories shape how we study or do research. Latin America has long captured the global imagination and people around the world have been inspired by its revolutionary struggles against imperialism, environmental destruction, foreign debt, and human rights abuses or by its folklore, literature, and popular culture, including salsa, football, telenovelas, and the novels of Isabel Allende or Gabriel García Márquez. Revolutionary leaders such as Che Guevara or Subcomandante Marcos or charismatic footballers such as Diego Maradona, Garrincha, or Pelé are global icons. So it is worth working out why we find ourselves drawn to these cultures and politics and why we might find ourselves in Bolivia, Mexico, or Nicaragua travelling, studying, researching or volunteering. In her autobiography, Nicaraguan writer Gioconda Belli (2001) describes Nicaragua as the country under her skin. After getting engaged in the solidarity movement in defence of the Nicaraguan revolution and visiting Nicaragua as a young woman in the early 1990s, Nicaragua too got under my skin. I had become interested in Nicaragua in the 1980s as an undergraduate student of Spanish, inspired by its revolutionary politics and the way the Sandinista Front for National Liberation (FSLN) had stood up to dictatorship and US imperialism and began to work on implementing models of healthcare, education and literacy, and agrarian reform that benefited the poor majority. I became a solidarity activist working with the Nicaragua Solidarity Campaign and the Central American human rights committees and campaigned to raise awareness of what was happening in Central America in the UK. My first few visits to Nicaragua were life-changing in so many ways. I participated in an environmental solidarity brigade in 1990 and had a passionate love affair with a Sandinista militant (Cupples, 2002). I was in Nicaragua in 1991 when I discovered I was unexpectedly pregnant. I went to a women's health clinic in Managua in search of a pregnancy test. I was first offered an abortion (even though abortion was and still is illegal in Nicaragua) and the women there only congratulated me when I confirmed that I wanted to continue with the pregnancy. I was in Santiago Atitlán, Guatemala in 1991 just after the villagers had thrown out the army, tired of the massacres and the intimidation, in Chiapas Mexico in 1993 just before the Zapatista rebellion and got to talk to Bishop Samuel Ruiz about the growing insurgency, and in 1997 I got to meet the Madres of the Plaza de Mayo in Argentina and observe their regular protest. In Nicaragua, I became good friends with many inspirational people who had fought in the revolution, as well as some that had been part of the counter-revolution. One of these was Sadie Rivas who had joined the revolutionary struggle at age 16 and two years later became part of the insurrección de los niños (the insurrection of the children) in Matagalpa in 1978 (see Castillo and Largaespada, 2010), so named because those participating were so young. According to Mónica Baltodano (2008), in taking over the army barracks in Matagalpa, Sadie "distinguished herself by her bravery. She was daring, bold. The men had no choice but to follow her in risky, near-suicidal attack operations." She also participated in a hunger strike against the dictatorship and then in the 1980s during the Contra war she lost her partner and father of her youngest child in a Contra ambush. Sadie was later expelled from the FSLN, as she was openly critical of the way that the verticalist and masculinist leadership had begun to betray their revolutionary principles, and began to work for an NGO focused on campesino empowerment, food security, and the rural economy. I was living with Sadie during my doctoral fieldwork in 1999, when she was tragically killed in a car accident on her way back to Matagalpa from Managua. During her wake I got a sense of the impact she had had on so many people and how widely she was loved. I didn't set out to become a disaster scholar, but Hurricane Mitch had hit my fieldsite in 1998 and took me into that literature and that politics. It was Hurricane Felix in 2007 that first took me to the Caribbean Coast of Nicaragua where I did some work with the Black Creole and Miskito survivors of the disaster and led me into a long-term research programme working with Black Creole and Miskito activists and broadcasters where I have gained a whole new perspective on the revolution and the workings of coloniality. More recently, my work with Black Creole Nicaraguans has taken me to Limón and Cahuita in Costa Rica, Bocas del Toro in Panama, the San Andrés archipelago, and mainland Colombia. Other activities have taken me to Chile, Mexico, and Ecuador and I got myself finally to Cuba in 2015. My scholarship with Latin America is entangled with my politics, my career, my life experiences, my relationships, and the emotions of love and loss. It is something that I feel as well as something that I theorize.

A decolonial approach also involves being properly attuned to the contemporary conjuncture. We live in a dangerous world characterized by growing inequalities, imminent ecological collapse, and societal fascism. Societal fascism is different from political fascism, because it "coexists easily with the democratic state" and it ejects whole populations from the social contract so that everyday life for many people on the planet becomes terrifyingly insecure and precarious (de Sousa Santos, 2014: 50). We are running out of time to save the planet from irreversible ecological collapse because in many places the reproduction of capital to produce more wealth for people who are already wealthy is more important than the reproduction of life. In Europe, the elite political class believes it is more politically expedient to let people drown in the Mediterranean than to save them and to resolve the forms of desperation that put them on the move. Similarly, many Mexicans and Central Americans die in the desert in an attempt to reach parts of the United States that used to be part of Mexico, fleeing conditions that result from US foreign and domestic policy. In 2020, Oxfam reported that by 2019 the world's 2,153 billionaires had more wealth than 4.6 billion people and that the seriously undertaxed 22 richest men in the world had more wealth than all of the women in Africa (Coffey et al., 2020). We could put an end to poverty and inequality in the world if we wanted but we choose not to do so, yet money is always found to fight wars, fortify, and militarize borders, and bail out bankers. We all need to develop the analytical capacities to do something about this state of affairs and also recognize that the struggle for the climate is entangled with the struggle for wealth redistribution and for gender, racial, and sexual liberation. So perhaps addressing the state of the world lies not in western Eurocentric modernity which has inflicted and continues to inflict colonial violence on people and the planet, but in a different set of intellectual traditions. We need then to turn to intellectuals and activists who write and speak from a decolonial locus of enunciation. This book requires you to engage with some of these in a way which will encourage you to reflect on your own subject positions, your privilege or relative lack of privilege, and your education. It constitutes an opportunity to learn about Latin American geographies and about development as a complex, contradictory, and contested concept, to learn from those actively engaged in defending life and territory, and to learn some theory that will help you to develop intellectually. In other words, it is a space from which to make sense of the world and your own life experiences, to develop your own politics, and determine your own interventions and modes of solidarity.

It is worth spending some time to think about why you are reading this book, what has motivated you personally, politically, or intellectually to study Latin American development and what you already bring to it and hope to gain from it. With this in mind, you should use your own biography as a resource to figure out these questions. The book contains many suggestions for further reading as well as films, documentaries, websites, and a playlist that will enhance your engagement and support your learning. There is a companion reader, the *Routledge Handbook of Latin American Development*, that will provide additional resources (Cupples et al., 2019b) and relevant chapters from the reader are suggested at the end of each chapter.

Please be warned that this book contains challenging material that deals with many human rights abuses in Latin America, including state-led massacres of Indigenous peoples, spectacular forms of misogyny and sexual violence, and the violent pollution of land and water by oil and other multinational companies. At the same time, it also seeks to be uplifting and provide a sense of hope, by focusing on inspirational collective movements to build a better world. It encourages reflection on how we might build solidarity with oppressed peoples and move beyond a world that excludes and destroys.

Chapter 2 is a decolonial history of Latin America to understand the world that was brought into being in 1492, the legacies of which are still with us today. Chapter 3 focuses primarily on the 20th and 21st centuries to understand why and how neoliberalism became the dominant economic model. Chapter 4 explores extractivism, a key colonial continuity, and discusses the courageous ways in which Latin Americans are standing up to multinational companies, often with tragic consequences. In Chapter 5, we explore the economic, political, and cultural relationships between Latin America and the rest of the world, including the United States and China. Chapter 6 discusses gender politics and dynamics through a decolonial lens, while Chapters 7 and 8 outline Indigenous and Afrodescendant politics and movements. Chapter 9 analyses the coloniality of disaster and how earthquakes and hurricanes are important sites of coloniality and decolonial resistance. In Chapter 10, we look at the Latin American city and analyse different forms of urban transformation and mobilization. We end with some concluding reflections on coloniality and decoloniality and how to work with the theoretical approaches covered in this book.

SUMMARY

- Development as we understand it today was brought into being after the Second World War. It constructed the so-called third world as a site of lack that required first world expertise and technology.
- It is clear that Europeans have as much or more to learn from Latin America than Latin Americans from Europe.
- Development is not a monolith and its meanings can be rearticulated.
- Latin America as a geocultural entity was created after the conquest of 1492.
- Exploring our biographies and personal motivations can help us to gain understanding.

DISCUSSION QUESTIONS

1 Why could it be considered problematic to say "America" when referring exclusively to the United States? How can we define Latin America? How do we decide which nations belong to Latin America?

2   Study the statistics in Table 1.1. Discuss the ways in which such development indicators are useful and ways in which they might be considered
limited or problematic.

3   How do colonial legacies shape contemporary Latin American development and cultural politics?

4   What do we mean when we say development was invented? What does
the invention of development have to do with geopolitics?

5   What is decolonization and why is it important in the contemporary university as well as in Latin America?

6   What are the basic principles that underpin a decolonial approach to
development? What might it mean to think from and with Latin America?

7   What do *you* bring to the study of development and decolonization in
Latin America? What motivates you to study this topic?

8   What do you think it might mean to engage with non–Eurocentric
knowledges and perspectives and what might your learning look like as
result? What are the implications of taking a political position that is on
the side of oppressed or subordinated people? How does that help you to
think through your own privilege or lack of privilege?

FURTHER READING

Routledge Handbook of Latin American Development Chapters 1 (Cupples et al.); 2 (Esquerra
Muelle); 5 (Ziai); 10 (Willis); 18 (Mato)

Latin America Readers. Duke University Press produces a series of country-specific (and
sometimes city-specific) Latin America readers (*The Colombia Reader, The Guatemala
Reader, The Peru Reader* etc) that are a useful source of geographically-specific material
and key debates.

Bhambra, G., Gebrial, D. and Nişancıoğlu, K. (eds) (2018). *Decolonising the University*.
London: Pluto Press.

Cupples, J. and Grosfoguel, R. (eds) (2018) *Unsettling Eurocentrism in the Westernized
University*. London: Routledge.

These two books are edited collections that seek to analyse the ways in which coloniality
is perpetuated in the contemporary university and the ways in which decolonial student movements and decolonial intellectual debates are seeking to forge a different
kind of university.

Escobar, A. (1995) *Encountering Development: The Making and Unmaking of the Third World*.
Princeton: Princeton University Press

Written by a US-based Colombian anthropologist, whose work has been central to Latin
American studies and development studies, this is an important book that reflects on
how development was brought into being in the postwar era and how it functions
discursively.

Galeano, E. (1973) *The Open Veins of Latin America: Five Centuries of the Pillage of a Continent*,
transl. C. Belfrage. New York: Monthly Review Press

Originally written in 1971 in Spanish by Uruguayan author, this book has become a classic and an absolute must-read for anybody interested in Latin American development and the consequences of the conquest.

Mignolo, W. D. (2005) *The Idea of Latin America*. Malden: Blackwell.
An intellectually stimulating critique of the idea of "Latin" America.

Useful websites

Global Social Theory
https://globalsocialtheory.org/
A tremendously useful pedagogical resource that provides concise and accessible summaries of many decolonial thinkers, topics, and concepts.

BBC Country Profiles
http://news.bbc.co.uk/2/hi/country_profiles/default.stm
The BBC has detailed profiles on all of the territories and countries in the world, with data on history, politics, economics, leaders, media and links to BBC audio and video archives.

World Bank DataBank
https://databank.worldbank.org/home
DataBank is an interactive analysis and visualisation tool that contains data and statistics on a range of development indicators.

Human Development Data Center
http://hdr.undp.org/en/data
Run by the United Nations Development Programme (UNDP), this website also contains data and statistics on a range of development indicators organized around the Human Development Index (HDI) and the Sustainable Development Goals

CHAPTER 2

A decolonial history of Latin America

INTRODUCTION

At its heart, a decolonial approach to Latin American development involves working with knowledges and worldviews that have been silenced and marginalized by Eurocentric dominance in knowledge production. It means taking the epistemic as well as the economic, political, cultural, and environmental consequences of colonialism into account and exploring how coloniality persists and continues to shape people's life opportunities, political processes, and development interventions today. For many of us trained in the westernized university, it means learning to speak from a different locus of enunciation, one that operates with relative exteriority from colonial and Eurocentric ways of knowing (Dussel, 1976; see also Grosfoguel, 2013). It means engaging with the geopolitics and body-politics of knowledge and adopting an attitude of epistemic disobedience (Dussel, 1976; Mignolo, 2011). It involves the interrogation and dismantling of binaries – first world/third world, north/south, nature/culture, man/woman, developed/underdeveloped, and mind/body. It means embracing pluriversalism and challenging anyone who claims their knowledge to be universal. Decolonial theory is focused on putting knowledge to work in support of Black and Indigenous sovereignty, autonomy, and self-determination; of campesino livelihoods and food security, and of gender and sexual liberation. While there are theoretical overlaps, a decolonial approach differs from a postcolonial one (see Box 2.5). This focus does not suggest that everything that takes place in Latin America today shaped by colonialism, but rather seeks to understand how contemporary oppressions, including racism, sexism, and homophobia, have colonial roots.

Of course, the region we now call Latin America did not begin with the arrival of Europeans. There is evidence of human settlement on the continent dating back many thousands of years and it is believed that the

DOI: 10.4324/9781003110453-2

first human migrants to settle on the continental land mass crossed the Bering land bridge some 20,000 years ago. Our concern here is, however, with the creation of Latin America which coincides with the creation of Europe. As Mignolo (2011: 66–67) argues "there would be no Europe without the discovery and conquest of America and the colonial matrix of power. That is why modernity/coloniality are two sides of the same coin. The colonial matrix is therefore a structure not only of management and control of the non-Euro-American world, but of the making of Europe itself and of defining the terms of the conversations in which the non-Euro-American world was brought in." 1492 thus ushered in a new kind of world, based on this colonial matrix of power (see Box 2.1). It made the modern-capitalist world system that is in place today possible and we are still living with its negative consequences. Formal colonialism might have come to an end with independence, but coloniality persists. Decolonial practice, activism, and research seek to both reveal and undo these negative consequences and to create a pluriverse, a world that according to the Zapatistas is "a world in which many worlds fit" (EZLN, 1996; see Chapter 7).

BOX 2.1 THE COLONIAL MATRIX OF POWER

Late Peruvian sociologist Aníbal Quijano coined the concept colonial matrix of power to refer to the ways that the conquest of America produced a global economic, political, and epistemic system in which the populations of the world are racially classified. The colonial matrix of power was established in America in the 15th century but now covers the entire planet. It works across four overlapping and mutually reinforcing registers of control that have been reified, or made to appear natural. These are the control of economy, the control of authority, the control of gender and sexuality, and the control of subjectivity and knowledge. These elements were enforced through various laws, such the *requerimiento*, through which Spain awarded itself the right to take control of Indigenous lands, and the *encomienda*, which gave European settlers the right to coerce native peoples to work on the lands they had stolen. The matrix was then a European invention created to justify and legitimize genocide, expropriation, and dispossession. As a result, coloniality shapes the division of labour and use of environmental resources, the power of state institutions and the military, the role played by the family and constructions of gender, and the education system. Furthermore, it has the effect of negating the knowledge production of Black and Indigenous peoples, campesinos, and women, and generates epistemic exclusion and violence. Coloniality is reproduced through this matrix and has survived the end of the formal structures of colonialism. Grosfoguel (2011) has extended Quijano's formulation to talk instead about a modern/colonial capitalist/patriarchal world-system underpinned by several entangled global hierarchies of power. These are an international division of labour, where coerced labour forms persist especially for racialized groups of people; a global racial hierarchy, which understands European peoples to be superior to non-European peoples; a global and binarized gender hierarchy that privileges men over women and European patriarchy over other forms of gender relations; a sexual hierarchy that privileges heterosexuality over homosexuality; a spiritual hierarchy that privileges Christianity over non-Western spiritualities; and an epistemic hierarchy that privileges Eurocentric knowledge over non-western and Indigenous knowledges. Mignolo argues that the colonial matrix of power is not something that can be observed from the outside and he encourages his students to "know your place in the colonial matrix of power, where you have been located and classified" because this will help you identify your political, ethical, and ideological options and make the decolonial turn in your studies, your work, and your activism (Mignolo and Hoffman, 2017). The colonial matrix of power is then a useful concept for understanding the forms of racial injustice and dehumanization that we see in the world today. Quijano and other decolonial scholars have, however, been subject to an important feminist and Indigenous critique by Maria Lugones and Silvia Rivera Cusicanqui for their insufficient understanding of the gendered, intersectional, and heteronormative dynamics that are part of the coloniality of power (see Chapter 6).

Sources: Asher, 2017; Grosfoguel, 2011; Lugones, 2007; Quijano, 2000, 2007; Mignolo, 2007; Mignolo and Hoffman, 2017; Rivera Cusicanqui, 2012

Decolonial theorists have made the radical assertation that modernity began in the 15th century – it did not begin with the European Enlightenment in the 17th century – and that there is no modernity without coloniality. If we are going to make such an assertion, we need to interrogate dominant colonial histories of Latin America. If we want to understand in decolonial terms the persistence of coloniality, the ongoing oppression and exploitation of racialized groups, and historical and contemporary forms of decolonial resistance, we need to understand its historical origins. We need, in other words, a decolonial history of Latin America. So let us start in the Iberian Peninsula and the territory once known as Al-Andalus. The following section draws extensively on the work of Ramón Grosfoguel and especially Grosfoguel (2013).

THE CONQUEST OF AL-ANDALUS AND THE CONQUEST OF AMERICA

The dominant history that is taught to Spanish schoolchildren is that in 711 Spain was invaded by the Moors – Berber Arabs from the North of Africa – and that this invasion was followed by a Christian Reconquest, a *reconquista*. It was a long-term process – the Moors were on the Iberian Peninsula for 700 years – but were gradually pushed back by Christian forces. The Muslim territory known as Al-Andalus was still quite large in 1260 but by 1400 had been reduced to the Kingdom of Granada. It is, however, a nonsense to talk about a Reconquest, as Spain didn't exist in 8th century. The Iberian Peninsula was subject to waves of invasions from different people including the Visigoths and Vandals and there was no unified nation-state that needed to be reconquered. Indeed, nation-states didn't exist at that time. What is now Spain was composed of a series of Christian kingdoms and a Muslim sultanate in which there was also a large Jewish population. It makes more sense to talk in terms of a conquest rather than a reconquest.

In the 14th and 15th centuries, Al-Andalus was a site of great learning and intellectual activity, unlike much of the rest of Europe that was mired in obscurantism and religious tyranny. At that time, the libraries of Al-Andalus were large compared with libraries in Christian Europe. It is estimated that the library of Córdoba had half a million books, that of Granada 250,000. In Christian Europe at that time no library possessed more than 1,000 books. Al-Andalus was a Muslim territory but one that was home to multiple identities and spiritualities. In Al-Andalus, Muslims, Jews, and Christians co-existed peacefully and there was no religious persecution.

The conquest of Al-Andalus was driven by the imperialist, nationalist, and religious ambitions of the Catholic monarchs, Fernando (1452–1516) and Isabel (1451–1504), who had already unified their kingdoms (Aragón and Castille) and wished to spread their rule into Al-Andalus, in order to create a large, unified territory with one identity and one religion, thus forging the beginnings of the European nation-state. The Catholic monarchs met with Genoan navigator Christopher Columbus who, given the intense imperial competition over trading routes between European powers, was determined to find an alternative route to Asia, primarily India, China, and Japan, which would strengthen Spain's position within Europe. They were highly in favour of this proposal and they codified their agreement in the Capitulations of Santa Fe but wanted to complete the conquest of Al-Andalus first before embarking on a new conquest.

A few days after Granada fell to the Christians on 2 January 1492, the Catholic monarchs gave Columbus authorization to set sail. The conquest of Al-Andalus was devastating for Jews and Muslims who found themselves subject to surveillance, forced conversion, and repression. Many were killed or expelled, others left for territories in Northern Africa where they could practise their religions without persecution. While the methods used were extremely cruel and driven by religious intolerance, the humanity of Jews and Muslims was not at this point in question.

Columbus arrived in the Americas on 12 October 1492. Despite his skills as a navigator, Columbus thought he was in India and this is why this region of the world is referred to as the West Indies and the Indigenous peoples as Indians.[1] After meeting the native inhabitants of the island, Columbus returned to his ship and wrote in his diary that these were people without religion. He didn't mean that they were atheists, but rather that they were people without a soul. The implications of Columbus' hasty judgement were profound. If the native inhabitants encountered on these voyages did not have souls, then their enslavement could be justified. Thus began the brutality of the conquest, in which the Indigenous peoples were subject to genocide, torture, disease, forced labour, and evangelization. It involved the theft of Indigenous land and resources, especially minerals such as gold and silver. Deadly pandemics imported from Europe including influenza, smallpox, and measles were commonplace (Lovell, 1992; see Chapter 9). It is important to remember at this point much of Christian Europe was an intellectual and scientific backwater, while the Indigenous peoples of the Americas had advanced astronomy, engineering, mathematics, surgery, medicine, agricultural technologies, and transport. Indigenous labour shortages led to the establishment of the transatlantic slave trade to replace the Indigenous peoples who had succumbed to ill treatment, overwork, and disease with African slaves. The conquest of Al-Andalus and the conquest of America were part of the same process, each with implications for the other. And the result was racism, the belief that some people are inferior to others or that they are not human at all.

Not all of the colonizers were comfortable with the atrocities committed. Spanish Dominican friar and landowner and first bishop of Chiapas, Fray Bartolomé de las Casas, who was there to convert the natives to Catholicism, spoke out against the cruelty he had witnessed (see Box 2.2). He described the journey of one European colonizer from the Kingdom of Venezuela to Peru as follows:

> On this dreadful journey he took with him a vast number of native bearers, shackled together and each weighed down by a load of three or four *arrobas*[2]. Whenever one of these poor wretches fainted from hunger or became too exhausted to carry on, they cut his head from his body at the point where the iron collar bound him to his companions, so as not to have to waste time unshackling him from the bearers to either side of him; his head would fall to one side and his decapitated body to the other, and his load would then be distributed among his fellows to their already heavy burdens. To set out in detail the vast

tracts of land he ravaged, the towns and villages he burned to the ground (for all the houses are built of thatch), the people he slaughtered and the atrocities he committed *en route*, in place after place, would be to produce an account that nobody would believe, only too terribly too though it would be. Other adventurers from Venezuela and Santa Marta were to pass that way on their holy Pilgrimage to the Golden shrine of Peru, and even they, seasoned campaigners that they were, were shocked and appalled by the grim trail of destruction blazed by this butcher over a distance of two hundred leagues or more, turning what had been a fertile and populated area into one vast scorched wasteland.

(de las Casas, 1992[1552]: 99–100)

BOX 2.2 THE VALLADOLID DEBATE

The debate over whether native Americans had a soul raged throughout the 16th century. If they did not have a soul, then it would not be a sin in the eyes of God to enslave them and deny them their political autonomy. In 1537, the Pope intervened and suggested that native Americans did have a soul but that it was an animal soul, which left the whole issue in a problematic limbo. In 1550, the King of Spain convened a debate in the Spanish city of Valladolid, between Bartolomé de las Casas and philosopher and theologian Juan Ginés de Sepúlveda. The debate also sought to pose the question of whether the Indians of the Americas were capable of self-government and what right the Europeans had to colonize and subjugate them. For Sepúlveda (1941[1547]), who defended colonization and slavery, it would "always be just and in conformity with natural law that such peoples be subjected to the empire of princes and nations that are more cultured and humane, so that by their virtues and the prudence of their laws, they abandon barbarism and are subdued by a more humane life and the cult of virtue" (cited in Blanco and Teixeira Delgado, 2019), while de las Casas believed the Indians did have souls but were at a backward stage of human development and that they needed to be peacefully converted to Christianity and assimilated. The Valladolid debate was fundamental with profound and long-term consequences. As Enrique Dussel (2014: 27–28) notes:

This was a moment at which Modernity could have changed its course. It failed to do so and its route was inflexibly fixed until the 21st century. The astonishment of the conquistadors that their every action was unjust and morally-lacking was such that they could not believe it. The discussion was lengthy. The Dominicans had the philosophical arguments; the colonizers their unjust and tyrannical habits. In the end, the latter prevailed permanently, and it was on their basis that Modern European Philosophy was established. From the 17th century on the right of the modern Europeans (and North Americans of the 20th century) to conquer the Planet would never again be discussed.

While Sepúlveda was expressing a biological racism in defence of atrocities, de las Casas was developing a cultural racism. De las Casas' views would also persist for centuries and have even more powerful and long-term consequences. During the European Enlightenment, when epistemic authority was passed from the Church to science, the two discourses became secularized. Sepúlveda's discourse was secularized in the natural sciences and emerged as the scientific racism that legitimized later colonial endeavours in Africa and Asia. While scientific racism is now widely discredited, it did not, as Angela Saini (2019) writes, completely go away and continues to be present in Eurocentric natural sciences. De las Casas' discourse was secularized in the social sciences and went onto to underpin dominant thinking in 20th century geography, anthropology, and development studies. The idea of barbaric peoples that required evangelization morphed over time into the idea of primitive peoples that required first civilization and then development. While the Indigenous peoples of the Americas were the first group to be racialized by Europeans, to be denied their humanity, the effect spread not only to the millions of Africans who would be kidnapped and transported to the Americas to work as slaves but also to Muslims and Jews and as a result formed the basis of the forms of Islamophobia and anti-Semitism that characterize global geopolitics today. As Grosfoguel (2013: 84) writes, the "medieval religious discriminatory discourses against Muslims and Jews were transformed into racist discrimination. The question was no longer whether the religiously discriminated population have the wrong God or wrong theology" but rather whether they could be considered human at all.

Sources: Grosfoguel, 2013; Dussel, 2014; Blanco and Teixeira Delgado, 2019

As the transatlantic slave trade intensified, the concern that some colonizers such as de las Casas expressed for the lives and well-being of Indigenous peoples did not extend to Black Africans. Some 12 million African slaves were transported to the Americas, 5 million of these to Brazil, and many did not survive the violence of the Middle Passage, the several month-long journey from Africa to America. Grosfoguel (2013) has shown how the entangled politics of the two conquests led to four genocides – the Muslims and Jews of Al-Andalus, the Indigenous peoples of the Americas, enslaved Africans, and

European witches. All of these genocides were accompanied by epistemicide, the destruction of knowledge, often through the use of fire. The Christians in Spain burned the great libraries of Córdoba and Granada, just as the Mayan codices, books written by the precolonial Maya, were destroyed by Spanish conquistadors and priests. Diego de Landa (1978 [1937/1566]: 82), who was responsible for the destruction of Mayan codices in the Yucatan, wrote:

> These people also used certain characters or letters, with which they wrote in their books about the antiquities and their sciences; with these, and with figures, and certain signs in the figures, they understood their matters, made them known, and taught them. We found a great number of books in these letters, and since they contained nothing but superstitions and falsehoods of the devil we burned them all, which they took most grievously, and which gave them great pain.

Only four Mayan codices survived the conquest. The knowledge of European witches was oral rather than written, hence the need to burn the witches themselves, who had called into question dominant ways of knowing. The destruction of the witches underscores that there were (and still are) European knowledges that are not Eurocentric.

Grosfoguel (2013) notes that the four genocides/epistemicides of the 15th and 16th centuries were followed by an important geopolitical and philosophical shift in the 17th century. At this moment, hegemony moves from Iberia to North West Europe, especially France, England, and Holland. René Descartes with his "I think therefore I am" formulation poses a profound challenge to Christian authority, as western man now replaces God as the source of knowledge. This proposal constitutes the beginning of modern scientific knowledge based on a separation of mind and body – the Cartesian dualism – and the assumption of a view from nowhere. This move also leads to the subalternization of southern Europe; the forms of racial classification invented by the Spanish and Portuguese in order to dispossess Indigenous Americans were now applied to them by northern Europeans.

Drawing on Enrique Dussel, Grosfoguel (2013) argues that such a position was only made possible as a result of the conquest. The *ego cogito* "I think therefore I am" was preceded by the *ego conquiro* "I conquer therefore I am," and these are connected by the logic of genocide/epistemicide, an *ego extermino* "I exterminate therefore I am." This formulation is still the basis of modern scientific thinking in the westernized university today that constructs itself as objective, rational, and universal and dismisses Other knowledges as local, particularistic, and inferior, often reducing them to superstition or sorcery. Many scientists today still claim that their knowledge is universal and objective and deny that it is rooted in social relations. Any decolonization of science needs to reckon with this past and its profound consequences for contemporary knowledge production. As Frantz Fanon (2004[1963]: 37) put it: "For the colonized subject, objectivity is always directed against him" (cited in Maldonado-Torres, 2016: 9).

After Columbus, many Spaniards travelled to Latin America in search of wealth and fame, some of them inspired by mythical tales, such as that of El Dorado, which promised great prosperity should it be found. The colonization of the continent was at times patchy and erratic. Many areas were barely settled and of little interest to colonizers because of a lack of mineral resources, while others were subject to destruction and dramatic transformation. In some places, the Indigenous populations were wiped out, in others they continued to survive and maintain their traditional ways of life. One of the most well-known conquistadors was Hernán Cortés, who moved into the great Aztec capital of Tenochtitlán, the site of present-day Mexico City and the seat of a vast empire ruled by the emperor, Moctezuma. Tenochtitlán was an impressive feat of civil engineering and urban planning. Built on a swamp, it possessed awe-inspiring temples, ceremonial buildings, gardens, markets, sports arenas, zoos, and sacrificial platforms, as well as efficient systems of water and sanitation. It had wide streets, avenues and a series of interconnecting canals adorned with *chinampas,* floating gardens in which crops and flowers were grown. Cortés and his men were overwhelmed by Tenochtitlán's beauty and magnificence. Cortés described it as the most beautiful city in the world and then he ordered his men to lay the entire city to waste. In one of his letters to Carlos V, Cortés (2018[1521]: 17) wrote, that "their fashion of living was almost the same as in Spain, with just as much harmony and order; and considering that these people were barbarous, so cut off from the knowledge of God, and other civilized peoples, it is admirable to see to what they attained in every respect." During the conquest of Mexico, a local woman, known variously as Marina, Malintzin or Malinche, became Cortés' mistress and interpreter and gave birth to the first Mexican (see Chapter 6). Cortés could not have achieved what he did without her.

COLONIAL ERA RESISTANCE, REBELLION, AND DECOLONIAL THOUGHT

It is important to recognize that the conquest was resisted right from the start. The native peoples never consented to the theft of their lands and the destruction of their temples, Africans never consented to their forced displacement and enslavement, and so the colonial period was characterized by multiple forms of resistance and rebellion. This is central to a decolonial reading of history. All of these rebellions contain and arise from different ways of knowing the world (see Box 2.3).

BOX 2.3 18TH-CENTURY ANTI-COLONIAL REBELLION IN THE ANDES

The colonial era was punctuated by constant Black and Indigenous uprisings and political struggle against European rule. In the Andean region in 1780–1781, in what is now Peru and Bolivia, Tupac Amaru II, Tupac Katari, and Tomás Katari courageously mobilized against their Spanish rulers and led anti-colonial revolutions in Cuzco, Northern Potosí, and Oruro and La Paz respectively. These rebellions challenged multiple manifestations of colonial power, including forced labour, government corruption, trade restrictions, and taxes. In the region around Cuzco, Tupac Amaru II and his mostly Quechua and Aymara troops were able to take control of a large region, while Tupac Katari was able to keep the Spanish forces from La Paz for several months. These initiatives were also supported by some *criollos*[3] and *mestizos*. While colonial elites managed to defend themselves in the large cities, smaller towns and rural areas were emptied of Spanish control. For two years, there was a serious possibility that colonial rule would be brought to an end. When Tupac Katari was eventually captured by Spanish forces, they responded with spectacular brutality. He was drawn and quartered and his severed limbs were publicly displayed to discourage further insubordination. He died promising to "return tomorrow as thousands." Tupac Amaru was tortured and sentenced to death and he was forced to watch the execution of his wife and son before being quartered and beheaded. The rebellions were characterized and indeed fragmented by the presence of different worldviews. As Rivera Cusicanqui (1991) writes, some of the leaders had internalized and circulated political concepts from the colonial world that made little sense to most native peoples. Reforms followed, but these were designed to calm tensions among the colonized while preserving the colonial order. When independence came a few decades later, many of the radical elements present in 1780–1781 were suppressed. There is no doubt, however, that these movements continue to shape revolutionary politics and political subjectivities in the Andes today. The turn-of-the-century struggles in El Alto and elsewhere drew inspiration from these rebellions (see Chapters 4 and 10) and when Evo Morales came to power in 2006, he framed his own struggle as one that builds on this anti-colonial past. In the current period, Andean rebellions persist and partially succeed, because of the survival of the collective forms of Indigenous governance, such as the *ayllu*. Veteran Aymara activist, organic intellectual, and founder of the Movimiento Indígena Túpac Katari, Felipe Quispe (1942–2021) once wrote:

> Tupac Katari is alive and will come back ... damn it (Tupac Katari vive y vuelve ...carajo.) The fire of truth of the oppressed and exploited will make this new Sodom and Gomorrah that is capitalist society cry and howl.
>
> (cited in Zibechi, 2021)
>
> Sources: Rivera Cusicanqui, 1991; Hylton and Thomson, 2007; Zibechi, 2021

Constantly, at the same time, even though Eurocentric thought was becoming hegemonic and was used to legitimize brutality, there were many Indigenous intellectuals and activists who were thinking and writing from the perspective of the colonized and producing their own decolonial scholarship. One especially important example of this from the colonial era is the work of Felipe Guamán Poma de Ayala, a native nobleman from the Inca empire who was born just after the Spanish conquest of Peru. He is the author of a famous handwritten manuscript *The First New Chronicle and Good Government* written around 1615 as an illustrated 1,200-page letter to the then King of Spain, Philip III. It appears that he worked for the colonial government as an interpreter and spoke Aymara, Quechua, Spanish, and Catalan. The letter describes Inca culture and society and the impacts of the conquest. It draws attention to the ways in which Peruvians were treated by the colonizers, the serious threats to Andean culture that colonization entailed, and how the colonizers worshipped gold and silver.

> Your Majesty, in your great goodness you have always charged your viceroys and prelates, when they came to Peru, to look after our Indians and show favor to them, but once they disembark from their ships and set foot on land, they forget your commands and turn against us. Our ancient idolatry and heresy was due only to ignorance of the true path. Our Indians, who may have been barbarous but were still good creatures, wept for their idols when these were broken up at the time of the conquest. But it is the Christians who still adore property, gold, and silver as their idols.
>
> (Guamán Poma de Ayala, 2005: 80–81)

The text reveals the author's ability to work with Christian European *and* Indigenous Andean literary and artistic traditions and constitutes a direct appeal to the king to put an end to colonial atrocities (Mignolo, 2011). It is not likely that the letter ever reached the king. It disappeared for nearly 300 years and was found in 1908 in the basement of the Royal Library of Denmark and was first published in Paris in 1936. Nobody knows for sure how it came into Danish possession. Since then, it has attracted much scholarly attention and as result of the drawings contained within and the manner in which it was compiled is considered by many to be both a work of art as well as a detailed Inca history and important critique of colonialism (see Guamán Poma de Ayala (2005[1978]); Adorno, 2000; Nash, 2012). Similarly, in the 18th century, Ottobah Cugoano was kidnapped in Ghana at the age of 13 and shipped to Grenada where he worked on a plantation. He was later purchased by a British slaveowner, Alexander Campbell, and given his freedom. In 1787, he wrote a book *Thoughts and Sentiments on the Evil and Wicked Traffic of the Slavery and Commerce of the Human Species* calling for the abolition of slavery, a book that the cover describes as "humbly submitted to the inhabitants of Great Britain." Both Guamán Poma de Ayala and Cugoano appropriated Christianity and used its discourses "to unveil the inhumanity of European ideals, visions, and self-fulfilling prophecies" (Mignolo, 2011: 134). They both think with and from different epistemic positions.

While high-profile rebellions and important intellectual productions were common during the colonial era, it is important to recognize the multiplicity of ways that colonialism and slavery were accommodated and resisted. Indigenous and Afrodescendant peoples also engaged in a range of practices of creative subversion, appropriation, hybridization, mimicry, and masquerade (Rowe and Shelling, 1991; Gates Jr., 2011; Hall and Schwarz, 2017). In Brazil, slaves developed capoeira, which is part dance and part martial arts training, so that they could train to defend themselves from torture and brutality without the slave owners realizing that was what they were doing (Gates Jr., 2011). Michel de Certeau (1984: 31–32) captures the creative subversion that colonialism engendered:

> The spectacular victory of Spanish colonization over the indigenous cultures was diverted from its intended ends by the uses made of it; even when they were subjected, indeed even when they accepted their subjection, the Indians often used the laws, practices, and representations that were imposed upon them by force or fascination to ends other than those of their conquerors; they made something else out of them; they subverted them from within – not by rejecting them or transforming them (although that occurred as well), but by many different ways of using them in the service of rules, customs or convictions foreign to the colonization they could not escape. They metamorphized the dominant order; they made it function in another register.

As Stuart Hall and Bill Schwarz (2017: 214–215) note with respect to the successful appropriation and creolization of cricket by the colonized in the Anglophone Caribbean, slavery was challenged through the "centuries-long practice of colonial mimicry and masquerade" an attitude characterized by a "*carnivalesque* unravelling of the master narrative from within."

INDEPENDENCE AND THE HAITIAN REVOLUTION

The colonial occupation of Latin America lasted 300 years and was characterized by ongoing Indigenous and African slave resistance to colonial rule. In addition, the Spanish population was increasingly a native-born one, and these Spaniards became known as *criollos* (creoles), in distinction from the *peninsulares*, or the settler population. Born in the New World, *criollos* had a different relationship to Spain and grew to resent the imposition of laws and taxes from afar. They felt such control was hindering their ability to trade and develop. At times, this resentment resulted in open insurrection and military conflict such as the Revolt of the Comuneros in Paraguay in 1721–1735 and in New Granada in 1780–1781. Such demographic changes and the concomitant transformation of attitudes contributed to the gradual unravelling of colonialism and the intensification of the demand for independence.

The desire for independence was further fuelled by the Napoleonic occupation of both Spain and Portugal, a move which c⁾used the Portuguese monarchy to relocate temporarily to Brazil and which undermined the allegiance felt by *criollos* in Spanish America towards Spanish colonial rule. The political crisis and power vacuums created in Iberia by the Napoleonic invasion led to the creation of juntas in Latin America, made up of individuals who wished to maintain colonial rule as well as those seeking independence. The Napoleonic invasion of the Iberian Peninsula also provoked fears that France would gain control of Spain's colonies in the Americas. The loss of political power was augmented by economic developments as Spain progressively lost its monopoly on trade with the Americas. It was struggling to provide adequate levels of supplies to the colonies and was coming under constant challenge from the British, French, and Dutch, who all had colonizing and trading interests in the region and had established colonies in North America, Latin America, and the Caribbean. In 1806, Britain invaded Buenos Aires, and in 1814 Guyana was taken from the Dutch by the British. In the absence of direct occupation and colonization, these nations also developed investment and trade linkages with the continent. By the 1800s, these challenges were joined by those from a newly independent United States.

The early 1800s saw the emergence of a number of independence leaders who waged bloody battles against Spanish troops. The most famous of these was Simón Bolívar, others include Miguel Hidalgo, Francisco de Miranda, Manuel Belgrano, José de San Martín, and Bernardo O'Higgins. Miguel Hidalgo, a *criollo* who had plotted against the Spanish, is credited with beginning the move for independence in New Spain with his Grito de Dolores, a cry from the town of Dolores. Most of the independence fighters were men but some women joined the struggle, such as Manuela Sáenz and Policarpa Salvarrieta (see Box 2.4).

BOX 2.4 SIMÓN BOLÍVAR AND MANUELA SÁENZ: THE FATHER OF SOUTH AMERICAN LIBERATION AND THE LIBERATOR OF THE LIBERATOR

Simón Bolívar (1783–1830) was born in Caracas, the descendant of Spanish aristocrats. The Bolívar family amassed a fortune in the New World in silver and copper mining and sugar plantations. Bolívar, who received part of his education in Spain, dreamed of and fought for Latin American independence and unity. In his famous Letter to Jamaica, he posits the idea of America not as a set of distinct nation-states but as a unitary whole. After a number of battles against the royalists, Bolívar became the president of Gran Colombia, one of the first independent republics of Latin America, composed of the contemporary nations of Colombia, Venezuela, Ecuador, and Panama. Along with José de San Martín, the liberator of Argentina, he also secured the independence of Peru. He also helped to liberate Bolivia, and as a consequence the new nation was named after him. Bolívar was ruthless in his drive for liberation and urged for a "war to the death" which would involve the killing of *peninsulares* or those born in Spain. Bolívar's struggle, while undoubtedly heroic and inspiring, suffered many setbacks. In many places, royalist control was re-established after independence was declared and Bolívar would have to retreat before launching a new set of attacks. Political instability and confusion were part of his presidency. Although he is regarded by many Latin Americans as a nationalist hero, he died a bitter man. He resigned his presidency and was planning to go into permanent exile in Europe but died of tuberculosis at the age of 47. There are statues of Bolívar all over Latin America (see Figure 2.1)

While Simón Bolívar is well known and his struggle is well documented, like that of other male liberators, we should also acknowledge the important role played by women in the struggle for independence. For example, Manuela Sáenz (see Figure 2.2), born in Quito in 1795 to an upper-class *criollo* family, is often excluded from historical accounts. When she is included, historians reduce her role to that of being Bolívar's lover and many accounts focus on her lack of feminine qualities (Murray, 2001). Yet she was also an important fighter and military strategist who disrupted many of the gender norms of the time. Together with Rosita Campuzano, she risked her life passing information about the advance of José de San Martín in Argentina to forces in Peru and Bolivia. She was a key participant in the Battle of Pichincha that secured the liberation of Ecuador in 1822 and in the Battle of Ayacucho in 1824–1825 that secured that of Peru and indeed the rest of South America. In 1828, she saved Bolívar's life from attempted assassination in the presidential palace in Bogotá. After his death in 1830, she was forced into exile from Colombia and Ecuador. She went to Jamaica for a short period

Figure 2.1 Statue of Simón Bolívar in Havana, Cuba

Source: Photo by author

but spent the remaining years of life in a small village in northern Peru and she died in 1856 during a yellow fever epidemic. She remained politically active through prolific correspondence with important political leaders across the region. Through these letters, as Sarah Chambers (2001: 256) writes, "she defined a discursive space within civil society similar to the salon. Her

Figure 2.2 Portrait of Manuela Sáenz

Source: Marcos Salas, in the public domain, via Wikimedia Commons

visits and letters also created both a practice and a discourse of politics based upon friendship." The crucial role she and other women played in the liberation of South America is increasingly recognized, especially in more recent years by supporters of the Bolivarian revolution.

Sources: Chambers, 2001; Murray, 2001; teleSUR, 2020

When thinking about the 18th and 19th century movements for independence, it is essential to pay close attention to the earth-shattering events that took place in Haiti, that the colonial matrix of power seeks to marginalize and erase from the history books. In discussions about slavery and colonialism, European powers such as Britain and France often like to stress the key role they played in the abolition of slavery. You've probably heard some of these debates in the media quite recently. Dominant colonial discourses – that still circulate today – emphasize that slavery was indeed wrong but was brought to an end by courageous abolitionist leaders such as William Wilberforce. Some people also continue to assert that colonialism brought all kinds of benefits to colonized places and people. For example, A YouGov survey conducted in the UK in 2019 showed that more than a third of British people believed the British empire was something to be proud of and that former colonies were better off as a result of having been colonized (Booth, 2020). Such narratives and beliefs are deeply racist and politically harmful as they seek not only to erase the brutalities and violence exacted by colonial systems and its multiple political, economic, and cultural legacies that still constrain people's opportunities today, but also to suggest that abolition was benevolently gifted by the colonizer to the colonized. They have also had the effect of erasing the most important historical event in the abolition of slavery. It was the rebellion of Haitians that led to the abolition of slavery and led to the creation of the first Black republic in the world. As Julia Gaffield (2020) writes:

> Neither the French nor the British were the first to abolish slavery. That honor instead goes to Haiti, the first nation to permanently ban slavery and the slave trade from the first day of its existence. The bold acts of Haitians to overthrow slavery and colonialism reverberated around the world, forcing slaveholding nations like Britain and France to come face to face with the contradictions of their own "enlightenment." Many would now like to forget this reckoning.

For Michel Trouillot (1995: 73), the Haitian revolution has never made its way into the history books because "it entered history with the peculiar characteristic of being unthinkable even as it happened." In other words, because the Haitian rebellion challenges colonial historical narratives, it simply gets ignored, an exclusion which is beginning to be rectified by some important recent scholarship (for an excellent coverage of the literature that now seeks to rectify this silence, see Salt, 2019).

Haiti, formerly known as Saint Domingue, was an extremely wealthy and profitable sugar- and coffee-exporting French colony. Slaveowners were brutal and many of them worked their slaves to death and then replaced them quickly (Salt, 2019). In 1791, in a rebellion that would last for 13 years, slaves rose up against their masters and declared the abolition of slavery. In 1794, in the middle of the French Revolution and as a direct result of the insurrection, the French government confirmed the abolition of slavery in all French colonies. Between 1799 and 1804, however, Napoleon Bonaparte tried to restore colonial order and went on to restore slavery in other French colonies in the Caribbean. He failed, however, to do so in Haiti. In 1801, revolutionary leader and former slave Toussaint L'Ouverture (see Figure 2.3) published the new constitution that enshrined the abolition of slavery.

On 1 January 1802, the revolutionary forces led by Jean-Jacques Dessalines victoriously declared Haiti's independence. This event had significant political repercussions, accelerating the abolition of slavery elsewhere in the Americas. The British government didn't pass the Abolition Act until 1833 (just after the 1831–1832 slave rebellion in Jamaica) and continued to hold slaves in the Caribbean until 1838. Furthermore, the government went on to compensate slave owners for the loss of their "property" to the tune of £20 million (equivalent to £17 billion in today's money) and it took the British taxpayer until 2015 to pay off the money borrowed to do so (Draper 2010; Olusoga, 2018).

Figure 2.3 Bronze bust of Toussaint L'Ouverture by Haitian sculptor Ludovic Booz and gifted to the French city of Bordeaux in 2004, on the bicentenary of Haiti's independence

Source: Jefunky, CC BY-SA 4.0, via Wikimedia Commons

In other words, slavery ended because of the decolonial resistance of the enslaved, not because of the actions of benevolent and forward-thinking parliamentarians. It underscores that political change comes about when people act outside the law. The Haitian rebellion was hugely influential throughout the Americas. It decisively shaped the way in which struggles for independence were waged and the responses of European powers to these struggles. The newly independent Haitians went to establish important trading and diplomatic relations with the Atlantic world (Gaffield, 2015) and used "whatever means were at their disposal to resist closure to and consumption of their independence and power" (Salt, 2019: 15). Haitians have been repeatedly punished for their rebellion and have been subject to colonial and imperial interventions ever since. After the Haitians declared their independence, not only did France retaliate militarily, but they also only recognized Haiti's independence in return for a large debt burden. Haiti accepted this debt burden in order for its independence to be recognized and to avoid being invaded by French military forces. In other words, France demanded reparations from its former colony, and it took the Haitians 122 years to pay back this debt. In 1915, Haiti was occupied by the US marines and remained a US protectorate until 1934. During this period, the United States purchased Haiti's outstanding debt from France and so the final payment was made to the United States. The United States went on to support the brutal Duvalier dictatorship and the reign of terror led by Papa Doc and Baby Doc. The first democratically elected president, Jean-Bertrand Aristide, was deposed in a US-back coup in 2004. This coup came after Aristide said he was seeking reparations of $22 billion from France (Engler and Fenton, 2006). The earthquake in 2010 killed 230,000 people and left half a million homeless. When Haitians desperately needed food and shelter, the United States sent in troops to keep (colonial) order on the streets of Port-au-Prince (see Chapter 9). Since then, many Haitians have been forced to migrate to other countries in search of economic opportunities and end up in many different places, not only the United States and the Dominican Republic, but also Brazil, Colombia, and Chile (McIlwaine and Ryburn, 2019), where they continue to experience racism and discrimination.

More than two centuries later, Haitians are still fighting for their liberation. In the European and Eurocentric media, when Haiti is acknowledged, it frequently appears as a site of dysfunctional despair, disaster, suffering, and abject poverty, rather than as a site where abolitionist and other political struggles triumphed (Dubois, 2012). As Erin Durban-Albrecht (2017: 201) puts it:

> Since Haitian independence in 1804, European and US imperialists have worked methodically to undermine the success of the Black Republic as a way to punish Haitians and their enslaved ancestors who fought for and claimed their freedom from white colonial domination.

These various imperialist campaigns— including the refusal to recognize the existence of Haiti, the implementing of trade embargos, the withholding of aid, and neoliberal restructurings—have cumulatively contributed to what some call Haiti's chronic "underdevelopment."

The history of Haiti and in particular its slave rebellion is inadequately taught in schools and universities around the world. Even though they were crucial historical events with long-term and decisive consequences for the entire world, there has been a tendency to deny they even happened. As Christelle Gomis (2018) writes, we can see how the erasure of the Haitian revolution from official histories is deeply embedded in Eurocentrism and its tendency to elevate European achievements and deny or ignore non-European ones. She notes how the colonization of Saint Domingue and the Haitian slave rebellion are as significant for French history and society as the French Revolution, but French schoolchildren learn a lot about the latter and nothing about the former. Indeed, white French revolutionaries in spite of all of their claims about liberty and equality were resurrecting chattel slavery and mobilizing the idea that liberty and equality did not apply to enslaved peoples. The erasure of Haiti from dominant historical narratives enables France to maintain the idea that racism is external to French civilization and that colonial violence belongs firmly in the past. Essentially, in spite of the emancipatory dimensions of the French Revolution, the French embraced and embellished modernity and buried the history of the conquest, and the French colonization of what is now Haiti. According to Diana Paton (2020), Britain also benefits from the denial of the Haitian rebellion as it

enables the fiction that Britain is and was a progressive outlier in relation to race and racism to be preserved. It helps the Prime Minister to claim that Britain is not racist, contrary to evidence. Erasing the Haitian revolution also maintains a fiction that historical change has come primarily through a process of gradual reform and the use of the 'proper channels.' These are the same 'proper channels' that Bristolians used for years to try to get the Colston statue removed, before taking matters into their own hands.

As Black Jamaican leader intellectual and founder of the UNIA (United Negro Improvement Association), Marcus Garvey (1925, see Chapter 8) wrote in an editorial "Toussaint L'Ouverture's brilliancy as a soldier and statesman outshone that of a Cromwell, Napoleon and Washington; hence, he is entitled to the highest place as a hero among men." A century later, this historical recognition is still in its infancy.

We could debate the extent to which Haiti, as a French Creole-speaking Caribbean island, should be considered to be part of contemporary Latin America, but either way the Haitian rebellion has had and continues to have dramatic continental consequences, that are at once political and epistemic. The Haitian revolution decisively shaped not only what became of the transatlantic slave trade, but also the historical struggles for independence and the contemporary struggles for Black liberation and for a decolonized education. Furthermore, Haiti along with the Dominican Republic make up the island of Hispaniola, where Columbus first made contact with native Americans, so it was indeed the place where the conquest began and the concept of race invented.

COLONIAL LEGACIES

The colonial influence on contemporary Latin America is more than apparent. Iberia gave Latin America languages, religion, and modes of organizing cities and agricultural production still in evidence today. While many Indigenous languages are still spoken, Spanish and Portuguese are the most widely spoken languages on the continent and Catholicism continues to be the dominant religion. All Latin American cities and many towns have impressive Catholic churches built by the Spaniards. The Spanish also implemented a particular mode of urban planning which is still in place today. Cities were based on a grid system, in which rectangular streets emanated from a central square where the most important buildings were located. The Spaniards also created a specific form of unequal land tenure, the *latifundio*, in which large landowners amassed vast quantities of productive land, displacing Indigenous peoples and converting them into labour. The *latifundio* and its counterpart, the *minifundio* or excessively small plot of land on which many plantation labourers and their families relied for basic subsistence, have plagued Latin America's development since the colonial period. Indeed, Latin America's mineral and agricultural wealth has also been the source of its poverty. The plantation and extractive economy established during the colonial era has been one of the most enduring and most destructive legacies of the colonial era.

The independence movements reshaped Latin America, but they didn't bring an end to the colonial matrix of power. Cuban revolutionary and independence fighter, José Martí, described the period 1810–1825 as Latin America's first independence, but it was an independence that remained incomplete given the semicolonial dependence created by the extractive economy. He believed therefore that Latin America required a second independence. In a much-celebrated essay entitled Our America (Martí, 1977[1891]), Martí warned of impending US imperialism against which Latin America should unify drawing on local Latin American and Indigenous knowledges. But after independence, Iberian colonialism gave way to US imperialism with equally consequential and tragic impacts (see Chapter 5) and externally imposed colonial administrations were replaced by internal forms of colonialism and coloniality (see Chapters 7 and 8), as political elites drew their inspiration from Europe and Eurocentric ways of knowing in their creation of independent polities (see Fanon, 2004[1963]). Many things have changed dramatically but there are also many colonial continuities that continue to oppress and destroy.

BOX 2.5 DECOLONIAL OR POSTCOLONIAL?

A key question often raised by students is what the difference is between a decolonial and postcolonial approach. While there are important overlaps and synergies between the MCD and postcolonial studies, in terms of attitudes, theorists, and research questions, there are also some important differences in emphasis. Both are concerned with the histories and legacies of colonialism and with the politics of knowledge production and are heavily influenced by theorists such as Aimé Césaire and Frantz Fanon. The two fields do, however, work with different understandings of modernity and have different historical starting points. While decolonial theorists start with the conquest of America in 1492 and understand modernity to be entangled with and inseparable from coloniality, postcolonial theorists such as Edward Said, Homi Bhabha, and Gayatri Spivak are concerned with the colonial histories of Palestine and India and tend to understand the internal critique of Europe that came with the Enlightenment as an emancipatory project. Decolonial scholars believe postcolonial scholars have neglected "the formative role of the conquest of the Americas and the racialized practices of settler colonialism in constituting Europe" (Asher, 2017: 519). Furthermore, postcolonial scholars tend to emphasize culture and neglect political economy, while decolonial scholars seek to work with both simultaneously (Grosfoguel, 2011). While postcolonial scholars often stress that the 'post' in postcolonial does not mean that they believe colonialism has ended, Mignolo argues that the use of the 'post' "keeps you trapped in unipolar time conceptions" trapping us "in a universal time that is owned by a particular civilization" conceptually closing down the possibility of multiple times and sidelining the question of epistemic diversity (Mignolo and Hoffman, 2017). While both postcolonial studies and decoloniality are intellectual movements, decoloniality has a much stronger activist component and emerges from and is connected to a range of Black, Indigenous, feminist, and campesino social movements. There are, however, important and productive dialogues taking place between decolonial and postcolonial theory that are likely to be transformative for both fields.

Sources: Bhambra, 2014; Asher, 2017; Grosfoguel, 2011; Mignolo and Hoffman, 2017

SUMMARY

- It is not possible to understand the conquest of America without also understanding the conquest of Al-Andalus.
- The conquest of America called the humanity of native Americans into question. If some people are less than human, their enslavement is justified. This is the basis of racism.
- Genocide was accompanied by epistemicide.
- The colonial era was characterized by extreme violence and brutality, as well as by ongoing rebellions and decolonial intellectual production.
- The Haitian slave rebellion (1791–1804) led to the abolition of slavery and the creation of the first Black republic. Despite the enormity of this event, it gets erased and ignored by Eurocentric histories.
- Slavery was bought to an end because of the resistance of the enslaved. It was not benevolently granted by European parliamentarians.
- Despite successful independence movements, contemporary Latin America continues to be shaped by its colonial past.

DISCUSSION QUESTIONS

1   What is meant by the colonial matrix of power? Where are you located in it?
2   What is the relationship between the conquest of Al-Andalus and the conquest of America?
3   What is the significance of the Valladolid debate and what relevance does it have for the contemporary period?
4   If modernity and coloniality are two sides of the same coin, what implications does it have for our understanding of global history?
5   Discuss why the history of Haiti is so infrequently taught in schools and universities?
6   How was colonialism resisted by Indigenous and Afrodescendant populations?
7   What are the similarities and differences between a decolonial approach and a postcolonial one?

NOTES

1   Columbus made a total of four voyages to Latin America but died believing he had reached Asia as was intended. It was not until Amerigo Vespucci, from whom the name America is derived, participated in a series of Portuguese voyages to colonize Brazil that it became recognized that a

new continental land mass had been discovered. The term *indio* (Indian) stuck, however, despite the geographical confusion from which it emerged, although not without contestation and rearticulation.

2 An arroba was about 25 pounds in weight.

3 The term *criollo* or Creole means different things in different parts of the continent. I use the Spanish term *criollo* to refer to people of Spanish descent who were born in Latin America. I use the English term Creole to refer to the English-speaking Afrodescendant population that you find on the Caribbean Coast of Nicaragua or Costa Rica and elsewhere in the Caribbean.

FURTHER READING

Routledge Handbook of Latin American Development chapters 1(Cupples et al.), 4 (Esguerra-Muelle)

Cultural Studies 21(2–3), 2007

This is a combined special issue of the journal *Cultural Studies* that provides a useful introduction to MCD and the decolonial option.

Restall, M. (2004) *Seven Myths of the Spanish Conquest*. Oxford: Oxford University Press

This book tackles some of the common misunderstandings about the Spanish conquest of Latin America.

Salt, K. (2019) *The Unfinished Revolution: Haiti, Black Sovereignty and Power in the Nineteenth-Century Atlantic World*. Liverpool: Liverpool University Press.

An excellent introduction to the Haitian Revolution and attempts to re-centre it in world history.

Williamson, E. (2009) *The Penguin History of Latin America*. London: Penguin

A concise yet comprehensive introduction to the history of Latin America.

Recommended films

The Uprising (2019), directed by Pravini Baboeram

An excellent introduction to the legacies of colonialism and decolonial resistance in Europe and beyond. It includes a number of leading decolonial scholars and activists.

Bolívar (2019), directed by Luis Alberto Restrepo, Andrés Beltrán, and Jaime Rayo

A 40-episode dramatization of the life and independence campaign of South American liberator Simón Bolívar.

Conquistadors: Fall of the Aztecs (2001), directed by David Wallace, written and narrated by Michael Wood

A PBS documentary focused on Hernan Cortés and the conquest of Mexico.

CHAPTER 3

Coloniality, capitalism, and neoliberalism

INTRODUCTION

During the 19th century, after most of Latin America had gained its independence from Spain, Portugal, and France, competing narratives about how the newly independent nations of Latin America should be organized economically, politically, and culturally were in circulation. Spain and Portugal left behind important legacies in terms of land tenure, commodity trade, modes of government, militaries, religion, languages, and political culture. These legacies have manifested themselves in bureaucratic authoritarianism, *caudillismo*, corporatism, clientelism, populism, and frequent military-led coups. The struggles for independence did little to undermine Iberian political traditions because they were waged largely, although certainly not exclusively, by white, conservative, and often aristocratic *criollos*. While these actors were opposed to colonial rule, they were of European descent and their aim was to create modern and independent nations like those in Europe. The dominance of Eurocentric thinking meant that even after independence, economic and political conditions did not improve much for Indigenous, Afrodescendant and poor mestizo populations. The new independent nations continued to be ruled by elites, and land and other kinds of wealth remained very unevenly distributed leaving large, dispossessed majorities whose living conditions were extremely poor. The failure to adequately address this state of affairs in the 19th century created prime conditions for social and political unrest and resistance.

This chapter explores the politics and economics of post–independence Latin America, including the role of commodity trade, import substitution industrialization, the debt crisis, and the implementation of neoliberal structural adjustment. It ends with a discussion and evaluation of the Pink Tide governments that came to power at the turn of the century as neoliberalism became ideologically exhausted.

DOI: 10.4324/9781003110453-3

THE 19TH CENTURY

While the Indigenous and Afrodescendant populations continued to fight for their liberation from oppression, discrimination, and internal coloni-alism, national politics in many Latin American countries was dominated by battles between elite groups, who divided into liberals and conservatives. There were fierce and unresolved struggles over the role of the Church, the armed forces, the landed elites, and international commerce in the making of modern nations. While most nations became republics after independ-ence, both Mexico and Brazil experimented for a short time with monarchies before becoming republics. The conservatives felt nostalgia for the Catholic monarchy and believed the power of the Church and the armed forces should be protected, while liberals on the other hand believed in the sovereignty of the people, representative government, and free markets (Williamson, 2009). Despite their disagreements, as Edwin Williamson (2009) notes, both liberals and conservatives developed dictatorial and authoritarian tendencies, feeling that the chaotic situation in which they found themselves demanded strong leadership. Thus after independence, the phenomenon of *caudillismo*, or strong man politics, emerged and became widespread throughout the continent. Williamson (2009: 237) defines a *caudillo* as "a charismatic leader who advanced his interests through a combination of military and political skills, and was able to build a network of clients by dispensing favours and patronage." Some of Latin America's most famous *caudillos* include Antonio López de Santa Anna of Mexico, Juan Manuel de Rosas of Argentina, and Diego Portales of Chile. Amidst authoritarianism, clientelism, and patriarchy, liberal ideals did make important inroads into continental political culture and the post-independence era also produced leaders such as Benito Juárez of Mexico who was in favour of democracy and social reform.

By the second half of the 19th century, some countries had achieved a degree of political and economic stability and there were visible signs of eco-nomic development and modernization in some parts of the continent. But the land tenure system set in place during the colonial period became intensi-fied in the 19th and 20th centuries as a result of the growth of export agricul-ture driven by both US multinationals and national elite capitalists. This was a system that condemned the large Indigenous and poor mestizo majority to landlessness, hunger, and illiteracy, leaving them with little option other than being an exploited labour force on plantations and large farms. It fuelled their sense of political outrage and encouraged political mobilization in favour of political and social change.

The ruling elites frequently resorted to military rule to protect the eco-nomic status quo and maintain order. Order-and-progress dictators such as Mexican Porfirio Díaz (1876–1911) often rigged elections and engaged in corrupt and anti-democratic practices but modernized the nation through investment in public infrastructure and urban planning and enabled industrialists, large landowners, and foreign investors to prosper.

International trade, which had fallen with the collapse of the Spanish empire, was expanding through increased exports and foreign investment in railways, mining, and plantation agriculture. Coffee, which had been brought to Latin America from Africa, began to boom owing to high demand in Europe. Brazil, Venezuela, Colombia, Costa Rica, and Nicaragua became large exporters of coffee. Venezuela and Ecuador exported cacao, while Mexico expanded its silver production. Peru and Chile provided the guano and nitrates used in European fertilizers, while Argentina became a leading exporter of both meat and wool. New export crops, such as bananas, rubber, copra, and chicle, began to be exported. Whole nations became reliant on one or two crops. Export crops tended to produce a boom-and-bust economy, producing great wealth at times for a minority. For the majority poor, converted into an exploited labour force, these commodities provided long hours of work and insecure income when they were booming and misery and hunger when their prices collapsed on international markets.

THE 20TH CENTURY

By 1920, Latin America was thoroughly integrated into the global economy. Despite some apparent stability after the turbulence of the independence struggles, the failure to address structural inequalities and in particular the question of unequal land tenure would continue to produce social tensions and would lay the pattern for the economic and political struggles of the 20th century and for the ongoing political violence that we see today.

The model of commodity trade has been particularly harmful, socially and environmentally. It relegated Latin America to a peripheral or dependent status, in which its economic role was that of producing raw materials to fuel the economic growth and industrialization of Europe and the United States. Consequently, Latin America was forced to use the foreign exchange earned to import industrial and manufactured products in a vicious cycle of ever deteriorating terms of trade. Traditional agricultural and mineral exports were joined by non-traditional agricultural exports (NTAEs) such as melons, oranges, mangetout, flowers, and biofuels produced from sugar or soy. Agriculture that is oriented towards exports rather than domestic consumption undermines food security and exacerbates landlessness.

As a result, over the course of the 20th century, economic misery and political violence coincided. The Mexican Revolution (1910–1920) put the question of land reform, workers' rights, and social justice firmly on the continental agenda. Socialist and revolutionary thought flourished and Latin American intellectuals such as Argentinean Ernesto Che Guevara and Peruvians Victor Raúl Haya de la Torre and José Carlos Mariátegui began to develop an endogenous socialism embedded in Latin American realities and Indigenous forms of governance and land tenure. Many Latin Americans – women, campesinos, workers, students, urban squatters, Indigenous and Afrodescendant peoples, LGBTQ+ activists – mobilized to fight for human dignity and improved living conditions, while those attempting to defend the colonial-capitalist quo fought back, often successfully, installing dictatorships and military governments across the continent and most of these had overt and covert US support (see Chapter 5).

DICTATORSHIPS AND GUERRILLA WARFARE

In the 20th century, military governments, dictatorships, and guerrilla armies formed all over Latin America. Alfredo Stroessner seized power in a coup Paraguay in 1954, curtailed civil liberties and during his 35 years in power presided over political kidnappings and torture. This is a pattern that would be repeated in Argentina, Brazil, Chile, and Uruguay, as well as in parts of Central America and the Caribbean, as military juntas came to power and killed and tortured political opponents and suspected guerrilla sympathizers. The Uruguayan government resorted to increasingly repressive measures in their attempt to contain a revolutionary movement known as the Tupamaros, and by 1972 the military had taken control, civil liberties were eliminated, and arrest and torture of suspected guerrillas became commonplace. Uruguay became famous for being the country with the world's highest percentage of political prisoners and over the next few decades half a million Uruguayans would be forced to leave the country, among them large numbers of teachers, professors, and artists (Salgado, 2003). In 1973, the socialist and democratic-ally elected president of Chile, Salvador Allende, was overthrown in a coup which had the support of the CIA and a military dictatorship led by General Augusto Pinochet was installed. Supporters of Allende, students, workers, trade unionists, were rounded up in the national stadium and many were killed. Democracy did not return to Chile until 1989 and in that time more than 2,000 Chileans were disappeared by state forces. A similar approach was adopted in Argentina in 1976, when a military junta seized power in a coup and Argentina would embark upon one of the darkest periods in its nation's history, the so-called Dirty War in which between 10,000 and 30,000 people were disappeared in a state-sponsored reign of terror, dubbed Operation Condor. In 1980 in El Salvador, people were outraged when popular bishop of San Salvador, Oscar Romero, who had denounced the human rights abuses, was gunned to death as he said mass.

All of the Latin American dictatorships and military governments of the 20th century deployed paramilitary death squads to disappear, torture and eliminate "subversives," using tactics that they had learned at the School of the Americas (see Chapter 5). In 1992, a human rights lawyer who was investigating the case of a client and political prisoner at a police station in Asunción, Paraguay discovered archives which documented the kidnapping, torture, and assassination of thousands of Latin Americans from Argentina, Bolivia, Brazil, Chile, Paraguay, and Uruguay as part of Operation Condor (McSherry, 2002).

In this context, guerrilla movements proliferated. Many Latin Americans took up arms and became revolutionary guerrilla fighters: the Tupamaros in Uruguay, the Montoneros in Argentina, the FSLN in Nicaragua, the FMLN in El Salvador, the Shining Path in Peru, the URNG in Guatemala, and the FARC and the ELN in Colombia. Some of these revolutionary movements triumphed, including the Cuban Revolution in 1959 and the Nicaraguan Revolution in 1979, although in both cases, revolutionary transformation was seriously hampered by US opposition (see Chapter 5). Some guerrilla groups, including the FARC of Colombia (see Box 3.1) or the Sendero Luminoso (Shining Path) of Peru, deployed terror tactics of their own. Sendero Luminoso was formed in 1980. Drawing on Maoist ideology, it aimed to implement a perfect communist system and convert Peru into a peasant-worker republic (Barton, 1997). Intense conflict between the guerrillas and the Peruvian military, centred particularly in the Ayacucho region, hit ordinary Peruvians hard. Many young men suspected of being guerrilla sympathizers were disappeared by the security forces, while the guerrillas also killed ordinary villagers it believed were supporting the government. Sendero's violent tactics disgusted many progressive Peruvians. Their strength declined substantially after 1992 when the organization's leader Abimael Guzmán was captured and by 2000 it appeared the struggle was all but over.

Although the vast majority of Latin American countries succumbed to military dictatorship or guerrilla warfare or both, some countries such as Costa Rica, Ecuador, and Venezuela did enjoy periods of democratic stability and civilian rule in the 20th century. It is important to understand the economic underpinnings of the forms of political violence and political activism that dominated 20th century Latin America and to note that while the varied economic policies discussed below were being implemented, much of Latin America was embroiled in state-led violence, civil wars, and armed conflict.

BOX 3.1 COLOMBIAN CONFLICT (1964–2016)

All of Latin America's 20th-century civil wars have colonial origins. Colombian journalist Julián Esteban Torres López (2020: 1) describes being "born into an armed conflict spanning centuries, whose roots are still entangled in the colonial mindset and situations we inherited after liberation from Spain" when land and resources remained in the hands of a wealthy elite. Many civil wars were long and deadly, but the Colombian conflict was the longest and one of the deadliest. "La Violencia" began in 1948 after the assassination of presidential candidate Jorge Eliécer Gaitán. It escalated in the 1960s, when groups of campesino farmers and dissident forces tired of Colombia's rampant inequalities and political repression and began to organize. Several left-wing guerrilla groups were formed, including the FARC, the ELN, the M-19, the EPL, and the MAQL. The FARC was the largest of these groups and at the peak of the conflict had more than 20,000 soldiers and many more supporters. All of them were opposed to the Colombian state and US imperialism and were fighting for various forms of liberation, especially agrarian reform, the distribution of wealth, and an end to state-led repression. They were opposed not only by the Colombian security forces but also by government- and cartel-backed paramilitary death squads such as the AUC, the MAS and the ACDEGAM, who carried out brutal massacres of civilians. Some of these paramilitaries received training and logistical support from US, Israeli, British, and Australian military instructors. The United States had advised the Colombian government to create paramilitary forces as part of its counterinsurgency drive and there is no doubt that the US-driven War on Drugs became entangled with the ways actors in the conflict engaged in and benefited from cocaine production (see Chapter 5). In his memoir, Colombian author Héctor Abad Faciolince (2006) writes movingly about his father, Héctor Abad Gómez, who was a doctor, university professor, and human rights activist who was murdered by paramilitaries in Medellín in August 1987. His book captures the impacts of this period on family and university life in Colombia, as well as the political barriers that stood in the way of peace. While by far most of the violence was committed by paramilitaries, the FARC and the ELN were also responsible for the murder of innocent civilians. They frequently engaged in kidnapping, drug trafficking, illegal mining, and acts of terrorism and support for them dramatically declined as a result. The Colombian conflict resulted in the deaths of more than 200,000 people, most of them civilians. It also led to the forced displacement of some eight million Colombians; the second largest number of people displaced by conflict anywhere in the world (see also Chapter 8). While it is important to recognize that weapons used in the conflict have been financed by the drugs trade, drugs are *not* the source of the conflict, which was already underway before the drugs trade started in the 1970s (Koopman, 2020). In 2012, after four decades of civil war, the Colombian government entered negotiations with the FARC in an attempt to end the conflict. In 2016, the agreement was put to a national referendum and rejected, although most people who lived in the zones of conflict voted in favour. Despite this rejection, the two parties signed a peace agreement in November of that year which brought the civil war to an end. While violence has decreased in Colombia since the peace deal, Colombia remains a violent country beset by extrajudicial killings (see Chapter 4 and Chapter 8) and some armed actors have still not demobilized.

Sources: Roldán, 2002; Abad Faciolince, 2006; Fuerzas Armadas Revolucionarias de Colombia and Ejército de Liberación Nacional, 2017; Koopman, 2020; Torres López, 2020

IMPORT SUBSTITUTION INDUSTRIALIZATION

The liberal free trade model ran into difficulties following the crash of the Wall Street stock exchange in 1929. The crash ended a speculative bubble and led to global economic depression and mass unemployment in much of the world, which lasted until the end of the Second World War. While the recession was a global one, it had very particular effects and consequences in Latin America. Commodity prices slumped and Latin America's export markets all but disappeared overnight and with them the foreign currency Latin America used to purchase imported manufactured goods. Poverty, misery, and social unrest increased. At this point, there was a clear sense that the free-market model had failed and some economic rethinking was urgently required. It became apparent to many that Latin American governments needed to protect their economies and produce for themselves many of the manufactured goods that were being imported.

In 1944 at the end of the Second World War, the institutions of global governance, the International Monetary Fund and the World Bank, were created at Bretton Woods to regulate and stabilize global economic relations. Their role would shift decisively in the 1980s, as we shall see. This was also the period in which Keynesianism came to the fore. British economist John Maynard Keynes, who had played a key role in the Bretton Woods conference, was in favour of private enterprise, but believed that states should actively use fiscal and monetary policy to stabilize and stimulate the economy. Governments should control the money supply and therefore inflation through the setting of interest rates. In times of depression or stagnation, they should spend public money to activate the economy. Keynesian thinking became influential in Latin America as it did throughout the world but was extended through geographically specific economic thought. Argentinian development economist and president of the UN Economic Commission for Latin America (ECLAC), Raúl Prebisch (1950), theorized that the free-market model which had dominated Latin America in the 19th century could not lead to economic independence for the region. He had observed that during the depression the prices of raw materials had fallen more dramatically than the prices of manufactured goods, a situation economists refer to as unequal terms of trade. He believed Latin American nations needed to undergo their own processes of industrialization and the state needed to play a central and decisive role in directing the economy away from the production of commodities such as minerals and agricultural goods. The new local industries would need to be protected from international competition to allow them to flourish (see Box 3.2).

Consequently, a new form of development came into being, known as Import Substitution Industrialization or ISI. It was a model in which states actively intervened to promote economic development rather than leaving it to the invisible hand of market. The governments used import tariffs to protect local industry from foreign competition and embarked upon the nationalization of key industries such as iron and steel production. Electricity, mining, telecommunications, transportation, and finance were all areas in which there was substantial state ownership and control. They also implemented price controls, maintained an overvalued exchange rate, and subsidized basic goods such as foodstuffs and utilities such as electricity to keep them affordable for the majority.

BOX 3.2 DEPENDENCY THEORY

ISI was inspired in part of by the scholarship of a group of Latin American economists who developed an anti-imperialist critique of orthodox economics known as dependency theory (DT). It was developed in a specific geopolitical context – that of the Cold War, the triumph of the Cuban Revolution, and the rise of the development industry, by Osvaldo Sunkel, Raúl Prebisch, Celso Furtado, Theotônio dos Santos, Ruy Mauro Marini, André Gunder Frank (a Belgian who worked at the University of Chile until the coup against Marxist president Salvador Allende sent him back to Europe), and Fernando Henrique Cardoso (who later became president of Brazil and abandoned his earlier thinking). Dependency theorists produced a trenchant critique of modernization thinking, the dominant approach to development that saw Latin American countries as backward and held back by tradition and that asserted that they needed to "catch up" with industrialized countries by following the same linear trajectory.

Not only did they draw attention to modernization's imperialistic underpinnings, but they showed how it was empirically flawed, as the attempt to integrate Latin American economies into the world economy through an emphasis on the colonial system of commodity trade was actually producing underdevelopment and dependency. They showed how the core was developing because of the way it exploited the periphery. DT has different variants and emphases. The structuralist variant led by the Economic

Commission for Latin America and the Caribbean was focused on ending dependency by reforming capitalism. Chilean Osvaldo Sunkel describes dependent development as follows:

Development and underdevelopment should therefore be understood as partial but interdependent structures, which form part of a single whole. The main difference between the two structures is that the developed one, due basically to its endogenous growth capacity, is the dominant structure, while the underdeveloped structure, due largely to the induced character of its dynamism, is a dependent one. This applies both to whole countries and to regions, social groups and activities within a single country
(Sunkel, 1973: 136, cited in Kay, 2018: 19).

The structuralist variant did not jettison the idea of development as progress. The Marxist variant was more radical and asserted that it was necessary to overthrow capitalism through socialist revolution in order to bring an end to dependence. Brazilian Theotônio dos Santos really felt that the choice was between socialism and fascism and asserted that:

the profound Latin American crisis cannot find a solution within capitalism. Either one advances in a revolutionary and decisive manner towards socialism ... or one appeals to fascist barbarism, the only alternative able to secure the conditions of political survival for capital'
(Dos Santos, 1978: 471, cited in Kay, 2019; 613-14)

The dependency theorists were not really decolonial scholars, but their work is important to the MCD as well as to world systems thinking, as it underscores the importance not only of taking colonial legacies seriously but of thinking in, with and from Latin America rather than importing theories generated by economists in different parts of the world.

Sources: Kay, 1989, 2018, 2019

Some of the results of ISI were impressive, particularly in the larger nations such as Mexico and Brazil. Gross Domestic Product (GDP) tripled and Latin American economies grew more rapidly than the first world economies of Europe and by the early 1960s, Brazil had become self-sufficient in the production of manufactured goods such as televisions, refrigerators, and electric stoves (Green, 2006). High import tariffs also encouraged transnational companies to initiate production within Latin America, so Brazil and Mexico became major producers of GM, Ford, Volkswagen, and Fiat cars, which used locally made parts and were sold locally (Franko, 2019). They did, however, expatriate profits. The standard of living also improved. Infant mortality fell and life expectancy increased. There was some public investment in literacy programmes, worker education, and housing. In Argentina and Chile, industrialization combined with populism facilitated the creation of organized labour.

ISI did not, however, prove to be sustainable over the long term. Protectionism allowed domestic industry to grow but also encouraged inefficiency and corruption. Many ISI products were expensive and of poor quality and the lack of external competition meant there was little incentive to innovate. The ISI model was good for the cities but devastating for the rural areas. Because government kept food prices artificially low and concentrated public investment in the cities, small farmers found making a living increasingly difficult. From the 1950s onwards, vast numbers of rural people migrated to Latin American cities, in search of employment and hoping for a better standard of living. As a result, shantytowns developed on the edges of most large Latin American cities (see Chapter 10). ISI created some wealth and industrial growth, but it was not evenly distributed and both misery and inequality continued to characterize the Latin American economy. ISI could not be fully implemented because Latin America still needed to import capital goods from Europe and to be a key exporter of primary products. Capital flight was also a serious problem, not just through the multinational repatriation of profits but also because wealthy Latin Americans preferred to keep their money in US bank accounts and currency overvaluation at home made it profitable for them to do so. While ISI was floundering in Latin America, in other parts of the world free-market thinking was starting to make a comeback. The Chicago School of Economics under the leadership of Milton Friedman and Arnold Harberger was staunchly advocating a return to free-market economics, arguing that state-led economic development had been a failure. These intellectuals had a strong influence in Latin America, especially in Chile in the 1970s who began to implement reforms (see Box 3.3). The rest of the continent would soon follow.

BOX 3.3 THE CHICAGO BOYS AND PINOCHET'S CHILE

In terms of neoliberal economics, Pinochet's Chile was an early reformer and implemented such policies before the debt crisis. It was the first country in Latin America to move away from the ISI development model. The military, led by Augusto Pinochet, had come to power in Chile in a coup which ousted democratically elected socialist president Salvador Allende. Allende killed himself in the presidential palace as the coup was unfolding. A number of Chileans had studied economics at the Chicago School of Economics with Milton Friedman and Arnold Harberger, and they later joined the faculty at Santiago's Catholic University and came to be known as the Chicago Boys. Their ideas were seen as marginal and obscure in the 1960s, but they subsequently became central to the transformation of the Chilean economy under Pinochet. Free-market scholars including Milton Friedman and Friedrich von Hayek made frequent visits to Chile in the 1970s and 1980s to advise on economic matters. Panizza (2009: 15) argues that the Chilean military supported free-market capitalism primarily for political rather than economic motives because they saw it "as the best option to impose an alternative social, political and economic order that would make impossible the return of the left." The Chicago Boys had substantial funding from the CIA with which to develop their programme, and as Naomi Klein (2007) writes, the economic shock therapy was soon accompanied by another kind of shock in the guise of state-led torture and brutality, as Pinochet began to kidnap and disappear opponents. Chile's impressive economic growth rates in the wake of reform led neoliberal proponents to refer to Chile as an economic miracle, but the country paid a very high price for this miracle. Between 1980 and 1990, inequality increased and employment became increasingly precarious.

Sources: Klein, 2007: Panizza, 2009; Ffrench Davies, 2010; Hutchison et al., 2014

THE DEBT CRISIS

The ISI model lasted in various forms until the early 1980s amidst growing dissatisfaction with its results and those of development more generally. The global economy changed dramatically in the 1970s in ways which would have long lasting effects in Latin America and would lead to the total dismantling of the ISI model. In Europe and the United States, Keynesianism had facilitated massive public investment in education and health care that improved standards of living. In 1973 and again in 1979, the oil-producing countries (OPEC) dramatically increased the price of oil, a move which produced stagflation (simultaneous stagnation and inflation) in oil importing nations such as the United States and the UK. Keynesian demand management had always proceeded to correct either a depressed or an inflating economy but in the 1970s both phenomena appeared simultaneously, a situation for which there were no Keynesian tools.

While western nations were stagflating, the OPEC countries were making massive profits. These profits, known as "petrodollars", were often deposited in commercial banks in both the United States and the UK. Given the economic situation in their countries, the bankers were desperate to recycle these petrodollars and they aggressively offered multimillion dollar loans to Latin America. At this time, many countries in Latin America were ruled by military dictatorships that spent the funds in unproductive ways – on steel plants that never produced any steel, on dams which destroyed the surrounding environment, or on weapons which they used against their own people. Some of the funds were, however, used to finance development, creating new industries and sources of employment, building roads, extending the provision of electricity, and developing telecommunications. Despite the scale of the borrowing, debts appeared sustainable to both lenders and borrowers, as global interest rates were low, indeed they were negative in real terms, and commodity prices were high.

This situation began to change after the oil crises and the rise to power of neoliberal global leaders such as Margaret Thatcher in the UK and Ronald Reagan in the United States. Both Thatcher and Reagan implemented restrictive monetarist policies to deal with the stagflation afflicting their economies, hiking up global interest rates. The oil crisis had also resulted in a fall in the price of commodities. Suddenly, the gap between what Latin American countries were able to earn from their exports and what they were required to pay to service their debts widened. This was the beginning of the Mexican default, when Mexico announced on a Friday that it had run out of foreign exchange and could not therefore make its debt service payments the following Monday, and the Latin American debt crisis.

The Mexican default had sent the first world bankers into a tailspin, but for many economists the debt crisis provided the catalyst needed for a move back to the free trade model that had been discredited in the 1930s. The IMF, with support from the commercial banks and the US Treasury, took advantage of the situation and underpinned by a growing body of economic theory coming out of prestigious schools used the debt crisis to demand a dramatic overhaul of the economies of Latin America in line with neoliberal thinking. They negotiated with each country individually to prevent the formation of a debtors' cartel. Countries were offered fresh loans with which they could meet their debt service requirements but in return were expected to comply with a stringent set of economic conditions, known as IMF conditionality, conditions which would bring the ISI model to an end (see Box 3.4). It amounted to a substantial reversal of economic policy from one based on inward-focused development and protectionism to one based on exports and liberalization. This approach became known as structural adjustment and over the course of the 1980s and early 1990s most Latin American countries fell subject to IMF conditionality. The support for such policies from the US government and powerful institutions based in Washington DC meant that the policy package became known as the Washington Consensus. In 1995, the IMF and the World Bank were joined by a third powerful multilateral organization, the World Trade Organization (WTO), which added further momentum to their economic prescriptions. While the WTO maintains it is a neutral site in which member states can resolve trade disputes, its ideological position nonetheless is "passionately against protectionism and just as profoundly for trade liberalization" (Peet, 2003: 160).

BOX 3.4 THE IMF RECIPE

Cut public spending
Promote exports
Eliminate government subsidies
Devalue currency
Privatize state-owned enterprises
Liberalize foreign trade and investment
Liberalize interest rates
Weaken labour protections
Guarantee private property rights

NEOLIBERAL STRUCTURAL ADJUSTMENT

While the IMF recipe (see Box 3.4) has a number of common features, it is important to recognize that the timing, rate, and the degree to which this set of policy prescriptions was implemented varied quite markedly from one country to another. It is important not to paint a picture of an all-powerful IMF imposing its will on weak developing nations who had no leverage or bargaining power. In some cases, such as in Mexico in the early years after default, as Ngaire Woods (2006) has noted, structuralist and nationalist positions within the government were resilient, cabinets were heterodox, and the IMF had to bargain hard to get an agreement in place. But after the election of Carlos Salinas de Gortari in 1988, Mexican officials became closely aligned with the policy prescriptions of both the IMF and the World Bank and a high level of cooperation developed. In other countries, government officials and the general public were frequently quite receptive to such policies from the start, given high inflation, persistent poverty, and the general state of national economies. Neoliberal supporters often tend to present their policies in a "there is no alternative" (TINA) mode, acknowledging that the medicine might be painful but it is the only possible option. In the midst of debt default, fiscal deficits, and spiralling inflation, it did seem at the time like there was no alternative. Bolivia, for example, was considered to be a basket case by the international community. The government, unable to secure sufficient revenue through taxation, simply printed money to pay for the things it needed. As a result, the country's inhabitants were confronting the daily reality of hyperinflation which reached a staggering 23,000% and meant that prices increased every hour. The heterodox stabilization measures implemented after the debt crisis were showing no signs of working.

After the 1985 election, young Harvard economics professor, Jeffrey Sachs, flew to Bolivia to draft a drastic economic plan with the Minister of

Planning, Gonzalo Sánchez de Lozada, which brought in free-market reforms, a plan guaranteed to gain IMF approval. Price controls were eliminated, public spending was slashed, and trade was liberalized. To get the harsh and controversial measures through with limited opposition, the plan was negotiated in secret and then approved by presidential decree 21060 and the president declared a state of siege so that workers could be fired and union leaders silenced. Hyperinflation was resolved and eventually the Bolivian economy returned to growth, and these positive economic indicators meant that Bolivia could be held up by the IMF and World Bank as a reform success. Yet the price paid for such "success" by Bolivians was extremely high (Green, 2003). The elimination of price controls put foodstuffs out of the reach of many ordinary Bolivians, malnutrition increased, and thousands of workers were laid off. The downsizing of the state-owned tin mining company brought economic misery to thousands of miners and their families.

Neoliberalization has been a complex, uneven, and at times contradictory process in Latin America. Francisco Panizza (2009: 10–11) maintains that it is useful to think of neoliberalism as "a relational construct, whose contested meaning is defined and redefined by the political struggles between its defenders and detractors, and to which specific policies are contingently articulated according to the history and contexts in which they are fought." It is useful therefore to gain insight into the general principles on which IMF conditionality is based but also to study its implementation in specific contexts and the ways neoliberal economic policy is negotiated by both policymakers and ordinary citizens.

While structural adjustments produced different impacts, in most countries, these were truly devastating and marked the beginning of what is understood in terms of development as Latin America's lost decade. Real wages fell and poverty and unemployment increased, as trade liberalization wiped out local businesses and local markets were flooded with cheaper imported goods. Privatization of state assets meant that wealthy national and foreign investors were able to make windfall profits while ordinary Latin Americans were pawning their belongings and struggling to put food on the table. People were also harmed by cuts to health care and education and the elimination of subsidies on food, fuel and transport dramatically increased the cost of tortillas, bread, milk, and bus fares. Many children joined the informal labour market or begged for coins at traffic intersections. Structural adjustment reoriented the banking sector towards the forms of agriculture that would generate foreign exchange depriving small farmers of bank credit that often resulted in their demise and in their losing their land to large agribusiness. For many years after the debt crisis, governments were spending more on debt service than they were on health care and education, but still ending up owing far more in 2004 (about US$750 billion) than they had in 1982 ($300 billion) (Franko, 2019).

The impacts were also particularly gendered as women became the shock absorbers of the crisis (Moser, 1993; Alarcón-González and McKinley,

1999, see Chapter 6). While some economic indicators showed improvement, structural adjustment didn't end poverty and in many cases did not even bring economic stability. Many countries went on to suffer economic collapse, including Mexico in 1994–1995, Argentina in 1998 and 2001–2002, and Brazil in 2002 (see Figure 3.1). Mexico and Argentina collapsed after they had been touted as structural adjustment success stories by the IMF and the World Bank. Over time, the picture was more mixed and in the larger economies some improvements in living standards were recorded. But structural adjustment never brought the trickle-down wealth that was promised and its continued failures led to a decisive rejection of the neoliberal model across the continent. In fact, influential economists including Joseph Stiglitz and Jeffrey Sachs who had supported the neoliberal policy regime began to shift their positions and publicly acknowledge that these policies had failed. There have been various US-IMF-World Bank-driven attempts to resolve the debt crisis, including the Baker Plan, the Brady Plan, and the Heavily Indebted Poor Countries Initiative (HIPC) as well some bilateral debt reduction initiatives. Many of these have been inadequate and have often involved yet more structural adjustment. The HIPC did however result in some debt relief for the poorest countries, including Honduras and Nicaragua, and did at

Figure 3.1 Economic collapse in Argentina in 2002. Middle-class savers protest at being locked out of their bank accounts

Source: BarcexEspañol, CC BY-SA 3.0, via Wikimedia Commons

least constitute official recognition that debt burdens were unsustainable and could not be repaid.

By the 1990s, many of the dictatorships were ideologically exhausted and morally discredited and many countries were able to transition to democracy. The benefits of these transitions cannot be understated. With the end of the Cold War, the anti-communist hysteria subsided and with it the state-led terror that led to the disappearance and torture of suspected subversives. In many places, however, democracy remained fragile. One of the reasons for this fragility was that the transition to democracy coincided with economic neoliberalism which exacerbated poverty and inequality. Over the next few years, democratically elected Latin American governments entered into free trade agreements with the United States in order to further neoliberalism through trade liberalization (see Chapter 5). In many ways, neoliberalism made it difficult for Latin Americans to reap the benefits of democracy.

COUNTER-HEGEMONIC STRUGGLE

The criticisms of structural adjustment from ordinary people, aid agencies, activists, and scholars both within and beyond Latin America began to mount. Organizations such as Jubilee 2000 called for the cancellation of third world debt, others such as 50 Years is Enough called for the abolition of the IMF and the World Bank. There were important anti-IMF, WTO, and World Bank protests in the Global North too, in Seattle in 1999 (see Figure 3.2), in Prague

Figure 3.2 WTO protestors in Seattle in November 1999 being doused in pepper spray

Source: Steve Kaiser, CC BY-SA 2.0, via Wikimedia Commons

in 2000, in Quebec City and Genoa in 2001, in Cancún in 2003, and in Gleneagles in 2005. The scholarly and activist literature which outlines the harm done to ordinary Latin Americans by the IMF and the World Bank is enormous! The political response to neoliberalism took many forms – from everyday coping strategies such as the creation of communal kitchens to feed families collectively to high profile political action led by groups such as the Landless Workers Movements (MST), a land reform movement that occupied unproductive land, or the Zapatistas (see Chapter 7). In Cochabamba, Bolivia, the people were able to prevent the World Bank-mandated privatization of water (see Chapter 4). In Argentina, workers occupied factories that had collapsed because of the crisis (see Box 3.5). In 2001, civil society organizations that wanted to create a different kind of global economy formed the World Social Forum (WSF) using the slogan "Another world is possible." They usually hold their annual meetings at the same time as the Global Economic Forum held in Davos (de Sousa Santos, 2004). Through the WSF, the "hegemonic globalization of neoliberalism" is called into question by the "counterhegemonic globalization of social movements" (de Sousa Santos, 2018a: 258). The WSF Charter of Principles established in Porto Alegre, Brazil, in 2001 emphasizes their role as an international, democratic, pluralistic, and horizontal space for economic deliberation that is fighting to create alternatives to capitalist globalization based on solidarity and respect for nature and human rights.[1] The WSF, along with the Zapatistas, the MST, and the Recovered Factories Movement, is an important site for the generation of epistemologies of the south (de Sousa Santos, 2014). The neoliberal project has been one in which decolonial thinking has flourished.

BOX 3.5 THE RECOVERED FACTORIES MOVEMENT

Latin Americans did not simply oppose neoliberalism but began to develop creative alternatives to it. The transition to democracy in Argentina meant the intensification of neoliberal structural adjustment, as the peso was pegged to the US dollar and state assets including electricity, telecommunications, and the state-funded pension scheme were all privatized. After a short period of economic stabilization, the economy went into free-fall. Consequently, the government was forced to default on $100 billion worth of debt, freeze bank accounts, and devalue the peso. Thousands of Argentines protested angrily in the streets at the situation into which they had been plunged. Many businesses and factories closed, unemployment soared, and millions of people fell below the poverty line. Some of the newly unemployed workers became active in the National Movement of Recovered Companies, in which workers who had lost their jobs occupied and took over the failed companies. Their slogan was Occupy, Resist, Produce! As the film *The Take* reveals, the workers took direct control of production and turned failed companies into successful worker-run cooperatives. The movement grew very quickly and was motivated by the workers' desire to keep their jobs and out of the recognition that there was no alternative, especially for older workers who were fearful of not being to find a new job in the midst of economic crisis and of losing their pensions and other benefits. They involved tense political struggles and long-term occupations to prevent eviction. There are now hundreds of recovered factories that are owned and run by the people who used to be waged employees. They can be found in a range of different economic sectors. High-profile examples include the Hotel Bauen and the Forja San Martín car parts factory, both in Buenos Aires, and the Zanon tile company in Nequén, which was renamed FaSinPat or Fábrica sin Patrones – Factory without Bosses. Many of them also have a horizontal management structure and everyone earns roughly the same regardless of skills. Sometimes they rotate jobs and decisions are taken democratically and collectively in assemblies. While these assemblies take up time, productivity goes up as people are less likely to waste time or resources and there isn't a surplus value that is taken by the capitalist owner. The companies are well articulated with local communities, neighbourhood assemblies, and other civil society organizations such as the *piqueteros* which enables them to build important modes of solidarity. Sometimes they donate goods or services to local hospitals, schools, nursing homes, and soup kitchens, blurring the distinction between production and consumption. They reveal the possibility of forms of economic transaction that are rooted in local communities and not based on the exploitation of workers or customers. Since being taken over by its workers in 2003, the Hotel Bauen (see Figure 3.2) became an important site for left-wing intellectuals and activists, although they never became free of the constant threat of eviction. Zibechi (2012) sees this struggle through the lens of territorialization. He notes that many of the factories "demonstrate a novel characteristic of the workers' movement: an incipient but growing territorial rootedness. The link between worker-run enterprises and neighbourhood assemblies reveals the ongoing support of the community and their articulation with broader social movements. In some cases, this is manifested by a factory's commitment to hire unemployed neighborhood residents to fill job openings. Thus, by maintaining community activism, rebuilding social ties, and tending toward the territorialization of the struggle, the factory-recovery movement seeks to address one of the main problems it faces: distribution or, to put it differently, relations with the market." (Zibechi, 2012: 133–134). Despite many ongoing challenges, as former owners resorted to the courts and other means in an attempt to regain ownership, there is no doubt that the recovered factories are an important feature of the Argentine economic landscape and that they defy the dominant colonial-capitalist narrative in all kinds of ways.

Sources: *The Take*, 2004; Ranis, 2006; Zibechi, 2012

In addition, to explicit opposition to structural adjustment, Latin Americans also mobilized in response to poverty, terror, and counter-insurgency politics. Although it was not supported by the Vatican, Catholics developed a specific brand of critical Catholicism known as liberation theology. They formed Christian Based Communities (CEBs) and used the Bible to make sense of and respond to the conditions of poverty and suffering. Peruvian liberation theologist Gustavo Gutierrez (1974) coined the term "preferential option for the poor", which embodies the idea that Christians have an obligation to be concerned about those poor, suffering, and marginalized. In Central America, liberation theology was practised by the Jesuits who spoke out against human

Figure 3.3 Hotel Bauen in Buenos Aires, occupied and run by the workers since 2003

Source: Photo by author

rights abuses and as a result they too became a target of the terror campaign. In 1989, the Salvadoran military murdered six Jesuit scholars along with their housekeeper and her 15-year-old daughter at the UCA, the Jesuit university in San Salvador. Many of those who fought for peace, justice, and human rights in El Salvador, Guatemala, and Argentina were women who organized as widows, mothers, and grandmothers of people who had been disappeared or tortured by state. They had a decisive impact on the Latin American political landscape and changed the way political struggle was understood (see Chapter 6). All over the Americas, people mobilized in their communities and around identities (as women, campesinos, shantydown dwellers, Indigenous peoples, environmentalists and so on) to lobby for land rights, clean water, electricity, garbage collection, public transport, credit, legal aid, education, and healthcare (see Chapters 4, 6, 7, 8 and 10).

THE PINK TIDE

In the 1980s and 1990s, the dictatorships and military governments began to fall. The new democratically elected governments embraced trade liberalization and neoliberal economic policies. Consequently, political freedoms were secured while economic misery for many continued and as a result many social movements mobilized against these policies. In the first decade of the new millennium, when it was clear to many that structural adjustment had failed, much of the continent took a decisive turn to the left via the ballot box, electing governments who campaigned on explicitly anti-neoliberal and often anti-IMF and anti-imperialist platforms and garnered the support of urban, rural, and Indigenous social movements. The election of these left-wing governments was referred to a Pink Tide (see Table 3.1). They are referred to as pink, rather than red, as they tend to soften and hybridize neoliberal policies rather than reject them entirely.

The election of Hugo Chávez in Venezuela in 1998 marked the beginning of this phenomenon. Chávez nationalized the oil industry and diverted funds into social programmes, including social housing and the subsidization of basic foodstuffs. Poverty fell and living standards of the poor majority improved. Chávez talked of a Bolivarian Revolution and 21st-century socialism and often set out to deliberately antagonize the United States. Chávez led the development of an alternative trading bloc based on south–south solidarity, ALBA (the Bolivarian Alternative for the Americas), to rival US trade agreements and the creation of a news channel TeleSUR to provide alternative media coverage and rival CNN. In 2002, he survived a 24-hour coup led by some members of his own armed forces. Although the country made much social progress, Venezuela was very reliant on high oil prices and the government did not go far enough in its attempts to diversify the economy (Enríquez and Page, 2019). When the price of oil collapsed, the economy was driven into the

Table 3.1 Latin America's Pink Tide governments

Venezuela	President Hugo Chávez	1999–2013	Hugo Chávez died in 2013 and was succeeded by Nicolás Maduro. Venezuela has been in an ongoing political and economic crisis since then.	The left under the PSUV remains in power, but Maduro's leadership has been heavily contested since 2019.
Brazil	President Luiz Inácio da Silva (Lula)	2002–2010	In 2011, Lula handed over power to Dilma Rousseff, also of the Workers Party. Rousseff was impeached and removed from office in 2016.	The right, led by Jair Bolsonaro, elected in 2018 is back in power.
Argentina	President Néstor Kirchener	2003–2007	Néstor Kirchener died in office and was succeeded by his wife, Cristina Fernández de Kirchener (2007–2015).	In 2015, the right led by Mauricio Macri returned to power for one term. The centre-left Justicialist Party returned to power in 2019, and Cristina Fernández became vice-president.
Uruguay	President José Mujica	2004–2015	José Mujica, former Tupamaros guerrilla fighter of the Frente Amplio, served two terms and then was succeeded in 2015 by Tabaré Vazquéz who served until 2020.	In 2020, the right returned to power when Uruguay elected Luis Lacalle Pou of the National Party as president.
Bolivia	President Evo Morales	2006–2019	In 2019, Evo Morales was deposed in a right-wing coup and was succeeded by interim president, Jeanine Añez who served for less than a year.	Morales' left-wing MAS party returned to power in elections, led by Luis Arce.
Ecuador	President Rafael Correa	2007–2016	Succeeded by Lenín Moreno of the same party, the PAIS Alliance Movement. Correa convicted of corruption fled to Belgium to escape justice.	In 2021, the right returned to power, when Ecuador elected conservative businessman Guillermo Lasso.
Paraguay	President Fernando Lugo	2008–2012	The election of President Lugo, former bishop and liberation theologist marked the end of 61 years of rule by the right-wing Colorado Party. He was impeached and deposed by the senate in 2012.	Lugo's vice-president completed Lugo's mandate. In 2013, the right-wing Colorado Party regained the presidency.

ground, generating a profound crisis. Chávez died in 2013 and was succeeded by Nicolás Maduro, whose mandate has been constantly questioned and prone to destabilization by forces both within and beyond Venezuela. Empty supermarket shelves, hyperinflation, hunger, blackouts, and unemployment sent many Venezuelans over the border into Colombia, Ecuador, and Peru in search of work. Venezuela remains an extremely polarized country, as it has been for many decades, in which the Maduro government has loyal supporters and fervent detractors.

In 2002, Brazil elected the leader of the Worker's Party, Luiz Inácio da Silva or Lula, to the presidency. He was more moderate than Chávez and didn't antagonize the United States in the same way. The Brazilian economy grew rapidly and the government invested in social programmes aimed at families such as Bolsa Família (Family Purse) and Fome Zero (Zero Hunger). While these programmes were quite welfarist and some would say paternalistic in their orientation, they made it easier for families to feed themselves and so nutrition improved as did school attendance, so they decisively contributed to the reduction of both poverty and inequality. During his time in office, Lula managed to pay off Brazil's debt to the IMF. Lula was succeeded by former guerrilla fighter, Dilma Rousseff in 2011, and later found himself in jail on corruption charges, although his conviction was later overturned (see Figure 3.4). Rousseff was also impeached by her own parliament and removed

Figure 3.4 Hugo Chávez and Dilma Rousseff in Brasilía in 2011

Source: Roberto Stuckert Filho, CC BY-SA 2.0, via Wikimedia Commons

from office, a move that allowed for the election of right-wing ultranationalist populist, Jair Bolsonaro in 2018, who has sought to undermine progressive policies.

Argentina also elected two Pink Tide presidents, Néstor Kirchener in 2003, who died in office and was succeeded by this wife, Cristina Fernández de Kirchener in 2007 and who was re-elected in 2011. With the exception of the pension system, Argentina did not re-nationalize the privatizations that had occurred under the government of Carlos Menem and they continued to support large national and multinational agribusiness (Enríquez and Page, 2019), but they defaulted on debt and began to invest in social programmes. In 2015, with Fernández facing accusations that she had artificially manipulated the currency, Argentina elected a right-wing businessman, Mauricio Macri, as president but he only survived for one term in office and the Justicialist Party with Cristina Fernández as vice-president returned to power. In Uruguay, Pink Tide president and former Tupamaros guerrilla fighter, José Mujica, elected in 2004, became known as the poorest president in Latin America as he rejected the salary and perks of the presidency (Tremlett, 2014). His government engaged in a range of redistributive policies, including price controls, confronted the corporate abuses of the tobacco giant, Philip Morris, and legalized marijuana, equal marriage, and abortion.

The Pink Tide continued to expand in 2006, with the election of Rafael Correa in Ecuador and Evo Morales in Bolivia (see Figure 3.5). Morales was an Indigenous coca farmer and union leader and his election followed strong Indigenous forms of political action especially over natural resources, including the Water War in Cochabamba (2000) and the Gas War in Alto (2003). He won elections with decisive popular majorities, in both 2006 and 2009. He presented quite a different personal style to other heads of state in Latin American, often preferring to wear his Indigenous dress, a brightly coloured *chompa* or jumper made of alpaca wool rather than the conventional dark suit and tie. When he came to power, Morales nationalized oil and natural gas. A close ally of Venezuela's Hugo Chávez, Morales became a strong advocate of Indigenous rights and a vocal critic of neoliberal economic policies, the role of the industrialized countries in accelerating dangerous climate change, and the attempt by the United States to eradicate coca crops as part of the "war on drugs" (see Figure 5.5). By 2009, things were clearly improving for ordinary Bolivians. Poverty and infant mortality levels fell, illiteracy was eradicated, a state pension scheme was introduced, and grants and scholarships were provided for students. Funds from the newly nationalized natural resources found their ways into social development programmes and provided a massive stimulus to the Bolivian economy, allowing for budget surpluses and high economic growth rates. One of Morales' key achievements was the promulgation of a new constitution in 2009, which puts Indigenous rights and the concept of *Buen Vivir* centre stage. It establishes the Bolivian state as both plurinational and secular, it makes special provision for the traditional

Figure 3.5 Evo Morales and Rafael Correa in Cochabamba in 2012 for the 42nd Assembly of the OAS

Source: Fernanda LeMarie, Cancillería Ecuador, CC BY-SA 2.0, via Wikimedia Commons

cultivation of coca, it limits land ownership to 5,000 hectares, it establishes Spanish and 36 Indigenous languages as official languages, and it prohibits both the establishment of foreign military bases on Bolivian soil and the privatization of oil and natural gas. In 2011, the Bolivian government passed the Law of Mother Earth, giving non–human living beings (nature) equal status with humans and enshrining the idea that nature has the right not only to exist but also "to not be affected by mega–infrastructure and development projects that affect the balance of ecosystems" (Vidal, 2011). Correa's government also embraced the Quechua/Aymara concept of *Buen Vivir* and engaged in radical forms of constitutional reform. But in both Bolivia and Ecuador, the extractivist model remained in place which resulted in contradictions and confrontations with the Indigenous movements that had brought them to power (see Chapters 4 and 7).

There are other countries in the region that also elected left–wing governments during this period, including Chile and Peru, but it is debatable the extent to which they are part of the Pink Tide as they don't follow the same pattern as the others. In Nicaragua, former revolutionary president Daniel Ortega returned to office in 2006 but has behaved remarkably like the dictator he ousted in 1979 in the way that he has done corrupt deals with big business, rigged elections, and committed serious human right abuses against campesinos, Indigenous peoples, women, students, and environmentalists who have opposed the repressive measures in place (Cupples and Glynn, 2018).

As Will Grant's (2021) in-depth study of recent populist Pink Tide and left-wing presidents reveals, the vast majority rework some of the more problematic aspects of their more right-wing predecessors. In his analysis of the presidencies of Fidel Castro, Hugo Chávez, Luiz Inácio "Lula" da Silva, Evo Morales, Rafael Correa, and Daniel Ortega, he shows that while important gains for Black, Indigenous, and poor populations can be observed in some places, much of it on the back of high oil prices that didn't last, their terms were characterized by corruption, authoritarianism, repression, and in the case of Venezuela economic collapse, and an attempt to cling onto power, rewriting or violating constitutions in order to do so. He characterizes the period from the election of Hugo Chávez in 1999 to the death of Fidel Castro in 2016 as "a period of outsized exuberant personalities who shook up the natural order of politics in their countries and shifted the continental balance of power" (p.2). It ended badly for all of them. Chávez and Castro passed away, Lula ended up in jail on corruption charges, Morales was forced to resign and go into exile, Correa also went into exile and was convicted in his absence of bribery and kidnapping, and Ortega, still in power at the time of writing, was behind a series of human rights abuses in 2018 that left hundreds dead.

The Pink Tide governments are all quite different and it is important to do geographically specific research to understand the local conjunctures that brought them into power. There is no doubt that they constituted a serious decolonial challenge to both capitalism and coloniality, while maintaining colonial-capitalist continuities, especially around extractivist economic practices. Some continued to work with the IMF or participate in trade agreements with the United States, many have displayed authoritarian tendencies and try to cling onto power, and some have criminalized Indigenous peoples who are trying to defend lands or livelihoods (see Chapter 4). Indigenous social movements have found themselves weakened, divided, and fragmented in this context (Zibechi, 2012). Some Pink Tide governments have combined anti-poverty programmes with commitment to harmful forms of extractivism, including mining and agribusiness, reducing poverty without redistributing wealth (Zibechi, 2012) and they haven't been free of the accusations of corruption that are frequently levelled at the right. But the authority of neoliberal capitalism has been dislodged and the Pink Tide is part of broader decolonial struggles as well as the growing epistemological challenge to colonial and neoliberal logics. As commodity prices fell, and with them the social spending that they had permitted, and authoritarian and extractivist practices increased, support for the Pink Tide governments waned and in some countries the right has been able to return to power. But as recent social mobilizations reveal, a range of left-wing, progressive, and anti-extractivist politics continues to have wide support. There is no doubt that neoliberalism continues in the region, and as the next chapter shows is seeking to intensify itself through extractivist practices, but perhaps in a much more hybrid and less dogmatic way than in the 1980s and 1990s.

SUMMARY

- Colonial development models such as commodity trade continued to persist after independence. These have been subject to critique, including by dependency theorists.
- The 20th century was characterized by a struggle for power between right-wing dictatorships and military governments, left-wing guerrilla armies, and a range of social movements. Dictatorships were responsible for the kidnapping, torture, and assassination of thousands of Latin Americans.
- After the depression of the 1930s, Latin American governments began to protect their economies through import substitution industrialization with mixed outcomes.
- The rise in oil prices in the 1970s was the origin of the 1980s debt crisis that opened the path to free-market economics in the form of neoliberal structural adjustment.
- Starting in 1998, a number of Pink Tide governments came to power elected on anti-neoliberal platforms.

DISCUSSION QUESTIONS

1   What is Dependency Theory (DT) and what does it enable us to understand about Latin American economic development? How does the structuralist variant differ from the Marxist or revolutionary variant? Is DT still relevant today? To what extent is it compatible with decolonial theory?
2   How did ISI give way to neoliberal structural adjustment? What is the relationship between global oil prices and the Latin American debt crisis?
3   Discuss the relationship between Latin American development, the IMF, and the World Bank. How do you feel about these two institutions?
4   What were the main impacts of neoliberal structural adjustment packages? Why do some people see them as a success and others as a failure?

5 Discuss why and how military government and dictatorships came to power throughout Latin America in the second half of the 20th century?

6 Watch the section of *The Shock Doctrine* that deals with Chile. How does Chile differ from other Latin American countries when it comes to neo-liberal structural adjustment?

7 Explain why there was so much resistance to neoliberal economic policies in Latin America?

8 What do you think about worker–run factories or businesses? Do you support such movements? Should factories be owned and managed by their workers? Does industrial production and commerce require exploit-ation in order to be successful?

9 How does the Recovered Factories Movement challenge the colonial-capitalist world order? What kinds of democracy does it engender? Can it be considered a decolonial movement?

10 How useful is the notion of Pink Tide to describe the changes that took place in the Latin American political landscape in the late 1990s and 2000s? How have these governments challenged the neoliberal status quo?

11 What kind of president was José Mujica? How was he different from other Pink Tide presidents?

12 What are the factors that have resulted in the current crisis in Venezuela?

13 Is the Pink Tide over? Is it over? What evidence is there to suggest that it is and what evidence is there to suggest that it isn't?

NOTE

1 https://fsm2016.org/en/sinformer/a–propos–du–forum–social–mondial/

FURTHER READING

Routledge Handbook of Latin American Development Chapters 1 (Kay); 7 (Enríquez and Page); 9 (Ruttenberg).

Calderón F. and Castells, M (2020) *The New Latin America*, translated by R. McGlazer. Cambridge: Polity Press.
An assessment of 21st century Latin America and the challenges of the post-Pink Tide conjuncture.

Ellner, S. (ed) (2019) *Latin America's Pink Tide: Breakthroughs and Shortcomings*. Lanham, MD: Rowman and Littlefield.
An edited collection that assesses nine Pink Tide or leftist governments in Latin America and some of the reasons for their inability to overcome established models of devel-opment and governance.

Franko, P. (2019) *The Puzzle of Latin American Economic Development.* 4th ed. Lanham, MD: Rowman and Littlefield.
A thorough and detailed introduction to the economic aspects of Latin American development.

Recommended films

Commanding Heights: The Battle for the World Economy (2002), directed by William Cran
A documentary series on the global economy including in-depth coverage of neoliberal reform in Latin America.
The Take (2004), directed by Avi Lewis and Naomi Klein
An inspiring film on the rise of the Recovered Factories Movement in Argentina.
The Shock Doctrine (2009), directed by Mat Whitecross and Michael Winterbottom
The film based on Naomi Klein's book. It deals with the economic shock therapy implemented in Chile after the assassination of Allende and the question of disaster capitalism.
Wide Open (2009), directed by Gonzalo Arijón (featuring Eduardo Galeano)
A nuanced account of the Pink Tide governments and the struggle for liberation in Latin America
South of the Border (2009), directed by Oliver Stone
A sympathetic portrayal of some of the left-wing presidents associated with Latin America's Pink Tide. It was made before these governments began to run into difficulties but is nonetheless a very good introduction.
The Chicago Boys (2015), directed by Carola Fuentes
A film about the men trained at the University of Chicago who were behind Pinochet's economic policy after the coup that ousted Allende. The first structural adjustment package in Latin America.
El Pepe, a Supreme Life (2018), directed by Emir Kusturica
An intimate documentary about former Uruguayan president, José Mujica.
The Edge of Democracy (2019), directed by Petra Costa
This film explores the rise and fall of both Lula and Dilma Rousseff in Brazil and the dramatic political events that followed.
Massacre in the Stadium: A Victor Jara Story (2019), directed by Bent-Jorgen Perlmutt
A documentary about the murder of Chilean folk singer Victor Jara in the Chile Stadium after the 1973 coup and the more than four-decade struggle for justice by his widow, Joan Jara.
Fondo, otra vez la misma receta (Fund, the same recipe again) (2020), directed by Alejandro Berkovich
A film that provides an overview of the politics and economics promoted by the International Monetary Fund (IMF) and the geopolitical role of the IMF in maintaining US hegemony and capitalist structures of power.

Useful websites

The Bretton Woods Project

The Bretton Woods Project was created in 1995 to monitor and critically evaluate the activities of the IMF and the World Bank and work towards promoting a fairer, more democratic and transparent global economic system rooted in the principles of social justice. The website is a platform for information, networking and advocacy.

www.brettonwoodsproject.org/

CHAPTER 4

Extractivism and ontological politics

INTRODUCTION

As we've already established, the colonization of América/the Americas involved three core elements that are all entangled: extractivism, evangelization, and racialization. By extractivism, we are referring to "activities which remove great quantities of natural resources that are not then processed (or are done so in a limited fashion) and that leave a country as exports" (Gudynas, 2010:1). It can therefore include "any form of economic activity that relies on or benefits from resources or relations that are external to it" (Mezzadra and Neilson, 2019: 134). Extractivism began in the 16th century as colonizers extracted gold, silver, and other minerals as well as crops such as maize, tomatoes, cacao, indigo, and chillies. Native crops and minerals were soon accompanied by imported crops like sugar, coffee, and wheat. Many Indigenous and African slaves and labourers died working in mines or on plantations, while Europeans grew rich, developed industries, and got to consume new items.

In the early 20th century, Latin America became a key site for oil exploration and production and European and US companies such as Shell, Exxon, BP, Mobil, Texaco, Chevron, and Gulf moved into Venezuela, Colombia, Ecuador, and Bolivia, expatriating vast profits and leaving behind a series of serious environmental and health disasters (Valdivia and Lyall, 2019, see Box 4.1). All of these extractivist activities, that are embedded in highly unstable boom and bust economies and unequal forms of land tenure, have involved exploitation and created a cycle of dispossession and dependency. They have directly contributed to a multitude of problems, including land and water pollution, deforestation, soil erosion, food insecurity and hunger, and landslides, flooding, and drought. Some economists and dependency theorists such as Prebisch (1950) understood this problematic as a *resource curse*, a situation where there is significant resource wealth in a particular

DOI: 10.4324/9781003110453-4

location but the local people, including those who work as miners or labourers, are impoverished, as profits end up elsewhere (see Papyrakis and Pellegrini, 2019).

Not only does extractivism have a long 500-year history, but it has intensified in recent decades, promoted by both the right and the left and increasingly involving state-multinational partnerships. The old forms of extractivism – oil and mineral extraction, logging, and plantation-based agri-business – remain and are now joined by new forms – soy and sugar for biofuel production, intensified beef production, palm oil production, shrimp farming, and the building of mega-dams to produce hydroelectric energy (see Boxes 4.1 and 4.4).

BOX 4.1 BIOFUELS: NOT AS GREEN AS THEY SOUND

The cultivation of biofuels from plants such as sugar and soybeans is posited as one of the "solutions" to climate change which overcomes oil dependence. Unfortunately for Latin America, turning foodstuffs into fuel exacerbates unequal land tenure patterns and undermines local food security. Jim Goodman (2007) refers to biofuels as the "biggest scam going" and indeed there is substantial evidence that the cultivation of biofuels in Latin America merely replicates the inequalities, hardships, and environmental damage produced by earlier forms of commodity trade. Biofuel production is often more profitable for large landowners than food production as they receive subsidies from the Clean Development Mechanism, an initiative developed by the United Nations Framework Convention for Climate Change which aims to reduce global carbon emissions (Bird, 2011). South America is currently gripped by soybean fever. The biofuel-driven "soyification" of Paraguay, for example, is having serious social and environmental consequences. GM soybeans (grown since 1996 when illegal GM seeds were brought from Argentina) now account for 18% of Paraguay's GDP (UNDP, 2018). US multinational GM, seed, and agro-chemical companies such as Monsanto, Cargill, and Archer Daniels Midland as well as some Brazilian companies are all gaining massive profits from Paraguayan soybean production and enjoying big tax advantages. Soybeans, and what Joel Correia (2019), drawing on Fogel, calls *sojizacion* or soy territorialization, have transformed Paraguay's rural landscape. He writes how soybeans are embedded in the neoliberalization of nature and are associated with extensive environmental degradation, especially deforestation, a reduction in food security, as well as dispossession and state-led violence against campesinos that includes extrajudicial killings. Many children and adults have been exposed to pesticide poisoning that results in respiratory illnesses, headaches, skin rashes, vomiting, and diarrhoea, as well as death. As Kregg Hetherington (2020) notes, *soy kills*.

Sources: Goodman, 2007; Howard and Dangl, 2007; Correia, 2019; Hetherington, 2020

While extractivist practices have different impacts in different places, the consequences of this intensification are terrifying for both people and our planet. As a result, they are generating the most courageous forms of resistance, led in particular by Indigenous and Afrodescendant peoples and by women. Many of the social movements and rebellions of the past 500 years could be described as struggles against extractivism. While these struggles clearly have environmental dimensions, it is best not to reduce them to the environmental as they are much broader in their orientation and involve the defence of land, livelihoods, territory, culture, health, and food sovereignty. They involve the defence of ways of life and ways of knowing. They are then quite different from the environmental struggles in the Global North.

Drivers behind the acceleration of the extractivist model over the past century include postwar industrial growth in Europe and the United States and neoliberal structural adjustment policies in the 1980s and 1990s that promoted export-oriented production. As a result, land is increasingly used to grow export crops rather than food for domestic consumption and multi-national mining companies have penetrated new territories. The growth of beef cattle farming in many parts of the continent fuelled by the fast food and the pet food industries has been especially deadly and is often closely linked to illegal settlement and deforestation on ancestral lands (see, for example, Cupples and Glynn, 2018; Rautner and Cuffe, 2020). More people have been displaced from their land, deforestation has intensified, and food security has declined.

BOX 4.2 CHEVRON VS. ECUADOR

Some forms of extractivism driven by the oil industry have been truly devastating. The US oil company Texaco (which in 2001 became Chevron) is accused of dumping more than 30 billion gallons of crude oil and toxic waste in the Lago Agrío region of Ecuador between 1972 and 1993 and causing unimaginable damage for the 30,000 people who lived in the region. Thousands of square kilometres were contaminated – rivers were turned black as pits produced spills, plants and crops failed to grow, and the Indigenous (Shuar, Kichwa, Cofán, Siona and Secoya) and campesino inhabitants began to display extraordinarily high levels of cancer, birth defects, and miscarriages, as well as respiratory, renal, skin and intestinal illnesses (see Figure 4.1). Birds and animals, both wild and domesticated, succumbed after drinking contaminated water or were burned to death after falling into furnaces or waste pits. Staple foods such as bush meat and fish disappeared or tasted really strange, and consumption resulted in diarrhoea and skin rashes. There were also many accidents caused by the overflow of waste from pits and by seismic testing. In a study that sought to capture the environmental, economic, and psychological impacts of oil production in the region, one Secoya inhabitant described what life was life before Texaco came to the region.

> We had a lovely, excellent life. We had the jungle and clean wildlife, there were many animals. We lived very comfortably, without stress or harm, the jungle was a paradise. In the year 1941 there were no white people here, there was only jungle. Our daily market was the jungle, there were a lot of fish in the rivers. We had extensive territories where we walked, we did not have borders. We did not know what diseases were, for example, there was no flu. We had our rituals and encounters with the powers of the

Figure 4.1 Oil pollution left by Chevron/Texaco in the Ecuadorian Amazon

Source: Julien Gomba, CC BY 2.0, via Wikimedia Commons

jungle. The shamans were the ones who took care of the jungle, took care of the animals, attracted the fauna. When I was a child, I had never met a white person. In the jungle there were many monkeys and wild pigs. There, at that time, we only used blowpipes, we didn't have guns.

(cited in Beristain, Páez Rovira and Fernández, 2009: 30, my translation)

A Shuar described what it was like afterwards

There is no price that can be paid to remedy the impact of the damage that Texaco has caused, as it is this that killed my family and they want to pay me money. I could take the money, but I will never get the lives of my family back. It also killed the jungle where we live and left our lands and our waters contaminated. For us the jungle is a pharmaceutical secret where there are sacred places of wisdom in every sense of the word … the damage is so great that it couldn't be recovered even if they paid millions of dollars. They can pay, yes, but they are also unable to pay as it is not possible to bring it back like it was. It is a violation of a virgin jungle, yes… but it is just like the violation of a human being, a woman, it is just like raping a woman. This is what they did to the jungle, they violated it with this disastrous destruction, with that oil exploration!

(cited in Beristain et al., 2009: 52, my translation)

In 1993, with the help of human rights lawyers, those affected embarked on a 25-year legal battle against Chevron. In February 2013, the Ecuadorian Supreme Court ordered Chevron to pay $9.5 billion in compensation. Chevron refused to comply with the ruling citing irregularities and took the case to the Permanent Court of Arbitration in the Hague. In 2018, they managed to overturn the verdict using the highly contested Investor–State Dispute Settlement (ISDS) mechanism and demanded payment from the Ecuadorian state (the sum demanded has not been disclosed). ISDS are "corporate courts" that provide large companies with unfair legal advantages that enable them to harm both people and planet and get away with it. These courts are undemocratic in the sense that they can override national legislation that was drafted to protect human rights or environmental protection and order governments to annul judgements made in their own courts. The lawyer that supported the affected peoples of the Amazon was placed under house arrest and later imprisoned. Ecuador filed to get the ruling annulled which was denied. In December 2020, Ecuador announced it would appeal the arbitration award. In the meantime, many of the Indigenous peoples in the region have lost their land, their livelihoods, and some of their culture, perhaps permanently, raising the question of how you can compensate someone for something that can no longer be recovered (see Cepek, 2012).

Without denying the severe harms caused by oil and the importance of Indigenous resistance to it, Michael Cepek's research with the Cofán people presents a more nuanced picture, noting that oil does not also appear "as an instantly apparent and overwhelming power" but that rather oil is approached with a mixture of "uncertainty, surprise, curiosity, fear, fatalism, anger, disappointment, humor, and desire."

Sources: Sawyer, 2004; Beristain et al., 2009; Cepek, 2012; Valdivia and Lyall, 2018

THE WATER WAR AND THE GAS WAR

As well as accelerating extractivism, structural adjustment also encouraged the privatization of state-owned enterprises. What was particularly shocking for many Latin Americans was when privatization spread from airlines, banks, the oil industry, and telecommunications into the privatization of water. Obviously, nobody can live without water so the idea that access to water should provide profits for private companies is really controversial. When they tried to privatize water in the Bolivian city of Cochabamba in 2000 as the result of a World Bank mandate, this really hit a nerve and people rose up in protest against the US-owned multinational company Bechtel that had been granted a 20-year concession by the Bolivian government. Prior to privatization, public water provision in Cochabamba was inadequate. Residents had therefore developed their own water harvesting strategies through the digging of communal wells or the installation of roof tanks to collect rainwater. After taking over the water supply, Bechtel hiked water rates by 200% and residents were told that they required a permit to take water from their own wells and tanks. The people of Cochabamba organized street protests on a massive scale and for three days Cochabamba was paralysed in a city-wide general strike. The Bolivian military violently repressed the protests, but the people refused to back down. Local organizations made up of campesinos and Indigenous peoples produced a counterproposal which rejected the commodification of water and moved to define water as "a social and ecological good that guarantees the well-being of the family and the collectivity and their social and economic development" (cited in Assies 2003: 16–17). In the end, Bechtel was forced to flee Bolivia and the water supply was re-nationalized. The water war was fought and won by a diverse coalition of political actors, who had mobilized alternative understandings of water based not on individual ownership but on collective Andean cosmologies (see Laurie et al., 2002; Assies, 2003; Perreault, 2005). This uprising became known as the water war and it became a source of global inspiration.

It was rapidly followed a couple of years later by the gas war centred in the city of El Alto. As Forrest Hylton and Sinclair Thomson (2007) write, in El Alto, there were a number of dissatisfactions, that included the militarization of Indigenous communities, the criminalization of activists, and opposition to the proposed FTAA, but the uprising was unleashed in response to a proposal to export Bolivian gas via Chile to California. It started when 1,000 Aymara activists initiated a hunger strike at a local radio station and called for blockades. They managed to cut off the entire department and tens of thousands of people turned out to protest. The activists called their operations Plan Añutaya named after a small animal that eats pests or Plan Abeja – an *abeja* is a bee, meaning that in places where they couldn't maintain a full blockade, they would sting strategically. There were many arrests. Aymara radio stations called for the president's resignation. The protestors then called a general strike – the forces became increasingly heterogeneous and were joined by many others, including coca growers and miners, and the protests spread to many other towns and cities. The police and military became more violent – on 12 October, 54 protestors were massacred. But the city of El Alto had been effectively occupied and so the protestors began to set up their own self-government. These struggles are decolonial struggles, they are not just anti-neoliberal ones – they draw on ancestral knowledges and Indigenous modes of governace such as the *ayllu* to imagine a different kind of world. The *ayllu* is an Aymara and Quechua concept that predates the conquest and involves participatory forms of democracy, autonomous forms of organization, and the frequent rotation of community leaders. It is a system that facilitates rebellion and the formation of effective micro-governments and it is deeply unsettling to Eurocentrism – USAID labelled one of the neighbourhood Indigenous organizations in El Alto a terrorist organization, hence the "important role of *Ayllu* to cracking capitalism in everyday life, and to gradually changing the habitus is clearly revealed" (Petropoulou, 2018: 18) (see also Chapter 10).

THE PINK TIDE AND NEO-EXTRACTIVISM

As we discussed in Chapter 4, the Pink Tide (PT) governments came to power promising not only a move away from the colonial–capitalist model that had dominated Latin America since the conquest but new rights for both Indigenous peoples and Nature. The governments of Bolivia and Ecuador institutionalized the Aymara and Quechua concept of *Buen Vivir* which embodies the idea of living in harmony with nature. *Buen Vivir* encompasses everything in the cosmos, the spiritual and the material, the human and the non-human, and has mobilized a radical critique of capitalism and consumerism. Furthermore, it has been enshrined in the constitutions of both Bolivian and Ecuador. Not only did these constitutions recognize Indigenous peoples' rights, they also recognized the rights of Nature. For example, articles 71–74 of the 2008 Ecuadorian Constitution recognize that Nature or Pachamama, understood as the site where life is reproduced, has its own set of rights. This means that rivers, lakes, mountains, birds, and insects have the right to exist and to flourish and to be restored when polluted or threatened with extinction. Any person, community, people, or nationality can call upon the state to enforce these rights. As Eduardo Gudynas (2015b) writes, while this is a radical position and constitutes a clear departure from the Eurocentrism that characterized previous constitutions in Ecuador, it has serious limitations as it reproduces the rights of humans to a healthy environment, meaning that the anthropocentrism of nature as an extension of property rights of humans remains intact. The challenge for Gudynas is how to break with anthropocentrism and move to a biocentric position where Nature has its own rights that are independent of human valuations and appropriations. Despite these limitations, such constitutional reforms constituted key sources of political optimism. They did not, however, bring an end to the extractivist model of development.

There were both important shifts and continuities in terms of extractivism when the PT governments came to power. What we saw in Venezuela, Bolivia, and Ecuador was termed neo-extractivism, as it was a form of "progressive" extractivism in which the state plays a greater role. Taking advantage of high commodity prices at the time, many PT governments embraced traditional extractivist practices, but the profits were used instead to fund social programmes that benefit the poor majority. This form of extractivism produces environmental damage but is perhaps socially more beneficial.

Extractivist activities are then supported by both right-wing and left-wing governments. Even forms of extractivism that are state-led and result in increased social spending still usually involve international forms of dependency and transnational forms of investment, increasingly from China as well as from the United States, Canada, and Europe. While Chinese investment sits more easily with the anti-imperialist rhetoric of PT governments, it cannot necessarily be celebrated as a less colonizing form of South-South cooperation, given the forms of dependency it engenders (see Chapter 5). Indeed, extractivism means that the PT governments have had to negotiate many contradictions and Indigenous peoples whose mobilizations brought them to power have often felt betrayed (see Box 4.3).

BOX 4.3 THE TIPNIS AFFAIR

One example where Indigenous peoples have felt betrayed is in the TIPNIS affair in Bolivia. In 2011, five years after the election of Evo Morales and two years after the passing of the new Constitution that enshrined the concept of

Buen Vivir, Indigenous peoples protested a government proposal to build a 200-mile highway through the Isiboro-Sécure National Park and Indigenous Territory (TIPNIS), a national park that straddles the departments of Beni and Cochabamba. TIPNIS has an area of 1.2 million hectares and is inhabited by the Mojeño-Ignaciano, Yuracaré, and Chimán Indigenous groups (McNeish, 2013). Although it has been a national park since 1965 and the Indigenous inhabitants possess legal title to their territories in this area, it has been subject to pressures from settlers, some of whom have arrived to plant coca or rear cattle in the area. The Morales government stated that the building of the road would connect Amazonian and Andean groups, facilitate trade, and bring prosperity (Fabricant and Postero, 2015; Ranta, 2016). While some Indigenous peoples were in favour of the development, many were opposed to it because they believed a road would encourage the incursion of loggers and coca farmers and be a threat to the existing human and nonhumans in the reserve. The police repressed the protest quite violently – they used tear gas and rubber bullets – and it did grave harm to the Morales government that had come to power on a platform of Indigenous rights. Two ministers were forced to resign, and Morales was forced to suspend the project and embark on a widely criticized consultation project. In her discussion of the case, Anna Laing (2019) argues that there is persistent logic of coloniality within the plurinational state but that Indigenous peoples continue to fight for decoloniality and for the delinking (see Mignolo, 2007) of the Bolivian state from its colonial foundation. There is no doubt that state-led neoextractivism undermines the commitments made to Indigenous autonomy and territorial self-determination. This example reveals the contradictory dynamics of Indigeneity in Bolivia (McNeish, 2013), the limitations of *Buen Vivir* as state policy (Ranta, 2016) as well as the resilience of capitalist and extractivist models of development.

Sources: McNeish, 2013; Fabricant and Postero, 2015; Ranta, 2016; Laing, 2019

BOX 4.4 COCA-CODO SINCLAIR: HYDROELECTRIC POWER AND THE POLITICS OF *BUEN VIVIR*

One key form of extractivist activity in Latin America is the building of hydroelectric dams to generate electricity. While hydroelectric power is often seen as a green form of energy, the dams themselves create serious ecological and social problems and often involve displacement of people and ecosystems. In 2019, I visited a recently constructed mega-dam called Coca-Codo Sinclair on the Coca River in the Ecuadorian Amazon funded

Figure 4.2 The Coca-Codo Sinclair hydroelectric dam in Ecuador

Source: Photo by author

by Chinese investment. One of eight mega-dams constructed by the Correa government, it is a huge undertaking and is spread over several kilometres (see Figure 4.2). The Ecuadorian government sees this mega-dam as part of its Plan Nacional para el Buen Vivir (PNBV). This plan promotes a redistribution of wealth and a gradual exit from an export-focused economy to one where the means of production are more democratically accessed. It was built using financing from a Chinese company called Sinohydro (see below) and the dam was supposed to provide 30% of Ecuador's energy needs. Driven by China's demand for crude oil, it is financed through a loan-for-oil agreement, which means that Ecuador has taken on unsustainable amounts of debt and commitments to China that undermine both domestic sovereignty and commitments to *Buen Vivir* (Casey and Krauss, 2018). To repay the $19 billion loan, China gets to keep 80% of Ecuador's oil and actually gets the oil at a discount and can then sell it for an additional profit. It's so difficult to pump enough oil to repay China that Ecuador is drilling deeper and deeper into the Amazon, causing deforestation and displacing Indigenous peoples whose lands are found on oil fields. The government has also had to implement a series of austerity measures, making massive cuts in public spending – and these underpinned the massive protests we saw in Ecuador in 2019. The dam has become a national scandal – several workers died in a tunnel collapse (Araujo, 2014) and a number of politicians that were involved in the dam's construction have been convicted of bribery (Casey and Krauss, 2018). In addition, the dam is located in an area that is exposed to high seismic and volcanic risk. The nearby El Reventador volcano, which has experienced about 20 different eruptive periods since 1541, has been in a heightened phase of activity with intermittent eruption since 2002. The dam has also resulted in the ecological transformation of the Coca River – there are visible signs of erosion and the San Rafael waterfall has disappeared (Pacheco, 2020). Small local business owners such as kayakers involved in developing ecotourism in the region told us in 2019 that they have seen their business collapse. So it is very hard to say that Ecuador in spite of having a left-wing government committed to *Buen Vivir* and Indigenous rights moved away from a colonial model of dependency and extractivism. It is also hard to argue that China is a better partner than the United States.

Sources: Escribano, 2013; Nathanson, 2017; Casey and Krauss, 2018

The promotion of extractivism by PT governments has brought them into conflict with the social movements that brought them to power and has called into question their radical commitment to *Buen Vivir* and the Rights of Nature. The government of Rafael Correa found itself frequently in conflict with CONAIE (see Box 7.2) and other Indigenous organizations. As social movements and Indigenous groups began to oppose Correa's extractivist policies, his government became more repressive and many Indigenous leaders, even those who supported Rafael Correa's bid for the presidency, were criminalized and imprisoned, accused of "terrorism and sabotage" (Picq, 2011; Zibechi, 2011, see Chapter 7).

Thea Riofrancos (2019, 2020) has argued that we need to distinguish between the left in power – the elected politicians who are confronting the might of transnational capital and having access to resource rents that could fuel development – and the left in resistance made up of social movements that had imagined and mobilized for a different kind of world beyond extractivism. The left in power talks about resource sovereignty but fails to avoid new forms of dependency such as with China (see Chapter 5). Furthermore, entrenched socio-economic inequalities are not erased. In boom times, there is increased social spending that helps to secure or retain popular support but there is no redistribution of wealth. In bust times, the government must then resort to austerity measures that exacerbate inequalities. In both boom and bust times, the government resorts to repression to defend its extractivist model.

EXTRACTIVISM AND THE LOGIC OF COLONIALITY

As a result of extractivism and other destructive capitalist practices, some scholars assert that we are living through the Anthropocene, an epoch in which human beings are the most potent geological force.

> ... what can be more eloquent of human geological force than the removal of mountains in a time-efficient search for minerals, the damming of large bodies of water to reroute rivers for hydroelectric commercial purposes, the transformation of rainforests into palm oil plantations or cattle grasslands and of deserts into land for industrialized agriculture?
>
> (Blaser and de la Cadena, 2018: 2)

But there is a colonial element to the Anthropocene which is sometimes overlooked. Extractivism is without doubt a colonial practice, one that "continues the practice of terra nullis ... by rendering empty the places it occupies and making absent the worlds that make those places" (Blaser and de la Cadena, 2018: 3). As Claudia Composto and Mina Lorena Navarro (2014: 13, my translation) write, this "new cycle of dispossession (...) is a problem that has a history behind it, that is at least five hundred years long." In 2011, the Mapuche people in Chile issued a declaration against US gold mining company, Meridian, comparing their plunder of gold to that committed by the Spanish 500 years earlier (Urkidi and Walter, 2011). The situation is now so urgent that people threatened by extractivism have no choice left but to protest and to do everything they can to block these activities.

There are multinational mining projects underway in many Latin American countries, including Argentina, Brazil, Chile, Colombia, Costa Rica, Ecuador, El Salvador, Guatemala, Honduras, Mexico, Panama, and Peru. All of them are inflicting huge damage on rural and coastal communities, including forced displacement and environmental devastation and an increase in drug trafficking and prostitution, and so there are mining–related social conflicts in all of them. It's worth understanding what hard rock mining does to water. To process the ore you need large quantities of water and slurries that contain cyanide or arsenic or other chemicals. The water that is pumped from below the water table often makes contact with oxidizing rock and become acidified, as well as being laced with cyanide or arsenic and that affects water safety and quality as well as water quantity. Subsistence farmers can find themselves deprived of water for irrigation and water supply for drinking is reduced or becomes unsafe. These effects can persist for decades after a mine is abandoned.

The conflicts are significant because most Latin American governments have ratified the 1989 ILO Convention 169 on the rights of Indigenous peoples and the 2007 UN Declaration of the Rights of Indigenous Peoples. ILO C169 came into force in 1991 and it has been ratified by most Latin American countries, including Colombia, Guatemala, Honduras, and Nicaragua. C169 mandates that any development that affects the lands or lives of Indigenous peoples must be accompanied by a process of Free, Prior and Informed Consent (FPIC). It applies to any activity that will take place on ancestral land or that will involve the use of resources and non-human entities, such as rivers, owned and managed by Indigenous peoples. This means the establishment of a non-coercive process of consultation in advance of any activities commencing and the need for those affected to be provided with sufficient information on the risks and potential benefits of any development. Despite the protection of international law, FPIC is frequently and routinely violated by multinational and national companies and their activities are endorsed and supported by states, as well as sometimes by international organizations such as the World Bank. Violations therefore go unpunished, and impunity and corruption prevail. Campesino, Indigenous, and Afrodescendant peoples have therefore been organizing and protesting, doing what they can to protect their own lands and prevent access by extractivist companies. In the process, they often pay for their activism with their own lives. For example, most of the territory of the Shuar people who live in the Morona Santiago Province in the Ecuadorian Amazon has been concessioned to large-scale mining, oil, and hydroelectric projects without consultation. As a result, the Shuar have been forced to engage in risky anti-mining activities in defence of their lands and their water resources and the state has responded with military raids, harassment, and criminalization and three Shuar were murdered by state forces during the PT government of Rafael Correa (Riofrancos, 2019).

The murder of environmental defenders such as the Shuar is a frequent occurrence. According to Global Witness (2021, see also Bebbington, 2019), 227 environmental defenders were murdered in 2020 and two thirds of these were in Latin America. The most dangerous countries in which to be an environmental defender were Colombia (with 65 killings), Mexico (30 killings), Brazil (20 killings), Honduras (17 killings), Guatemala (13 killings), Nicaragua (12 killings) and Peru (6 killings). A large proportion of the people murdered are Indigenous or Afrodescendant activists. In the vast majority of cases, these murders take place in a context of impunity; it is very rare for perpetrators to be brought to justice. As Anthony Bebbington (2019) writes, as a rapidly transnationalizing extractivism expands into new areas, increasing numbers of actors are prepared to resort to violence in order to gain access to resource rents.

It is important to say and remember the names of those killed; they include Costa Rican Bribri leader **Sergio Rojas Ortiz** for his opposition to illegal settlement on ancestral lands, who was murdered shortly after he went to denounce the death threats he had received; **Christian Javá Ríos**, who had opposed oil extraction in the Peruvian Amazon; Mexican Tarahumara defenders **Otilia Martínez Cruz** and her son, **Gregorio Chaparro Cruz**, who were fighting to prevent illegal deforestation on their land; Nahua leader **Leonel Díaz Urbano**, who had fought against the construction of a hydro-electric plant in Zacapoaxtla in Mexico; Wajãpi member, **Emyra Wajãpi**, who was stabbed by armed goldminers and his body dumped in the river in the Brazilian Amazon; and Garífuna leader **Mirna Suazo** from Honduras, who had opposed the creation without FPIC of two hydroelectric plants on the Masca river. There are many more.

Despite the dangers involved in activism and the corruption and impunity that surrounds political violence in Latin America, many environmental defenders believe that there is no alternative, as if they lose their land or their water supply is contaminated, their community and their children and grand-children have no future. Environmental defenders are murdered because of the threat their activism poses to capitalist accumulation. Their collective action that involves blockades, occupations, appeals to international law, and cre-ative use of media tools often works to disrupt the aspirations of extractivists. The mining companies have substantial resources to hire lawyers and security guards and they often count on the support of the state. Nonetheless, the power of community resistance is substantial and often achieves large, small, or partial victories. Let us consider two examples in depth, one from Honduras and the other from Guatemala. In these cases, the environmental defenders mobilize a decolonial and anti-capitalist philosophy by drawing on ancestral knowledges and non-Eurocentric ontologies.

THE LIFE OF BERTA CÁCERES

The fragility of democracy in Honduras was underscored in June 2009, when the democratically elected president of Honduras, Manuel Zelaya, was over-thrown in a coup that had the support of Hilary Clinton and the US State Department (Lakhani, 2020). Although a member of the political elite and not strictly speaking a PT president, Zelaya was advancing a progressive agenda that included support for Indigenous land rights and LGTBQ+ rights that "deeply upset the Honduran and transnational capitalist, military and reli-gious elites" (Méndez, 2018: 11). The military clamped down on protests in support of the deposed president and Zelaya was never returned to power. In the period since the coup, extractivist projects such as mining and hydroelec-tric dams have increased exponentially and have been vociferously opposed by campesino, Indigenous and Afro-Honduran groups whose lands, lives, livelihoods, and health are threatened by such activities. As Sharlene Mollet (2017) writes, post-coup Honduras is characterized by intimidation, phys-ical assault, death threats, unlawful detention, and assassinations; Indigenous and Afro-Honduran environmental leaders, along with journalists, lawyers, trade unionists, and LGBTQ+ people are especially targeted. Honduras has become one of the most dangerous places in the world to be an activist, as the government uses repressive measures to fight "an intense counterinsurgency campaign on behalf of transnational capital" (Méndez, 2018: 11).

One of the most high-profile assassinations in recent years is that of Honduran Lenca activist, Berta Cáceres, who was murdered in her own home in March 2016. She was the co-founder of an organization called COPINH (Council of Popular and Indigenous Organizations) that was formed to defend the political and cultural rights of the Lenca people. COPINH's work had an intersectional focus and mobilized around predatory capitalism, US imperialism, racism and discrimination, and gender and sexual equality. Berta gained international visibility in 2011 when she helped to organize a protest against the proposed Agua Zarca hydroelectric dam, run by a company called DESA with investment from Chinese company Sinohydro, the same company that funded the Coca–Codo Sinclair dam in Ecuador mentioned above. DESA wanted to dam the Gualcarque River on Lenca territory and so the Lenca people led by Berta and others began to organize a series of protests, occupations, and road blocks to prevent the dam construction going ahead (for details of these events, see Lakhani, 2020). She worked closely with other Indigenous and Afro organizations in Honduras and internationally, making connections for example between Lenca opposition to the Agua Zarca dam and the Sioux opposition to the Dakota Access Pipeline in the Standing Rock protest. In 2015, Berta was awarded the highly prestigious Goldman Environmental Prize for her activism, evidence of the international recognition she had gained.

The COPINH defence of the Gualcarque River can't be understood in the Eurocentric language of environmentalism which sets up a binary between the human and nonhuman and nature and culture. It is not simply about avoiding environmental damage to the river's ecosystem but it is also driven by a Lenca cosmovision that sees the river as a sacred and sentient being that has its own political agency. Berta articulated this cosmovision repeatedly in interviews (see also Box 4.5).

> When we started the fight for Río Blanco, I would go into the river,
> I would talk to the river and I could feel what the river was telling me.
> I knew it was going to be difficult. But I also knew we were going to
> triumph, because the river told me so.
>
> (Frente Juvenil 2016, cited in Méndez, 2018:20)

> This mountain region has a strong relationship with the Lenca people,
> the forests are alive, the mountains are alive. This is a live river that is
> threatened by the construction of six hydroelectric dams … From the
> Lenca cosmovision, water is a fundamental element, just like land is
> part of balance and creation, the spirits live in the water. That is why it
> is crucial to respect and care for the water as a being just like us. This
> explains why a community has so much strength to defend a river.
>
> (Friends of the Earth 2017, cited in Méndez, 2018:21)

BOX 4.5 BERTA'S SPEECH ON RECEIVING THE GOLDMAN PRIZE

In our worldview, we are beings who come from the earth, the water, and the corn. The Lenca people are ancestral guardians of the rivers, who are in turn protected by the spirits of young girls who teach us that giving our lives in various ways is to give life for the good of humanity and this planet. COPINH, walking with other peoples for their emancipation, ratifies its commitment to continue defending water, rivers, and our shared resources and those of nature, along with our rights as peoples. Let us wake up! Let us wake up, humanity! We are running out of time! Our consciences will be shaken through contemplation of our own self-destruction wrought by predatory capitalism, racism, and patriarchy. The Gualcarque River has called upon us, as have other gravely threatened rivers. We must respond. Mother Earth, militarized, fenced-in, poisoned, where basic rights are systematically violated demands us to take action. Let us build societies capable of co-existing in a fair and dignified way that protects life. Let's get together and continue with hope defending and caring for the blood of the earth and the spirits. I dedicate this prize to all rebellions, to my mother, the Lenca people, the Río Blanco, to COPINH and all of the martyrs who died defending natural resources.

Source: Reproduced in COPINH (2015), my translation

The coloniality at work in the proposed construction of the dam and especially in the intimidation and assassination of Indigenous and Afrodescendant environmental defenders is clear. For the World Bank, Sinohydro, and DESA, the river is seen in instrumental terms as a source of electricity and a source of profit for the investors and Lenca territory in colonial terms as *terra nullis*. The murder of people who seek to defend territorial rights are then justified in the same terms as the genocide that followed the conquest was. Berta's comrade and Garífuna leader of the Fraternal Black Organization of Honduras (OFRANEH), Miriam Miranda, is very clear that the mestizo Eurocentric Honduran state wants to eliminate her people. Miriam Miranda has been repeatedly intimidated for her activism, she's been physically assaulted and unlawfully detained (Mollet, 2017). As she told Amy Goodman on Democracy Now (2020), there is a "well-crafted plan by the Honduran state to exterminate the Garífuna community … because we are a community in constant fight for ancestral rights."

In order to understand the colonial violence that underpins these struggles, how it is that somebody like Berta Cáceres can be murdered with impunity, it is useful here to draw on Enrique Dussel's (1993: 75) account of the myths of modernity which he summarizes as follows:

(1) Modern (European) civilization understands itself as the most developed, the superior, civilization. (2) This sense of superiority obliges it, in the form of a categorical imperative, as it were, to "develop" (civilize, uplift, educate) the more primitive, barbarous, underdeveloped civilizations. (3) The path of such development should be that followed by Europe in its own development out of antiquity and the Middle Ages. **(4) Where the barbarian or the primitive opposes the civilizing process, the praxis of modernity must, in the last instance, have recourse to the violence necessary to remove the obstacles to modernization.** (5) This violence, which produces, in many different ways, victims, takes on an almost ritualistic character: the civilizing hero invests his victims (the colonized, the slave, the woman, the ecological destruction of the earth, etc.) with the character of being participants in a process of redemptive sacrifice. (6) From the point of view of modernity, the barbarian or primitive is in a state of guilt (for, among other things, opposing the civilizing process). This allows modernity to present itself not only as innocent but also as a force that will emancipate or redeem its victims from their guilt. (7) Given this "civilizing" and redemptive character of modernity, the suffering and sacrifices (the costs) of modernization imposed on "immature" peoples, enslaved races, the "weaker" sex, et cetera, are inevitable and necessary.

So according to myth number 4, when Indigenous or Afrodescendant people become obstacles to (innocent modernizing) development, their elimination is both necessary and justifiable. This colonial understanding appeared in the views of World Bank Group president Jim Yong Kim, who in a keynote address given in 2016 described Berta's death as an "incident," stating that "you cannot do the work we're trying to do and not have some of these 'incidents' happen" (cited in Méndez, 2018:13). Indeed, this is a colonial discourse that gets recycled constantly in relation to extractivism. In Peru in 2007, then President Alan García accused Indigenous people who opposed extractivist projects as obstacles to progress, describing their position as resulting from "demagogy and deception that say that these lands cannot be touched because they are holy objects" (García Pérez, 2007, cited in Stensrud, 2019: 159).

They killed Berta Cáceres in an attempt to end the opposition to the dam, but it is clear that Berta did not die but lives on in all kinds of ways (see Figure 4.3). The Agua Zarca dam has still not been built and in that sense Berta's struggle has triumphed. Not only are there hundreds of Lenca people who continue the struggle, but she is a global source of inspiration who appears in other sites – underscoring the connectedness of decolonial struggles across the globe.

Figure 4.3 Tribute to Berta Cáceres in Guatemala City, International Women's
          Day 2019

Source: Photo by author

LA PUYA, GUATEMALA

In 1996, after the peace accords were signed in Guatemala bringing the civil
war to an end (see Chapter 5), the Guatemalan government began to encourage
foreign direct investment in polluting extractivist projects such as strip mining
and hydroelectric dams. This meant, as Diane Nelson (2019: 130) writes, that
Guatemalans "just barely recovering from the violence found entire mountains
leveled and water sources contaminated and rivers threatened with damming
or being sluiced away." The government of former president Otto Pérez
Molina (later jailed on corruption charges) granted hundreds of concessions
to multinational mining companies from Canada and the United States. The
people here knew that aside from dangerous and poorly paid employment
for local people, multinational mining projects tend to bring few benefits
to local communities and they often lead to violence, including community
displacement and environmental devastation, including the pollution of local
water resources. While mining companies sometimes provide a health centre,
a school, or electricity supply in the communities in which they operate under
the euphemizing guise of corporate social responsibility, these are services that
many of us would consider to be the responsibility of the state and to which
communities are entitled regardless of multinational investments. The spread
of extractivist projects means that Indigenous and low-income ladino communities who had to fight to defend their lives, lands, and livelihoods during

the civil war from racist and imperialist aggression have now had to organize to defend them from the scourge of mining. Although Guatemala signed ILO C169, Guatemalan environmental defenders are frequently subject to physical assault, harassment, intimidation, extrajudicial eviction, defamation, raids, death threats, and assassinations (UDEFEGUA, 2020).

In 2012, two Guatemalan communities, San José del Golfo and San Pedro Ayampuc, mobilized in resistance to gold mining in their community. They called their movement Peaceful Resistance La Puya (La Puya means cattle prod) and they rose up in opposition to a mining company called El Tambor originally owned by Canadian company Radius Gold Inc and US company Kappes, Cassidy, and Associates (KCA), and operated through a Guatemalan subsidiary known as Exmingua. This company came to mine on their land without consent or consultation. When company officials started to arrive in the community, they did so without informing local people that a gold mine was planned. They also secretly built a tunnel to carry out the mine tailings. As soon as community members realized what was happening, they began to organize. They set up a road blockade to block access to the mining site "in defense of life and water" and suffered three violent evictions by riot police. This roadblock was highly organized; people organized rotas and others provided food to keep those on the blockade fed. They engaged in several years of peaceful resistance in which they had to face riot police, intimidation, physical attacks, and harassment. Many community members have been seriously injured in police attacks. In June 2012, one protestor, Yolanda "Yoli" Oque Veliz, was ambushed by two masked gunmen and shot multiple times (Pedersen, 2014). After this attempted murder, Radius pulled out of the venture. Although it was hard, the resistance was seen by the community members as beautiful because of the way that the community came together and developed this strong sense of peaceful solidarity. They played music and sang songs, they prayed, they shared food, they lay on the ground and linked arms, they celebrated their culture. They used their cell phones to document harassment and intimidation from the company executives and raise global awareness of their struggle. Aggressive company executives constantly tried to provoke them into a violent response, but they maintained their peaceful resistance. They also developed an extraordinarily high degree of legal literacy and awareness of the rights that are enshrined in international law.

Alexandra Pedersen (2014), who did long-term research with the activists at La Puya, sees Canadian and US mining companies as modern-day conquistadors. Extractivism has a 500-year history in Guatemala: it began with the conquest, so the Mayan people are "well acquainted with the colonial mining technology as a form of 'development'" (Pedersen, 2014; 194). Just as in the days of the conquest, these companies try to buy support from community members through offering them trinkets, *espejitos*, and sadly in conditions of poverty some people accept. This sows division in communities in struggle and it is a very common strategy.

Despite these divisions, the people at La Puya have waged a successful defence against extractivism. In the process, they have negotiated Eurocentric legal frameworks (as well as mechanisms that come out of environmental sciences such as environmental impact assessments) and have engaged in a kind of border thinking or a form of transmodernity where Eurocentric law is harnessed in defence of ancestral knowledges and ways of being. They successfully defended their right to be consulted and in 2015 successfully got a court injunction to suspend all activities at the site. The company continued its activities in spite of the injunction and in February 2016, the Guatemalan Supreme Court ruled to provisionally suspend the mining license due to lack of prior consultation. One of the company employees was jailed for threatening and harassing protestors and journalists. All mining activity stopped in 2017. At the time of writing, they were still fighting for the entire concession to be cancelled definitively.

ONTOLOGICAL POLITICS

A decolonial analysis of extractivism in Latin America reveals that the present social and ecological crisis is a crisis in the dominant model of civilization, produced by coloniality, capitalism, productivism, patriarchy, xenophobia, plutocracy, and anthropocentrism working together. As these two cases show, activists are defending land and territory but doing so with a non-Eurocentric understanding of the social and natural world and as such they work to dismantle the dualist ontologies of the Eurocentric modern project. When Berta Cáceres talks about the river talking to her and being home to spirits, it is important not to see these as Indigenous beliefs to be either dismissed or respected as long as they do "not express an epistemic alternative to scientific paradigms" (see de la Cadena, 2010:349) but rather as knowledges that reveal a quite distinct relationship to the world from the one held by many Europeans, scientists, and multinational mining companies. Supernatural beings, earth-beings, or water spirits are usually deemed non-admissible in both modern polities and the environmental sciences and state actors and scientists often try to exclude them from both legislation and scientific scholarship, but as outlined below cannot always do successfully. This conflict has been framed by a number of decolonial scholars, including Mario Blaser, Marisol de la Cadena and Fabiana Li, as ontological politics. Indeed, as Juan Javier Rivera Andía and Cecilie Vindal Ødegaard (2019: 1, 19) write, "extractivism actualises questions of ontological difference."

Marisol de la Cadena is a Peruvian non-Indigenous anthropologist who did fieldwork in Cuzco in Peru in the mid-2000s when multinational extractivist practices were in ascendancy. The Indigenous inhabitants, her research participants, were opposed to a multinational mining concession awarded to mine Ausangate, a sacred mountain in Peru where an earth-being or Apu resides, and they defeated it through the activation of an ontological politics. She writes:

> The event was that of a mountain (perhaps replete of gold) that was also an earth-being (or an earth-being, also a mountain) participating in a political contest where one reality was more powerful than the other. The human participants in the conflict were environmentalists, Quechua indigenous-mestizos, and engineers working for a mining corporation. An alliance between the first two defeated the golden aspirations of the corporate engineers. The *mountain* won, the mining company lost; but to earn this victory, the earth-being was made invisible, its political presence recalled by the alliance that also defended it.
>
> (de la Cadena, 2014)

In Eurocentric thought, the earth-being would be reduced to a belief (to be either tolerated or dismissed by western academics) but in the Andean cosmovision, the earth-being is a fully fledged actor and so constitutes a challenge to what is deemed to exist and to be part of the analysis. What is at work here is a different and a plural kind of politics that comes into view, that can be very hard for non-Indigenous peoples to grasp. Indeed, de la Cadena expresses her inability to understand everything her Indigenous participants tell her. These ontological politics emerge, according to de la Cadena (2010: 346) "not because they are enacted by bodies marked by gender, race, ethnicity, or sexuality demanding rights, or by environmentalists representing nature, but because they bring earth-beings to the political and force into visibility the antagonism that proscribed their worlds." They involve practices that the Peruvian state cannot recognize as legitimate, because to do so "would require its transformation, even its undoing as a modern state" (de la Cadena 2015: 14). Even so, the state and the companies cannot completely dismiss them either because of the way they constitute themselves as a political force. We are not talking about different understandings of nature but about the existence of different worlds. Earth-beings, water spirits, supernatural spirits all become part of the political and challenge Eurocentric binary thinking. As Mario Blaser (2014: 55) writes:

> political ontology cannot be concerned with a supposedly external and independent reality (to be uncovered or depicted accurately); rather it must concern itself with reality-making, including its own participation with reality-making.
>
> (Blaser, 2014: 55)

Fabiana Li's (2015) ethnographic work on mining in Peru also analyses the ontological conflicts over different worlds. Cerro Quilish, also an Apu, was targeted for excavation by a gold mining company, Minera Yanacocha, who had other mining projects in the region. The defence of the mountain was led by rural Indigenous peoples and campesinos directly affected by mining, but was also supported by many other people, including students, NGOs, religious organizations, and urban market vendors and transport workers. It was clear that the rural protestors (who according to Li do not tend to refer to themselves as Indigenous) saw Cerro Quilish not a resource that could generate profits but rather as a source of water and as a sentient being that had not consented to being mined and furthermore could demand things from the mining company if mining were to go ahead. Cerro Quilish is alive but is not necessarily always benevolent. Li goes on to show how these distinct understandings of the mountain could not be captured within Eurocentric notions of corporate social responsibility that tried to maintain that Cerro Quilish could be mined responsibly and all that was required was techno-logical management along with the provision of compensation or some form of conventional development to the affected communities. She writes:

> The mountain was not just an economic resource to be defended, but the embodiment of life itself. By calling Cerro Quilish an Apu, the protestors suggested that it was a living entity, and furthermore, that other lives (both human and nonhuman) depended on its existence.
>
> (Li, 2015: 110)

This understanding of Cerro Quilish as an Apu could not simply be dismissed as Indigenous belief or superstition because "Cerro Quilish, as mountain, ani-mate being, water, and livelihood, entered into the sphere of politics and became a powerful force with which corporations and the state had to con-tend" (p.111) and ultimately led to the company abandoning its plans to mine Cerro Quilish. In this process, scientific knowledges on the importance of the mountain as an aquifer or as a site of biodiversity converged, albeit uneasily, with understandings of the mountain as an Apu, while a Catholic priest who supported the anti-mining campaign also embraced Cerro Quilish as an Apu. The company shifted its focus on water management in order to clean up its image, which "had the effect of destabilizing Cerro Quilish's multiplicity and enabling a singular reality (water) to take hold" (p.142). Such an effect can, however, only be temporary.

Adopting this ontological position means recognizing that "nothing, exists prior to the relations that constitute it (Escobar, 2016: 18). This rela-tional approach, according to Escobar challenges the Eurocentric worldview that separates humans from nature and that seeks to reduce everything in the jungle or forest to "nature" and nature into a resource that can be possessed and sold for a profit. It means recognizing that mountains, lakes, glaciers, and rivers are alive, sentient beings that have rights. It means recognizing the existence of multiple worlds, of a pluriverse. It means decentring ourselves as humans so that we can develop non-instrumental relations with non-humans. It also means that we can't separate the meanings that we apply to things or the stories that we tell about things from the things themselves (see Wynter, 2003; Wynter and McKittrick, 2015). Orlando Fals Borda (2009), a revolutionary Colombian sociologist coined the term *sentipensar* (thinking-feeling) to describe the way his campesino participants brought together reason, emotion, love, body, and heart in order to participate in harmonious interactions with the entities (humans, animals, sentient and non-sentient objects, ancestors, spirits, collective memory) that make up the world. Environmental defenders can be similarly thought as *sentipensantes*, as their activism draws on ancestrality and territorial connected-ness to generate non-dualistic knowledges, committed to life rather than cap-italist accumulation. As these movements gain momentum, we can see how Eurocentric thought is unable to solve the problems that it has created and that other ways of being, knowing, thinking, and feeling are required.

TERRITORY/TERRITORIO

Ontologies are enacted in place or more specifically in territories. Increasingly, resistance to extractivism in Latin America is articulated in terms of the defence of territory or *territorio*. *Territorio* is not simply a piece of land but is seen instead in the words of Maristella Svampa (2019: 27) "as a place of reappropriation and creation of new social relations," something that is both symbolic as well as material and explicitly connected to notions of autonomy and self-determination. Territory is then a collective space in which to repro-duce life. Talking about territory rather than land resignifies the meaning of conservation or environmental protection and is at odds with neoliberal or modernizing development visions. Territories are not, however, static; they are always in process. Escobar (2015:35) defines territory from this decolonial perspective as follows:

> The "territory" is the space — at the same time biophysical and epistemic — where life is enacted according to a particular ontology, where life becomes "world." In relational ontologies, humans and non-humans (the organic, the non-organic, and the supernatural or spiritual) are an integral part of these worlds in their multiple interrelations
> (my translation)

Territorial imaginaries are important in undermining colonial-capitalist visions of development. They challenge visions that see the sites of proposed extractivist activity as empty, and put forward spaces laden with cultural, pol-itical, or spiritual meaning inhabited by rights-bearing humans and non-humans. In the film *Defensoras-es*, the protestors at La Puya are particularly good at articulating what is meant by territory and how it encompasses quite different elements not shared by those who want to mine for gold for profit.

As Svampa (2019: 4) cogently puts it, what is going on in these forms of political action is a contestation of the concept of "development", as they reveal that there are "other ways of building society and inhabiting the world."

What is at stake in the extractivist drive is of course life itself and it is therefore important to understand how a colonial–capitalist worldview which is driving the destruction of life itself became so hegemonic despite substantial evidence of the irreversible and devastating harms it brings. It is now clear that Eurocentric environmental science alone cannot solve the ecological crisis. While there are many Eurocentric critiques of capitalist environmental destruction, it could be that it is their inability to embrace other worlds that limits their impact. In the United States and Europe, activists are turning to degrowth, commoning, ecofeminism, and local food sovereignty and we need to try and build alliances between these visions and those at work in territorial movements in Latin America.

Activism is clearly making a difference. In 2017, Colombia's constitutional court awarded rights to the Atrato River in the northwestern Pacific region of Colombia, an area known as Chocó, and the state is now obliged to uphold the ruling and protect and restore the river (Vargas-Chaves et al., 2020). After much mining-related conflict, violence, and displacement, an alliance of disparate actors was able to put sufficient pressure on the government to ban all metal mining in El Salvador (Riofrancos, 2019). In March 2018, after a long struggle, an international environmental human rights treaty – the Regional Agreement on Access to Information, Public Participation and Justice in Environmental Matters in Latin America and the Caribbean in Latin America – was approved. Known as the Escazú Treaty, it obliges signatories to protect both environmental resources *and* the lives of environmental defenders. At least 11 Latin American governments needed to ratify it for it to come into force and following pressure from civil society organizations 12 governments had done so by April 2021. Once ratified, it will become another legal mechanism at the disposal of protestors.

SUMMARY

- Extractivism has a 500-year history in Latin America but has intensified in recent decades.
- The Pink Tide governments continued to invest in extractivist activities in order to generate funds to invest in social programmes.
- Environmental defenders who oppose extractivism are often intimidated, criminalized, and murdered because of the threat their activism poses to capitalist accumulation.
- In struggles against extractivism, earth-beings constitute themselves as political forces that cannot be either admitted or dismissed by Eurocentric nation-states.

- While the situation is bleak in many ways, as extractivist practices and political violence continue all over the continent, collective resistance does sometimes succeed in disrupting the aspirations of extractivists.
- Extractivism and opposition to extractivism are evidence of the existence of different worlds and incompatible worldviews.
- The concept of territory/territorio as a space in which to defend life is often mobilized by environmental defenders.

DISCUSSION QUESTIONS

1   What are the continuities and discontinuities between the extractivist model put in place during the colonial era and the one in place today?
2   Describe some of the ways in which Indigenous and Afrodescendant people are fighting against extractivism? Why is it so dangerous to be an environmental defender in much of Latin America today?
3   Who was Berta Cáceres and why was she deemed to be a threat?
4   What do environmental defenders mean by territory and how is it different from land?
5   What is the ontological turn and how does it challenge Eurocentric thinking on environmental politics?
6   What are the features and dynamics of resistance to gold mining in La Puya, Guatemala?
7   Do you think it is possible to form alliances between different worlds, such as between first world environmentalists and Indigenous groups, to save the world from ecological collapse?
8   If we accept the epistemological and ontological position that mountains and rivers are sentient beings and political agents, how does that shape how we research what in Eurocentric terms we might call environmental struggles?

FURTHER READING

Routledge Handbook of Latin American Development Chapters 27 (Bebbington), 33 (Gudynas), 34 (Ojeda), 35 (Fletcher), 36 (Perreault), 37 (Boelens), 38 (Finley-Brook and Ramos), 39 (Valdivia and Lyall), 40 (Bee)

Blaser. M. and M. de la Cadena, M. (2018) (eds) *A World of Many Worlds*. Durham, NC: Duke University Press.

de la Cadena, M. (2015) *Earth Beings. Ecologies of Practice Across Andean Worlds*. Durham, NC: Duke University Press.

Li, F. (2015) *Unearthing Conflict: Corporate Mining, Activism, and Expertise in Peru*. Durham, NC: Duke University Press.

These three books provide an excellent theoretical introduction to the ontological turn in Latin American environmental politics.

Sawyer, S. (2005) *Crude Chronicles: Indigenous Politics, Multinational Oil, and Neoliberalism in Ecuador*. Durham, NC: Duke University Press.

This book charts the destructive environmental consequences of multinational oil extraction and neoliberalism in Ecuador and the mobilizations by Indigenous peoples against this state of affairs.

Recommended films

Berta Vive (2016), directed by Katia Lara

Worth Dying For? (2017), directed by Nicky Milne

La Voz Lenca No Se Calla (2013), directed by Nina Kreuzinger & Andrea Lammers (in Spanish with German subtitles).

Three excellent films about the anti-extractivist struggles of Berta Cáceres and the Lenca and Garífuna people in Honduras.

Crude (2009), directed by Joe Berlinger.

A film about the lawsuit in Ecuador against Chevron for the contamination it caused in the Amazon.

Daughter of the Lake (2015), directed by Ernesto Cabellos Damián

An Indigenous woman in Peru communicates with water spirits to prevent a mining company destroying the lake in their community.

Defensoras-es. La Puya un ejemplo de defensa del territorio (2015), directed by Álvaro Revenga

An excellent short film on La Puya resistance in Guatemala that includes an introduction on the growth of extractivist practices in the world.

Land of Corn (2015), directed by Bianca Bauer

A PBI documentary about four environmental defenders in Colombia, Guatemala, Honduras, and Mexico.

Soyalism (2019), directed by Stefano Liberti and Enrico Parenti

A film that documents the destruction being wrought by soy and other agribusiness investment.

When Two Worlds Collide (2016), directed by Heidi Brandenburg and Mathew Orzel

A feature length documentary on how Indigenous and capitalist worldviews clash in the Peruvian Amazon and on the Indigenous activism to oppose extractivism.

Useful websites

Eduardo Gudynas

The personal website of Uruguayan researcher Eduardo Gudynas with access to the full text of many of his publications (in Spanish, English, and Portuguese), videos and blog posts.

https://gudynas.com/

Global Witness
Global Witness campaigns are focused on resource conflicts, land, and human rights abuses committed against environmental defenders. They aim to raise awareness and bring those responsible for abuses of power to account.
www.globalwitness.org

Observatorio de Conflictos Mineros de América Latina (OCMAL)
Information and maps on mining-related conflicts in Latin America (in Spanish)
www.conflictosmineros.net/

World Resources Institute (WRI)
The WRI's focus is on global social and environmental challenges (food, forests, water, energy, climate, the ocean, and cities) and contains useful information, publications, and data about Latin America.
www.wri.org

CHAPTER 5

Latin America in the world

INTRODUCTION

The creation of Latin America is often described, perhaps somewhat euphem-
istically, as an encounter between three cultures: the native American, the
European, and the African. These three cultures were and are of course internally diverse and heterogeneous. As we have emphasized, this "encounter" was
driven by genocidal, extractivist, and epistemicidal logics, but it also resulted
in diverse forms of syncretism, hybridization, and cultural exchange in which
the distinct groups influenced and shaped each other. These are processes
that are ongoing. The European legacies of the conquest can be observed in
Latin America's languages, religious practices, political culture, land tenure,
and development models, but it is important to note that these legacies have
been embraced, accommodated, resisted, and Indigenized and Africanized
in culturally specific ways. Contemporary Spanish, Portuguese, and French
cultures are also decisively shaped by their colonial histories and centuries of
involvement in Latin America. Europe and the European Union continue to
have a decisive presence in Latin America, in aid, trade, FDI, and forms of
academic exchange, as does European and Eurocentric thought which formed
the basis of Latin American political institutions and universities. In fact, the
conquest and its aftermath unleashed a process of globalization that had and
continues to have multifaceted economic, geopolitical, and cultural impacts.
Latin America has also been constituted not only through its relationship with
Europe and Africa, but with other parts of the world, including Canada and
the United States, the Soviet Union and Russia, and parts of Asia and the
Middle East, including Iran, China, Taiwan, and South Korea. A range of
global actors, including multilateral organizations such as the United Nations,
the World Bank and Inter-American Development Bank, foreign investors
and multinational companies, and NGOs are all present in Latin America –
and their top-down presence has engendered a range of bottom-up responses

DOI: 10.4324/9781003110453-5

from communities and social movements, whose lives are affected by it. Many foreigners come to Latin America as migrants, tourists, investors, journalists, researchers, activists, artists, and athletes. Some come to learn, contribute, and show solidarity; others to extract, exploit, or sell weapons.

Latin America and Latin Americans are of course also present in other parts of the world. Outside of the continent, people consume Latin American products from coffee to telenovelas. Latin American migrants, refugees, and students have established communities outside of the continent and there are large migrant communities in many parts of the United States and Europe. Given that Latin America is composed of genetic, cultural, linguistic, and political influences of diverse origins, deciding what is internal and what is external is not straightforward (Gardini, 2021) but there are some notable economic, political, and cultural trends that are important to understand and that all have implications for processes of decolonization. This chapter explores these dynamics.

LATIN AMERICA AND THE UNITED STATES

It is impossible to understand the making of Latin America and the development project without a thorough consideration of the role of the United States. Latin America's geographic position to the south of the United States compounds the multiple legacies of colonialism. In the early 1800s, the United States came to believe that it was its manifest destiny to expand its dominion across the continent (Skidmore and Smith, 2005: 367). As Latin America was fighting for independence, the United States was expanding both to the South and the West. In 1803, in the wake of the Louisiana purchase, the United States doubled the size of its territory. In 1823, US President James Monroe stated clearly that any attempts to further colonize the Americas by European powers would be viewed as acts of aggression against the United States, a policy which came to be known as the Monroe Doctrine. In the middle of the 19th century, Cuba and Puerto Rico were described by the US Secretary of State as "natural appendages of the North American continent" and the United States tried to purchase Cuba from Spain (Sweig, 2009: 4). In 1846, ten years after the declaration of Texan independence and one year after Texas had joined the Union, the United States invaded Mexico and took one third of its territory. The present-day states of Texas, New Mexico, California, Arizona, and part of Utah were all part of post-independence Mexico. In 1898, the United States seized Puerto Rico and still occupies this nation today. The inability to do the same with Cuba led the US Congress to pass the Platt Amendment in 1902 in order to curtail Cuban sovereignty. The Platt Amendment stationed US troops in Cuba until 1933, giving the United States the right to intervene in the name of "good government." Cuba's first president, Tomás Estrada Palma, leased an area at

Guantánamo Bay to the US administration which was converted into a naval base. The dream of Cuban independence that independence leaders such as José Martí had fought for was thwarted (Sweig, 2009). Although it had no formal diplomatic relations with Cuba between 1959 and 2015, the United States still retains control of the naval base at Guantánamo Bay and after 2002 started to use it as a military prison to incarcerate prisoners captured as part of the US War on Terror.

In 1903, the United States also signed a treaty with Panama. Panama had been part of Colombia, but the United States helped it to become an independent nation to facilitate the furthering of its own interests in the region. Panamanian "independence" and the treaty gave the United States permission to build the Panama Canal and retain sovereign rights over the canal zone until 1999. In 1904 the "Roosevelt Corollary" to the Monroe Doctrine appeared, in which it was made clear that the United States would be prepared to resort to military intervention in Latin America if deemed necessary (Carmagnani, 2011). It was hardly a measure of last resort. Although US foreign policy towards Latin America has taken different forms over the years, it has frequently involved overt and covert political and military intervention designed to protect the economic and geopolitical interests of the United States. Between the 1846 invasion of Mexico and the present time, the United States has intervened militarily throughout Latin America and the Caribbean dozens of times (see Box 5.1 for a selective list). Some countries, especially in Central America and the Caribbean, have been subject to repeated invasions and occupations.

BOX 5.1 US MILITARY INTERVENTIONS IN LATIN AMERICA

1846 The United States invades Mexico and takes one third of Mexican territory

1855 An adventurer (*filibustero*) from Tennessee, William Walker, installs himself as president of Nicaragua

1894 US marines land in Panama

1898 US troops occupy Cuba and Puerto Rico

1903 US troops occupy Panama to gain land for the construction of Panama Canal

1904 US troops sent to Dominican Republic to protect US financial interests

1905 US troops sent to Honduras

1908 US troops sent to Panama to intervene in elections

1910 US Marines occupy Nicaragua

1912 US Marines occupy Cuba to crush a rebellion by sugar workers

1912–1925, 1926–1933 US Marines occupy Nicaragua

1915 US Marines occupy Haiti, it becomes a US protectorate until 1934

1916–1924 US Marines occupy the Dominican Republic, it becomes a US protectorate until 1941

1917 US troops invade Mexico in a failed pursuit of Pancho Villa

1917–1933 US troops occupy Cuba

1925 US troops sent to Panama

1932 US sends troops to El Salvador to suppress a suspected communist uprising led by Farabundo Martí

1934 Nicaraguan nationalist fighter, Augusto César Sandino, is assassinated with US support

1946 The United States opens the School of the Americas in Panama to train military officers from the Americas in counterinsurgency techniques, including torture

1950 The United States suppresses a movement for Puerto Rican independence

1954 CIA overthrows Jacobo Arbenz, democratically elected president of Guatemala, who had begun to implement a land reform which threatened the interests of the United Fruit Company (UFCO)

1960 Covert actions against Cuba begin

1961 The United States invades the Bay of Pigs in Cuba

1961 The CIA overthrows José María Velasco Ibarra, the democratically elected president of Ecuador

1962 The CIA backs a coup to overthrow Juan Bosch, the democratically elected president of the Dominican Republic

1964 The United States backs a coup to overthrow the elected president of Brazil, João Goulart, after he proposes agrarian reform and the nationalization of the oil industry

1965 The United States invades the Dominican Republic to prevent the return of Juan Bosch to power

1966 The United States invades Bolivia to carry out a campaign of counterinsurgency, which culminates with the assassination of Che Guevara in 1967

1973 The United States backs a coup to overthrow the elected president of Chile, Salvador Allende which brings the Pinochet dictatorship to power

1980s The United States financially supports the Nicaraguan Contras and Salvadoran and Guatemalan death squads and creates US military bases in Honduras near the Nicaraguan border

1983 The United States invades the Caribbean island of Grenada

1984 The CIA mines Nicaragua's harbours

1989 The United States invades Panama to topple President Manuel Noriega, a former CIA ally

2000 US provides aid for Plan Colombia to fund counternarcotics operations
      and military targeting of FARC guerrillas
2002 The United States converts its naval base at Guantánamo Bay in
      Cuba into a military prison for suspected terror suspects from Iraq and
      Afghanistan
2004 The United States assists in a coup to oust President Aristide of Haiti
2009 The United States supports a coup to oust President Zelaya of Honduras

US intervention military intervention has clearly been motivated by intersecting political and economic interests. The United States has acted to uphold the role played by Latin America during the colonial period as an exporter of raw materials as this model has provided highly profitable economic opportunities for US multinationals such as the United Fruit Company and has ensured a supply of cheap coffee, cotton, sugar, and bananas to US consumers. Direct military control of the Americas was, however, proving complicated for the United States and by the 20th century, dictatorship was seen as a more effective and less costly means of controlling its back yard. Consequently, the United States helped a number of dictatorships to come to power in Latin America in the 1930s. These include Anastasio Somoza of Nicaragua, Fulgencio Batista of Cuba, Maximiliano Hernández Martínez of El Salvador, Jorge Ubico of Guatemala, Tiburcio Carías Andino of Honduras, Rafael Trujillo of the Dominican Republic, and Gétulio Vargas of Brazil. The installation of these dictatorships coincided with the so-called Good Neighbor Policy established by President Franklin D. Roosevelt in 1933. The United States withdrew its troops from Haiti and Nicaragua and repealed the Platt Amendment.

Despite these moves, Latin America's US-backed dictators were brutal. They repressed any political opposition, and did not hesitate to torture, disappear, and incarcerate their opponents. Batista allowed US oil exploration, protected and promoted US corporations on the island and turned Havana into a gambling and prostitution mecca for wealthy US Americans. Ordinary Cubans endured miserable living conditions while Batista and his associates enriched themselves at their expense and the United States rich and famous partied in their capital. It is estimated that around 20,000 Cubans were murdered by Batista's forces.

In Nicaragua, the United States helped to create a National Guard designed to maintain social order within Nicaragua without the deployment of US troops. US military occupation had bred local resentment and fuelled an armed nationalist resistance led by Augusto César Sandino. Anastasio Somoza García, the son of a coffee oligarch, became head of the National Guard and later declared himself president. Sandino was persuaded to come down from the mountains and engage in peace talks and was duly assassinated by Somoza's guard. The Somoza family ruled Nicaragua until 1979. With US support, they tolerated no dissent and they became extremely wealthy through embezzlement of state funds, land grabbing, agro-exports such as coffee and cotton, and the granting of concessions to foreign companies.

Under President John F. Kennedy, however, there were signs that relations with Latin America might begin to move in a more progressive direction, motivated by a recognition that it was necessary to improve people's living conditions to prevent them turning to revolution. The Alliance for Progress thus involved a limited increase in aid to the region and encouragement to US entrepreneurs to invest in the region. Kennedy was assassinated in 1963, and his successor Lyndon B. Johnson authorized military interventions in the Dominican Republic, Bolivia, and Brazil. The Alliance for Progress policy was formally ended by President Nixon in 1969.

As the Cold War got underway, US foreign policy turned towards preventing the countries of Latin America embracing communist ideologies, a goal which turned into an obsession with tragic consequences. In 1946, the United States established a military academy, the School of the Americas, in the canal zone of Panama and over the next few decades would controversially train Latin American soldiers in counterinsurgency techniques, including torture (Gill, 2004). Its graduates include some of the continent's most notorious human rights abusers, including Roberto D'Aubuisson of El Salvador and Efraín Ríos Montt of Guatemala. In 1954 the CIA overthrew Guatemalan president, Jacobo Arbenz, after he initiated an agrarian reform which threatened US economic interests in the region.

With the failing war in Vietnam, counterinsurgency was therefore never far from the mind of US administration officials and its anti-communist obsession intensified. The Cold War might have been a struggle between two opposing superpowers, the United States and the Soviet Union, but in the 1970s and 1980s, it was played out in Central America, with tragic consequences for thousands of ordinary people.

US IMPERIALISM AND REVOLUTION

The brutality of the Batista regime was the catalyst for the Cuban Revolution, led by Fidel Castro and Ernesto "Che" Guevara. Fidel Castro was jailed for two years for his involvement in an attack on the Moncada army barracks. On his release, he worked with Cubans in Cuba and in exile, organizing the 26th July Movement and deploying troops to the Sierra Maestra. Over the next few years, Castro and Guevara launched a number of attacks on the Batista government. Batista underestimated the threat posed by the guerrilla forces and as a result continued to act repressively towards the Cuban people, which merely had the effect of generating more popular support for the revolution. The Cuban Revolution triumphed on 1 January 1959, with Batista fleeing the

country. Once in power, the new revolutionary government began to implement socialist reforms and nationalize US assets. It also began to invest heavily
in health care and education with impressive results. The United States was,
however, opposed to the Cuban experiment and fearful that it would export
its revolution to other parts of the continent did everything it could to undermine it. In 1961, it attempted to invade Cuba but failed. The botched Bay
of Pigs invasion was quickly followed by the Cuban missile crisis in which
the US administration discovered that the Cuban government had allowed
the Soviet Union to install missiles on its territory only 50 miles from the
coast of Florida. The Soviet Union withdrew its missiles in return for a guarantee that the United States would not invade Cuba. The United States did
not attempt another invasion but imposed a trade embargo (see Figure 5.1)
and the CIA has been implicated in a number of assassination attempts on
the life of Fidel Castro. Che Guevara left Cuba in 1965 in an attempt to
export revolution, travelling first to the Congo-Kinshasa and then to Bolivia,
where he was assassinated in 1967 by Bolivian state forces in collusion with
the CIA. Until 1989, Cuba was very dependent on its reciprocal trading and

Figure 5.1 Opposition to the US-imposed trade embargo in Havana, Cuba

Note: It reads: "Blockade, the longest genocide in history"

Source: Photo by author

diplomatic relations with the Soviet Union, exchanging sugar for oil. US actions have undoubtedly caused hardship on the island (as have the repressive actions taken by the Cuban government against political dissidents) but the Cuban revolution has hung on for more than 50 years, in some ways against all odds. Fidel Castro died in 2016 and his death led to mourning on the island as well as celebration by Cuban ex-pats in Miami. It is important to acknowledge that the Cuban revolution has played an important humanitarian role in the world, sending Cuban doctors to provide medical assistance to countries dealing with disasters or epidemics, including Central America after Hurricanes Mitch (1998) and Stan (2005), Haiti after the deadly earthquake of 2010, and Sierra Leone, Liberia, and Equatorial Guinea during the Ebola epidemic (2014–2015).

In the early 1960s in Nicaragua, a political and military organization, the Sandinista Front for National Liberation (FSLN), was created to fight against the Somoza dictatorship. Many of its members and sympathizers were tortured and imprisoned by the regime, but nonetheless support for the guerrillas grew. In 1972, a massive earthquake levelled Nicaragua's capital city, Managua, and killed more than 10,000 people. When Somoza embezzled the aid funds, support for the FSLN grew, including from middle class and conservative sectors of the population. In 1978, 50,000 Nicaraguans demonstrated in the street after the National Guard assassinated Pedro Joaquín Chamorro, editor of the conservative daily newspaper, *La Prensa*. As the regime's brutality

Figure 5.2 Barricades created during the insurrection of León, Nicaraguan Revolution

Source: Dora María Téllez, public domain, via Wikimedia Commons

grew more indiscriminate, the resolve of the revolutionaries to triumph strengthened. They managed to gain control of all of Nicaragua's towns and cities, taking the capital Managua on 19 July 1979.

As in Cuba, the triumph of the Nicaraguan revolution in 1979 paved the way for a period of political transformation. The Sandinista government put emphasis on the mixed economy, non-alignment, and political pluralism and implemented a series of social and economic redistributive measures aimed at improving conditions for Nicaragua's poor majority. The main gains of the early years of the revolution were in health, education, and agrarian reform. But they were, however, to be short lived, as the Reagan government began to fund and equip a counter-revolutionary army known as the Contra, which operated out of military bases in Honduras and Costa Rica and imposed a trade embargo. Initially the Contra was composed of former members of Somoza's guard, but as time went on, its ranks were increasingly filled by campesinos and Indigenous Miskitos disaffected with the revolution as a result of Sandinista mismanagement and cultural insensitivity.

The demands of the Contra war meant that resources had to be diverted out of social spending and into defence. In 1983, the government introduced military conscription and young men were often forcibly recruited by the Sandinistas to fight the Contra. While between 30,000 and 50,000 Nicaraguans had died in the war against Somoza, a further 30,000 were killed in the Contra war of the 1980s. The schools, health centres and co-operatives that had been created by the revolution became Contra targets and were frequently bombed and destroyed. Many of those killed were teachers, farmers, health workers, or environmentalists. The Contras, despite US backing, were never able to defeat the Sandinistas military, but by 1990, Nicaragua was economically devastated and emotionally exhausted. In the elections that year, Nicaraguan voted for the US-backed coalition led by Violeta Barrios de Chamorro, the widow of slain newspaper editor Joaquín Chamorro.

The successful revolutions in Cuba and Nicaragua had tragic consequences for El Salvador and Guatemala, whose guerrilla armies, the FMLN and URNG, were never able to triumph, because of brutal US intervention and financial support for armies that disappeared, tortured, and murdered thousands of so-called subversives. The Guatemalan civil war lasted 36 years. With US support, General Ríos Montt led a scorched-earth counter-insurgency campaign in which thousands of poor Guatemalans, most of them Indigenous Mayans, were murdered by their own government (see Chapter 7).

While some guerrilla movements, including the FARC of Colombia and Sendero Luminoso of Peru continued to operate for many years after the end of the Cold War, most guerrilla movements based on Marxist-Leninist principles became increasingly discredited. Revolutionary rhetoric began to give away to a plurality of discourses grounded in identity politics and diverse grassroots and civil society organizations. Some guerrilla movements became political parties or demobilized or signed peace agreements.

FREE TRADE AND REGIONAL INTEGRATION

After the Cold War came to an end, and the Soviet threat to the United States disappeared, US foreign policy shifted to the securing of Free Trade Agreements, as well as on fighting the War on Drugs (see Box 5.2). The North American Free Trade Agreement (NAFTA) between Canada, Mexico, and the United States came into force on 1 January 1994. NAFTA involved the gradual elimination of tariffs and non-tariff barriers between the three nations. Further trade liberalization had a deleterious effect on Mexican food security and on the livelihoods of those that grew crops for domestic consumption. Opposition to NAFTA was central to the Zapatista uprising in Chiapas (see Chapter 7). In 2005, the United States failed to secure a broad hemispheric agreement in place between the United States and Latin America known as the Free Trade Area of the Americas (FTAA), but in 2007 did manage to achieve the Central American Free Trade Agreement (CAFTA).

At the same time, Latin American nations have been producing trade agreements of their own which promote mutual intra-continental trade and economic cooperation. Some of these have long histories that predate the rise of neoliberalism. Indeed, many Latin American leaders have dreamed of and have been striving for regional integration since the time of Simón Bolívar. In South America, such agreements include Mercosur between Argentina, Brazil, Paraguay, Uruguay and Venezuela and the Andean Community, which comprises Bolivia, Colombia, Ecuador, and Peru, with Argentina, Brazil, Chile, Paraguay, and Uruguay as associate members. In 2008, Mercosur and the Andean Community merged to form UNASUR. In Central America, the nations have created the Central American Integration System (SICA). In 2004, Venezuela's president, Hugo Chávez, created ALBA, or Bolivarian Alliance for the Peoples of our America, an anti-neoliberal alternative to the FTAA. Initially an agreement between Cuba and Venezuela to exchange petroleum for doctors, it expanded to include eight member states including Bolivia, Ecuador, Nicaragua and three Caribbean nations, Antigua and Barbuda, Dominica, and Saint Vincent and the Grenadines. In defiance of the Washington Consensus and its conditionality, Argentina, Bolivia, Brazil, Ecuador, Paraguay, Uruguay, and Venezuela created the Bank of the South in 2009 with the intention of replacing the IMF and the World Bank as a regional lender. The regional economic landscape is a dynamic one with a number of divergent tendencies at work. It has, however, one in which the leverage of the IMF and the World Bank has been reduced.

In the post 9/11 world, US intervention in Latin America continued to decline, in part because the United States was focused on the War on Terror and in fighting wars in Iraq and Afghanistan and in part because of the rise of China (see below), but the question of Latin American migration to the United States has been a central political concern for many decades and continues to be so.

BOX 5.2 THE WAR ON DRUGS

One ongoing mode of US intervention in Latin America has been through the so-called War on Drugs. The drugs trade is a transnational one that connects countries, infrastructures, and bodies across the Americas. Indeed "a web of financiers, producers, suppliers, traffickers, money launderers, *capos* (drug lords) and runners" collaborate and cooperate "across time and space" in order to produce, move and sell drugs (Ballvé and McSweeney, 2020). Traffickers use speedboats, submarines, planes, and "mules," people who swallow several small bags of cocaine and then travel by plane to the United States or Europe, to get the drugs to their destination (Moldano, 2017). Over several decades, different US governments have implemented supply-side and militarized approaches that seek to prevent drug production in Latin America before it reaches the US. These include the Andean Initiative (1989), Plan Colombia (2000), and the Mérida Initiative (2008), all of which failed to curtail the drug trade and its associated violence. There are close linkages between organized crime, violence, and drug trafficking. Thousands of people have died in drug-related violence, with more than 200,000 drug-trade related deaths since 2006 in Mexico alone. During the Colombian civil war, drug dealers, paramilitaries, and guerrillas all collaborated. Plan Colombia involved US military training along with the aerial spraying and manual eradication of coca crops and it became an inseparable part of the civil war. Until 2015, the crops were sprayed with glyphosate, a herbicide produced by Monsanto, deemed by the WHO to be a possible carcinogen (Lyons, 2016). Coca eradication causes severe hardship and often also eliminates food crops, trees, and coca that is grown for traditional purposes as well as that destined for the production of cocaine. While some politicians claimed Plan Colombia to be a success, it failed on many levels. It didn't decrease coca production, as destruction of a coca harvest would lead either to its immediate replacement or to increased production elsewhere (the balloon effect). Traffickers are very good at frequently changing their routes and sites of production to evade detection and arrest (see Cupples, 2012). At times, areas cleared were deforested or taken over by extractivist activities such as mining or agri-business and so the strategy also contributed to the exacerbation of racialized forms of social exclusion and environmental destruction (Asher, 2020). Attempts to try and deal with the drug crisis are usually thwarted because the profits are so extraordinarily high and because the economy draws in the state, the capitalist class, the banking sector, and the poor. As Fernando Calderón and Manuel Castells (2020) write, when the drug economy first emerged, it had a number of key features. It was and is a transnational and exported-oriented one, involving primarily the United States, Colombia, Peru, Bolivia, Mexico, and the countries of Central America, with national origins. The original cartels in both Colombia and Mexico were created by local people, some of whom including Joaquín El Chapo Guzmán, Pablo Escobar, and Griselda Blanco went on to gain celebrity status. It involved the frequent use of hitmen (*sicarios*) and criminal gangs that deployed spectacular forms of violence. Criminal groups enjoyed the corrupt support and complicity of the police and other state institutions, so were able to operate with impunity. Drugs profits do not remain on the black market; they are laundered, invested in property and other assets, moved to financial havens such as Panama, and deposited in banks around the world. Dawn Paley (2014) has called this situation drug war capitalism as the War on Drugs produces benefits for transnational corporations while ordinary people are displaced and dispossessed. In Central America it is often gang members who participate in drug trafficking. While it is not on the same scale as Colombian cartels in the past or Mexican cartels now, there is evidence that Central American gang leaders are buying up "apartments, car washes, used car dealerships, discos, bars and restaurants in an attempt to launder proceeds and conceal their drug, kidnap, car theft and extortion" (Dudley, 2010: 23) and Central America is central to the routes that drugs take on their way north.

While the coca leaf provides the raw material for cocaine, coca is a mild stimulant that has been consumed and cultivated by Aymara and Quechua populations for centuries and the livelihoods of many rural Bolivians depend on it. It is highly nutritious, has a range of spiritual, ceremonial, and medicinal properties and is frequently safely used in other products including tea, flour, and toothpaste (Harten, 2011). When Evo Morales came to power, he expelled the US Drugs Enforcement Administration (DEA), describing the DEA as "an instrument the US uses to blackmail those countries who don't comply with imperialism and capitalism" (BBC, 2011; Farthing and Kohl, 2014). There is no doubt that the War on Drugs has furthered US foreign policy objectives in the region, has resulted in serious human rights abuses and campesino dispossession, and has enabled further extractivism led by transnational companies, while doing very little to curb drug consumption (Correa-Cabrera, 2019). It is not an anomaly but is central to neoliberalism and is a key part of the colonial matrix of power.

Sources: Harten, 2011; Paley, 2014; Lyons, 2016; Correa-Cabrera, 2019; Calderón and Castells, 2020

LATIN AMERICANS IN THE UNITED STATES

A constant feature of the past century has been Latin American migration to the United States, and this migration has been fuelled by US-led policies in the region, which have exacerbated hardship and misery. In 2019, the US Census Bureau estimated that people of Latin American descent (referred to collectively as Latin Americans, Latinos/Latinas/Latinx, or Hispanics, see Box 6.2) reached 60 million and that they made up 18% of the US population, up from 16% in 2010. While the majority (more than 30 million) are Mexicans, immigration from other Latin American countries is growing more rapidly. Such dramatic demographic changes and the presence of Latin Americans in all walks of life are having significant economic, political, and cultural implications for both the United States and the countries of Latin America. While there is plenty of evidence that Latin Americans are well integrated into US culture, Latin American culture, including language, food, sport, cultural activities, and religious celebrations, is also ubiquitous in the United States, especially in cities such as Los Angeles and Miami. There are many Latinx and Latin Americans who are prominent across the mediascape and in public life. In addition, millions more, some of whom have no legal status, are making massive and often invisibilized contributions to the US economy and everyday social reproduction by cleaning homes, caring for children, tending gardens, working in restaurants and factories, and picking fruit and vegetables. In addition to their decisive contribution to the US economy, they also send almost US$60 billion worth of remittances back to their families in Latin America, enabling them to improve their access to health care, send their children to school and invest in homes, infrastructure, and local businesses. In other words, remittances go some way towards compensating for the harms caused by US foreign policy and allow some family members to remain in their countries of origin. The massive presence of Latinx in the United States has transformed US culture in a range of ways, including the popularity and globalization of salsa music (see Box 5.3), television (see Box 5.4), and the quinceañera celebration (Valentina González, 2019). It is clear, for example, that US companies can no longer afford to ignore Latinx consumers and US politicians can no longer afford to ignore Latinx voters.

BOX 5.3 HÉCTOR LAVOE AND THE ORIGINS OF SALSA

Salsa is a cosmopolitan musical genre that has its origin in Latin American migrations to the United States in the 1960s and 1970s. Héctor Lavoe was born Héctor Pérez Martínez in Ponce in Puerto Rico (see Figure 5.3) and developed a keen interest in music and performance. In 1963, he moved to

Figure 5.3 A mural dedicated to Héctor Lavoe in Ponce, Puerto Rico

Source: Carmelo Ruro, CC BY-SA 2.0, via Wikimedia Commons

New York where he collaborated with other Latin American artists including
Willie Colón, Johnny Pacheco, and the Fania All Stars. The collaboration
between Lavoe and Colón was instrumental in the establishment of salsa as a
musical genre, although some Cuban musicians such as Damaso Pérez Prado
argue that salsa is a contemporary version of the Cuban son and did not
originate in New York (Aparicio, 1998; Taylor, 2005). Salsa combines a range
of Afro-Cuban and Puerto Rican styles including son, mambo, and bomba,
and also incorporates elements of Latin jazz. It was especially promoted and
popularized by Fania Records, which signed up most of the famous salsa
singers of the time, including Celia Cruz, Rubén Blades, Ray Barreto, and
Bobby Valentín. Lavoe and Colón recorded many hit songs together including
El Malo, La Murga, El día de mi suerte, and Aguanile. Aguanile incorporates
elements of Santería to which Lavoe turned as a result of his struggles with
depression and drug addiction. His drug use frequently resulted in his inability
to act professionally and led to the end of the collaboration with Colón.
Lavoe went on to have a successful solo career that included the global hit
Periódico de ayer. Lavoe survived a suicide attempt not long after the tragic
death of his teenage son and having tested as HIV positive as a result of
intravenous drug use. He died of AIDS in 1993, aged 46. Salsa is then a Latin

American music genre that drew on pan-Latin American traditions to create a distinctive musical style in New York and that then circulated back in Latin America, leading to further new fusions and styles. Lyrics are politically as well as romantically motivated and often refer to poverty, migration, the violence of everyday life, and the legacies of US imperialism. They can therefore be politically radical but sometimes they are merely deeply sentimental or problematically reproduce machismo or construct Black bodies as exoticized sites of consumption (Aparicio, 1998; M. Q. Sawyer, 2005). Salsa continues to evolve in new directions. While the Buena Vista Social Club was having huge global success at the turn of the new millennium with their revival of the pre-revolutionary, pre-New York salsa son sound fused with guajira and danzón, late Afro-Colombian salsero, Joe Arroyo, was combining salsa rhythms with cumbia and vallenato (Taylor, 2005) and injecting his music with a strong decolonial aesthetic. His song La Rebelión dismantles the racist and patriarchal objectification of the Black woman's body in salsa and elsewhere and draws attention to the legacies of slavery and oppression in Colombia (M.Q Sawyer, 2005). It is important to acknowledge that the most popular (US and Latin) American musical genres – salsa, merengue, cumbia, tango, jazz, and blues – that revolutionized world music in the 20th century all have African origins (Fernández, 2006). Salsa connects Latin America to the African diaspora and is a site of cultural struggle over the meanings of Blackness (Sawyer, 2005).

Sources: Boggs, 1992; Aparicio, 1998; Sawyer, 2005; Fernández, 2006

BOX 5.4 THE "LATINIZATION" OF THE US MEDIASCAPE

Whereas US imperialism works in economically and politically negative ways, undermining struggles for decolonization, the cultural sphere does not operate in quite the same way. While Latin Americans have creatively and selectively borrowed and adapted foreign elements for centuries, there is an ongoing tendency to see Latin American cultures as vulnerable to mass media in general and to US cultural products in particular, a phenomenon often referred to as cultural imperialism (see, for example, the highly influential text on Disney movies by Chileans Dorfman and Mattelart, 1971) This thesis, which sees US American culture as all powerful and able to eclipse and destroy the cultures of the periphery, has, however, been convincingly challenged by scholarship in cultural studies which suggests that it is highly problematic to assume that a cultural product has ideological effects simply because it circulates in a given site. Such an assumption allows no scope for the fact that audiences can be quite heterogeneous in the ways in which they read and interpret texts and that they also often construct oppositional meanings out of popular media texts (Rowe and Shelling, 1991; García Canclini, 2001; Martín-Barbero, 2004). In other words, all forms of popular culture, whether produced by Indigenous groups in rural communities or by multinational media corporations, are used by people in ways which *make sense to them*. Furthermore, Latin America is not merely on the receiving end of globalization, but is also a key contributor to processes of globalization and exports culture along with coffee, gold, and bananas. Salsa dancing is a popular pastime in Cape Town, London and Auckland; Bosnians and Russians are big fans of Latin American telenovelas; Gabriel García Márquez, Isabel Allende, and Mario Vargas Llosa have been translated into dozens of languages and for decades Latin America has produced some of the world's most famous footballers. Latin American media texts, themes, content, and actors are also transforming the US mediascape. The "Latinization" of the US mediascape can be seen in successful television series such as *Ugly Betty* (2006–2010) (see Rivero, 2003) and *Jane the Virgin* (2014–2019). *Jane the Virgin* is a remake of a 2002 Venezuelan telenovela, *Juana La Virgen*, and the protagonist is a young woman of Venezuelan descent living in Miami who plans to save her virginity until marriage but becomes pregnant when she is accidentally artificially inseminated. The show both celebrates and satirizes the telenovela aesthetic. Its bilingual, matrilineal, and multigenerational cast creatively deal with multiple representations of Latinidad and particularly Latina womanhood, as well as the cultural and linguistic complexities and realities of Latinx people in the United States (see Piñon, 2017; Melgarejo and Bucholtz, 2020). Both *Ugly Betty* and *Jane the Virgin* also raise important issues for the many US-born Latinx and Latin Americans, whose parents or grandparents (Betty's father Ignacio and Jane's grandmother Alba) may have migrated illegally to the United States and remain undocumented.

Sources: Rivero, 2003; Piñon, 2017; Melgarejo and Bucholtz, 2020

Migrating to the United States from Latin America is a dangerous and difficult endeavour. Many come on foot across the desert with the help of coyotes and thousands succumb to sunburn, dehydration, and hypothermia. If they survive the journey, they might then face detention, deportation, surveillance, or discrimination. But they still migrate. In the past two decades, there has been strident opposition to Latin American migration to the United States from some politicians and sectors of the population. In 2010 and 2011, Arizona, Georgia, Alabama, and South Carolina all passed laws targeting illegal immigrants and the US Immigration and Customs Enforcement (ICE) began carrying out frequent terrifying raids in farms and factories to find and arrest those working illegally. On the US-Mexican border, both the US Border

Patrol and self-organized vigilante groups, such as the United Constitutional Patriots (UCP), started to intimidate and detain migrants, while humanitarian organizations such as Humane Borders and No More Deaths provided water and other forms of assistance. Humane Borders (2021) have documented 7,000 migrant deaths in the border desert since 1998, with 2020 being the deadliest year.

Under the Obama administration, the DREAM (Development Relief and Education for Alien Minors) Act was introduced to provide a route to citizenship for the many Latin Americans who were brought to the United States by their parents while they were still children. Obama also worked to try and end the 50-year blockade of Cuba and normalize relations between the two nations. The embargo was not brought to an end, but the US Embassy in Havana was reopened, as was the Cuban Embassy in Washington, and restrictions on travel and the sending of remittances were eased. Many of these measures were reversed when Trump came to power, although the diplomatic ties remained. Despite positive moves towards Cuba, three million people, most of them Mexicans and Central Americans, were deported under the Obama administration (Guerrero, 2021). The effects were devastating. Families, some of them with long-term roots in the United States were separated and many deportees went on to suffer from serious mental health problems (Guerrero, 2015). After 2016, the situation worsened when President Donald Trump came to power wielding a violent anti-immigrant rhetoric. In his electoral campaign, he referred to Mexicans as rapists, drug dealers, and criminals and promised to build a border wall that Mexico would pay for.

While Latin American migration to the United States, especially from Mexico, had started to slow down around 2010–2011, it was at this time that Central American migrants began to migrate collectively in caravans composed of several hundred people. In many cases, minors started to travel alone, fleeing violence and insecurity at home, or trying to reunite with parents or other family members living in the US. In 2018, a caravan of 7,000 people departed from Tegucigalpa in Honduras (Figure 1.2). It attracted widespread media attention and produced a moral panic within the United States fuelled by the Trump administration. President Trump threatened to close the US–Mexican border and stop aid to Honduras and Guatemalans, and many were forced to return to their places of origin. Those that did make it into the United States often received quite brutal treatment. Parents were forcibly separated from their children by US border guards and horrific scenes appeared of unaccompanied children locked in cages covered in foil blankets in detention centres. Since 2018, ACLU lawyers have been trying to reunite children with their parents, but three years after the crackdown they had still failed to find the parents of 545 children (Pilkington, 2020).

The US–Mexican border has become a key site of cultural and political struggle. As Jill Fleuriet and Mari Castellano (2020) write, dominant border imaginaries construct Mexico and Central America as a security threat to the

US. The border and the borderlands in which the border is located produce, however, a diverse array of imaginaries and counter-imaginaries. As Mexican and Latinx immigration activists and artists have made clear "We didn't cross the border but the border crossed us" (see McCaughan, 2020), emphasizing that before 1848 Mexico included much of the present day United States, including California. Not all people of Mexican descent in the United States are migrants or descendants of migrants, some are descendants of Mexicans who were native to the areas taken by the United States in the Mexican–American war. Consequently, Gloria Anzaldúa (1999) described the US–Mexican border as a "1,950-mile-long open wound, dividing a pueblo, a culture." Furthermore, Spanish and not English was the first European language to be spoken in the United States.

CHINA AND LATIN AMERICA

In the past two decades, China's involvement in Latin America has intensified and now China is an important investor in and trading partner with many Latin American nations. Other Asian countries such as Taiwan and South Korea are also important investors in the region. But Asia's involvement with Latin America is not new and began in the colonial era. In the 16th century, the Manila–Acapulco trading route between Mexico and the Philippines was established and it lasted for 250 years. Mexico exported silver to Asia and imported Chinese and other Asian goods including spices and porcelain via Manila. The trade ceased in 1815 just prior to the independence of Mexico.

In the 19th century, there was substantial Chinese migration to Central America, Cuba, and Peru and people of Chinese descent are an important ethnic minority in the region. The first Chinese migrants arrived in Panama as indentured labourers in 1854 to work on the construction of the railways. Belize, Costa Rica, Honduras, Mexico, and Nicaragua went on to also receive Chinese migrants where they also worked on railways as well as in agriculture, mining, and domestic service. They were subject to horrific working conditions and racism, they often suffered ill health because of their working conditions, and suicide was a frequent response (Robinson, 2010). Some Chinese migrated to Latin American countries from the United States after the Chinese Exclusion Act was passed in 1882. Chinese people moved around the region, establishing themselves in a number of places, but especially in Bilwi and Bluefields in Nicaragua, in Colón and Panama City in Panama, and in Limón in Costa Rica and by the early 20th century, there was an established presence of Chinese merchants and stores that became central to local economies. The Chinese intermarried with Afrodescendant, Indigenous, and mestizo people. The Nicaraguan Caribbean port town of Bilwi/Puerto Cabezas has a well-established Chinese population that works in commerce, local government, and the Moravian Church and there are many Chinese surnames such as Lau, Lee, Siu, and Sujo among the populations (Pineda, 2001). Many of the Chinese merchants in Bocas del Toro in Panama are much more recent migrants and less well integrated with local populations.

In the 20th century, Chinese socialist ideologies such as Maoism also found their way into Latin America and inspired revolutionary movements such as Peru's Sendero Luminoso. Maoism was of course designed for Chinese social conditions but was seen as useful to mobilize the revolutionary consciousness of rural and peasant populations in Peru. The leader, Abimael Guzmán, studied Maoism in China in the 1960s prior to creating Sendero Luminoso in Ayacucho in 1967. Mao's militarism formed the basis of Sendero's violent actions as they confronted and sought to annihilate the dominant political order in Peru (Pinkoski, 2012). In the 1960s, Mexico and Argentina sent grain to China to help the nation overcome the famine induced by Mao's Great Leap Forward and in the 1980s China supported Argentina's claim on the Malvinas/Falkland Islands (Wise, 2021).

The Chinese Communist Party also supported Castro's Cuba initially. In 1960, they established diplomatic relations, Che Guevara visited China, and Cuba ended diplomatic ties with Taiwan. Fidel Castro, Che Guevara, and Raúl Castro were frequent visitors to the Chinese Embassy in Havana, where they often dropped in to eat Chinese food, but relations would soon break down as a result of the Sino-Soviet split and disagreements over Marxist-Leninist doctrine (Cheng, 2007). Cuba was fully implicated in this split because of the Cuban Missile Crisis, denounced by China who asserted that the USSR had betrayed Marxist-Leninist principles. As a result, Cuba and China found themselves on opposing sides and Cuba would then have very close relationships with the Soviet Union until 1991. After the collapse of the Soviet Union, trade between Cuba and China began to increase and today China is Cuba's second largest trading partner.

The nature of Chinese involvement in Latin America changed dramatically in the 21st century. China's recent economic rise is well documented, and it coincides with the decline of neoliberalism in Latin America and the rise of the Pink Tide governments who were trying to reduce US influence in the region (Ganchev, 2020). During the Latin America's lost decade in the 1980s, when Latin American economies were in free fall, the Chinese economy grew rapidly. China joined the WTO in 2001, strengthening its presence in and influence on global markets.

China has turned to Latin America to try and satisfy its demand for commodities and has developed strong bilateral trading and diplomatic relations with many other countries in the region, including Argentina, Brazil, Costa Rica, Chile, Ecuador, Peru, and Venezuela. In 2009, Brazil and China joined forces with India, Russia and later South Africa to form BRICS, an entity of powerful "emerging economies" that collaborates to foment mutually beneficial economic cooperation.

Chinese Foreign Direct Investment (FDI) remains on the whole fairly limited and tends to be concentrated in oil and minerals, consumer electronics, and telecommunications, including in 5G and satellite technology. Brazil still attracts the lion's share in the region. The Chinese Great Wall Industries Corporation (CGWIC) has developed the Simon Bolívar satellite in Venezuela and the Túpac Katari satellite in Bolivia (Marcella, 2012), both somewhat ironically named after independence fighters.

As Rhys Jenkins (2009) notes, there is a similar level of Latin American FDI in China. Brazil has invested in aircraft manufacturing and information technology, Argentina has invested in iron and steel production, and Mexican food companies such as Bimbo and Maseca have set up plants and distribution networks in China (Rosales and Kuwayama, 2012). So the China–Latin America relationship does involve some diversification away from commodities. But as we saw in Chapter 4, Chinese investment in extractivist projects including hydroelectric dams has been associated with human rights abuses, environmental destruction, and unsustainable debt burdens.

Trade between China and Latin America increased from $17 billion in 2002 to almost $315 billion in 2019 (Sullivan and Lum, 2020). The main Latin American exports to China are crude oil (Colombia, Ecuador, Venezuela), copper (Chile, Mexico, Peru), iron ore (Brazil, Venezuela), soybeans and soya oil (Argentina, Brazil, Uruguay), nickel (Cuba), tin (Bolivia), zinc (Honduras), cotton (Paraguay), frozen meat (Panama), shrimp (Ecuador), sugar (Guatemala), and fishmeal (Chile and Peru). While the vast majority of exports are primary products, Costa Rica and Mexico both manufacture and export integrated circuits to China. The main imports from China tend to be industrial and manufactured goods, including clothing, textiles and footwear and high-tech goods such as motor vehicles and parts, electrical machinery, televisions, cameras, as well as a diverse range of general consumer products.

China is also an important lender to Latin America. Since 2005, China's lending has increased to more than $140 billion, $62 billion of which it has lent to Venezuela in return for oil supplies and iron ore (Page, 2016) that Venezuela has struggled to provide (Segovia, 2021). Indeed, as Carol Wise (2021: 52) writes, because of their indebtedness to China, leaders in both Venezuela and Ecuador "have already sown the seeds for an entirely new kind of debt crisis—one based on sovereign-to-sovereign lending, in which the Bretton Woods institutions will hold little sway." In 2021, Ecuador and Venezuela were both seeking debt relief from China, so the situation is serious, but according to Wise (2021: 56) "is certainly preferable to the austerity-based debt carousel imposed by the Bretton Woods institutions in the 1980s."

As Barbara Hogenboom (2019) writes, China's involvement in Latin America is complicated. As a form of South–South cooperation it reduces the region's dependency on the United States and Europe and undermines the power of the Washington Consensus in the region. But there are least three negative impacts. First, it swaps one form of foreign dependency with another. Second, it encourages Latin American governments to continue to engage in extractivism, so the social and environmental harms associated with these activities continue. Third, the import of cheaper Chinese clothing, shoes, and manufactured goods has a negative impact on local production and employment in these areas.

Since 2017, several Latin American countries have joined China's Belt and Road Initiative (BRI), a project which involves the reactivation and extension of the Old Silk Road and could lead to Chinese investment in ports and other infrastructure that support trade (Koop, 2019). BRI projects might increase trade and prosperity for some, but they are likely to extend Latin America debt burdens, could lead to serious environmental damage, and could involve further incursions into Indigenous or Afrodescendant territories (Koop, 2019; Zhang, 2019).

It is important to avoid simplistic generalizations about China's involvement in Latin America. Given the destructive forms of extractivism promoted by China in the continent, it is clearly not a decolonial form of South–South cooperation. But it is also important to avoid Orientalist tropes that posit China as a dangerous and predatory actor whose economic rise is a threat to the world. But China's involvement is heterogeneous, there is no single clearly defined China strategy within the region, and it involves many kinds of engagement that Latin American leaders have pursued and welcomed.

RELATIONS WITH THE REST OF THE WORLD

While relations between Latin America and Europe, the United States, and China are perhaps the most significant in terms of development trajectories and consequences, Latin America has cultural, political, and economic links with many other countries. Canada is part of NAFTA and Canadian companies have been involved, often in destructive ways, in many mining projects (see Chapter 4).

As we've noted above, the geopolitics of the Cold War meant that the Soviet Union played a key role in supporting and providing weapons and military aid to countries such as Cuba and Nicaragua, which were dealing with US imperialist aggression. After the collapse of the Soviet Union, Russia largely withdrew from Latin America and countries like Cuba who had been very reliant on the Soviet Union had to go it alone, but now has a growing presence. Apart from its participation in the BRICS initiative, Russia's main goals in the region involve infrastructure projects, that often fail to materialize; weapons sales, especially to Venezuela; support for governments that are hostile to the US; and the seeking of support for Russian intervention in Georgia, Crimea, and Ukraine (Miles, 2021). In the past few years, for example, bilateral relations between Russia and Nicaragua have grown. Russian ministers including Putin have visited Nicaragua and Russia has supplied the Nicaraguan army with training, field hospitals, and weaponry, including tanks and armoured vehicles, anti-aircraft units, missile and patrol boats, and military planes. In return, Nicaragua has allowed Russia to instal a Russian Global Navigation Satellite System that strengthens its intelligence capabilities in the region. During the 2018 anti-government uprisings, in which a number of Russian weapons, including AK assault weapons and Dragunov sniper rifles, were used on protesting Nicaraguans, the Russian government sought to defend Ortega from criticism, by calling for non-interference in Nicaragua's domestic affairs. Nicaragua is one of a small handful of countries that has supported Russia's expansionist ambitions, recognizing, for example, the independence of Abkhazia and South Ossetia, considered by Georgia to be under Russian occupation (see Chaguaceda, 2019; Miles, 2021).

Iran has also developed relations with the Latin American nations that are hostile to the United States, including Venezuela, Bolivia, Nicaragua as well as with Brazil prior to the election of Jair Bolsonaro. Penny Watson (2021) believes that there are several possible reasons for the recent growth in Iran–Latin America relations – one is as an expression of mutually beneficial South–South cooperation, another is a means to extract resources such as uranium and lithium to develop nuclear weapons, and another is to have a base from which to prevent US military action against Iran. Venezuela and Iran have particularly strong collaborations in oil extraction (that date back to the 1960s and the creation of OPEC), nuclear technology, and finance. In 2020 in the midst of ongoing food shortages and much to the dismay of the United States, Iran opened a supermarket, Megasis, in eastern Caracas, full of Iranian products (Olmo, 2021) and in 2021, Iran entered into phase three of clinical trials of Soberana-2, a Covid-19 vaccine developed in Cuba (Tehran Times, 2021).

SUMMARY

- Latin America is both subject to and contributes to processes of globalization and these processes have economic, geopolitical, and cultural effects that are negative, positive, and contradictory.
- Driven by economic interests and obsessive anti-communism, the United States has intervened in Latin America many times, often with tragic and devastating consequences.
- There is a large population of Latin American descent in the United States that has transformed the cultural as well as the economic landscape in really important ways.
- China has become an important trading partner with and investor in Latin America. China's role in many ways is not dissimilar to the role played by the United States and Europe in earlier periods.

- Russia and Iran have developed relations with Latin American countries that are hostile to the United States.

DISCUSSION QUESTIONS

1   Why does the United States repeatedly intervene in Latin American domestic affairs?
2   Is China a better economic partner than the United States?
3   What is the "Latinization" of the United States and how does it disrupt the cultural imperialism thesis?
4   What are the geopolitical implications of relations between Latin America and Russia and Iran?
5   How is the US–Mexican border brought into being politically and culturally?

FURTHER READING

Routledge Handbook of Latin American Development Chapters 11 (Correa-Cabrera), 12 (McIlwaine and Ryburn), 13 (Malamud), 14 (Weeks), 15 (Hogenboom), 16 (Ayuso)

Anzaldúa, G. (1999) *Borderlands/La Frontera: The New Mestiza.* San Francisco: Aunt Lute Books.

A classic text that explores the mestiza consciousness of people of Latin American people who reside in the borderlands.

Gardini, G. L. (2021) *External Powers in Latin America: Geopolitics between Neo-extractivism and South-South Cooperation.* London: Routledge.

A detailed compilation that assesses the geopolitical dynamics of Latin America's international relations

Recommended films

Carla's Song (1996), directed by Ken Loach

A feature film set in Glasgow and Nicaragua about the relationship between a Glaswegian bus driver and a Nicaraguan woman fleeing the US-backed Contra war.

Cold War (1998), produced by Pat Mitchell and Jeremy Isaacs

A 24-episode TV series broadcast on CNN. Episodes 10 (Cuba) and 18 (Backyard) deal with Latin America.

Harvest of Empire (2013), directed by Peter Getzels and Eduardo López

A film that takes a look at Latin American migration to the United States and its links to US imperialism.

School of Assassins (1994), directed by Robert Richter

A short documentary about the School of the Americas and the involvement of its graduates in massacres and human rights violations throughout Latin America

The War on Cuba (2020), directed by Reed Lindsay

A three-part documentary series that explores the effects of U.S. sanctions on Cuban people.

A Three Minute Hug (2018), directed by Everardo González

A short documentary captures the violence of the US-Mexico border and its impacts on families who find themselves on different sides.

CHAPTER 6

The coloniality of gender and sexuality

INTRODUCTION

Gender is central to development theory and practice, as well as to many
social movement struggles. It wasn't always so. When development emerged
as an industry and organizing paradigm in the second half of the 20th century, development theory was largely gender blind. The differential impacts
of the development process on women and men were ignored. Women in
particular were excluded or viewed in essentialist ways as mothers rather than
as heterogeneous social actors in their own right. We now have a detailed
understanding of how the development process is gendered, how inequalities
of gender are an obstacle to development, how processes of colonialism, structural adjustment, and extractivism exacerbate gender-based disadvantages, and
how gender functions as a vehicle for political mobilization. More work needs
to be done, however, on how to situate debates on gender and development
within a decolonial framework.

Over the past four or five decades, Latin America has had vibrant feminist
and women's movements that have put gender-specific issues onto national
and international political agendas. For example, for a long time domestic violence was an endemic social problem in part because it was seen as a burden to
be tolerated privately and in silence, but women's organizing and awareness-
raising have brought the issue out into the open and now many Latin American
countries have legislation to deal with the issue. More and more Latin American
women are being elected to public office. Latin America has already had several female presidents, including Violeta Barrios de Chamorro of Nicaragua,
Cristina Fernández de Kirchener of Argentina, Michelle Bachelet of Chile,
Laura Chinchilla of Costa Rica, and Dilma Rousseff of Brazil, which have
had a disruptive effect on the entrenched masculinism of Latin American political culture. Other areas are proving more difficult to challenge. Abortion is
decriminalized only in Argentina, Cuba, and Uruguay and in two Mexican

DOI: 10.4324/9781003110453-6

states, and remains highly restricted or illegal elsewhere, meaning that women who are desperate to end unwanted pregnancies are forced to take grave risks with clandestine abortion providers. Femicide and misogyny remain serious issues and appear to be on the rise (see below). Many Latin American countries remain deeply homophobic and transphobic, but there are some tentative moves towards dismantling homophobic laws and attitudes.

A key barrier to gender and sexual liberation is the historical tendency to understand cultural identities in essentialist ways – as something natural, biological, and fixed – and to organize them hierarchically. Hierarchies are useful to systems of power because they facilitate the justification of unequal treatment based on a person's gender, race, or sexual orientation and thus they give rise to sexism, racism, homophobia, and transphobia. While many people maintain that gender is something that is fixed and natural, much feminist theory across the social sciences and humanities sees gender and sexuality as socially constructed and contingent. This means that these categories are not biologically given and can be therefore changed. This is an anti-essentialist approach. Change is, however, often difficult as gender is embedded in structural systems of power, including patriarchy, white supremacy, ableism, and the colonial matrix of power. Those that benefit from these systems work hard to keep them in place and often have the discursive and material resources to do so.

Identity categories do not act in isolation but intersect with one another. Gender, race, and class are mutually constituted and gender identities are fractured by race and class and by racism and poverty. In Latin America the mutual constitution of gender, race, and class emerges in categories such as domestic service (Radcliffe, 1999). It is therefore useful to approach gender and other identity categories with an intersectional lens and to see them as a process rather than as a thing. In other words, gender is what you do, rather than what you are (Butler, 1990).

This chapter deals with a number of key areas of Latin American development in which the question of gender and sexuality is particularly salient, including modes of organizing in the face of neoliberalism and gender-based violence. But first we need some theory on the coloniality of gender.

GENDER AND THE CONQUEST

The conquest was a thoroughly gendered affair. The conquistadors and the first European settlers tended to be men and colonialism proceeded through the plunder of (feminized) land and resources and the rape of Indigenous women by European men. These two processes are related – both can be captured by the Spanish world *rapiña*. According to Rita Segato, the word "rape has the same etymology as rapiña, looting or predatory behavior in Spanish" (in Uribe-Uran 2013: 16) and she notes that "with conquered territories came the insemination of women's bodies. Soldiers raped the women of conquered territories as if women's bodies were extensions of those territories" (p.18). Furthermore, the Europeans also brought with them their own patriarchal and binary gender order that involved the imposition of compulsory heterosexuality and the ideology of the nuclear family. This gender order was also racialized as the colonizers attached deviant sexualities to Black and Indigenous men and women, framing them as hypersexual beings while white women were seen as sexually passive. And even though white men routinely raped Black and Indigenous women, colonial discourses constructed Black men as sexual predators that are dangerous to white women, a colonial legacy that persists today.

A 16th-century painting by a Flemish painter, Jan van der Straet, captures this gendered framing of the conquest (see Figure 6.1). He represents the "discovery" of America as an eroticized and exoticized encounter between Amerigo Vespucci, as European colonizer and civilizer, and an Indigenous woman. Vespucci is fully dressed, armoured, and holds a crucifix, while she is almost naked and holds out her hand in an inviting and sexualized manner. There is a cannibalistic scene in the background, to the left are his ships and astrolabe. De Certeau (1988: xxv–xxvi) describes this encounter as follows:

Figure 6.1 The allegory of America. Amerigo Vespucci arrives in the "new world"

Source: Jan van der Straet (Stradanus), public domain, via Wikimedia Commons

An inaugural scene: after a moment of stupor, on this threshold dotted with colonnades of trees, the conqueror will write the body of the other and trace there his own history. From her he will make a historied body-a blazon-of his labors and phantasms. She will be "Latin" America.

This erotic and warlike scene has almost mythic value. It represents the beginning of a new function of writing in the West. Jan Van der Straet's staging of the disembarkment surely depicts Vespucci's surprise as he faces this world, the first to grasp clearly that she is a *nuova terra* not yet existing on maps-an unknown body destined to bear the name, Amerigo, of its inventer. But what is really initiated here is a colonization of the body by the discourse of power. This is writing that conquers. It will use the New World as if it were a blank, "savage" page on which Western desire will be written. It will transform the space of the other into a field of expansion for a system of production.

Another example of the gendering of colonialism is the treatment and subsequent interpretation of La Malinche, mentioned in Chapter 1. Malinche or Malitzin was a Nahua slave, who was hired by Cortés as his translator. They became lovers and had a son together, the first mestizo. Consequently, the mestizo Mexican nation has its origin in the sexual relationship between Malinche and Cortés. Her political significance for both Indigenous Mexicans and Spanish colonizers during the colonial period is indisputable. Malinche is discussed in *conquistador* and chronicler Bernal Díaz del Castillo's (1973[1576]) memoirs of the conquest, where she is referred to respectfully as Doña Marina. Malinche and Cortés appear together here in this image, which was published in the late 16th-century codex History of Tlaxcala (see Figure 6.2) and in this text Malinche, who displayed impressive linguistic and diplomatic skills, is clearly admired. Her fundamental role in the conquest, however, means that she has often been viewed as a traitor. She is also sometimes pejoratively referred to as *la chingada*, the raped one, and Mexicans as a result are all *hijos de la chingada*, children of the raped one, a perspective popularized by Octavio Paz (1950) that requires feminist deconstruction (see Medina, 2009). These are not then the only meanings that get attached to her and there is also a decolonial feminist reading of Malinche available to us. As Damián Baca (2010) writes, many Chicana and Chicano scholars have sought to underscore Malinche's agency and resistance and the way that she engaged in specific and strategic forms of border thinking and border crossing, between Mesoamerican and European worldviews. Drawing on Anzaldúa's work on the new mestiza that sees such border crossings as subversive, Baca suggests that rather than seeing Malinche as having contributed to the destruction of Aztec cosmologies, she can alternatively be understood as keeping them alive as she engages in what he calls "decolonial rhetorics in motion" that disrupt and subvert colonial meta-narratives and the colonial invention of America

Figure 6.2 A scene from the lienzo de Tlaxcala, a 16th-century painting showing La Malinche and Hernán Cortés

Source: Diego Muñoz Camargo, c. 1585, in the public domain, via Wikimedia Commons

by European powers. Anzaldúa (1999) rejects the binary thinking that characterizes Guadalupe as saintly and Malinche as traitor as forms of patriarchal thinking, noting that it was not Malinche who betrayed her people, but her people who betrayed her. This lens can help us to appreciate the ways in which contemporary Indigenous Latin Americans also engage in subversive border crossings, engaging with Eurocentric colonial and patriarchal power structures and forms of governance to exert a decolonial agency on colonial situations that are not of their own making.

THE COLONIALITY OF GENDER

These are some of the elements that combine to create what we call the coloniality of gender. Gender was central to Quijano's (2000) coloniality of power but has been extensively criticized most notably by María Lugones

(2007, 2010, 2016). Lugones' work builds on the postcolonial feminist critique led by feminists of colour that charges western feminism for failing to pay attention to racism and colonialism (e.g. hooks, 1984; Mohanty, 1991) and is important because it explicitly places "gender in relation to the genocidal logic of the coloniality of power" (Mendoza, 2016: 118). Lugones (2010, 2016) takes on Quijano for his failure to acknowledge that binary and heteronormative gender is a colonial imposition and that this gender order did not exist in precolonial societies. Instead, Quijano assumes that "control of sex is a dispute among men, about men's control of resources who are thought to be female" (Lugones, 2016: 19), thus naturalizing heterosexuality in cultures that did not pathologize homosexuality (Mendoza, 2016). Quijano's colonial matrix of power remains rooted in Eurocentrism and is profoundly sexist.

Lugones (2010) opened a debate over the degree to which precolonial societies had a notion of gender. She draws on the work of Nigerian scholar Oyeronke Oyewumi to argue that Latin American societies did not have gender before the conquest. Irene Silverblatt (1987), Rita Segato (2010, 2016), Silvia Rivera Cusicanqui (2004), and Julieta Paredes (2008) all believe, however, that there was a concept of gender and gendered difference, but that it didn't take the form that came with colonization. Segato (2010) talks in terms of precolonial low-intensity patriarchies that were transformed by colonialism into high intensity ones. Rivera Cusicanqui (2004) discusses the egalitarian and reciprocal gender system that existed prior to colonization in the Andean regime that was weakened not so much by the conquest but rather by processes of nation-state building after independence. Silverblatt's (1987) work on the pre-Incan worlds of precolonial Peru refers to the "gender parallelism" that was in place prior to the conquest. She believes that while there was a cosmological understanding of gender difference, both women and men could be political leaders and exert control over resources. These gender formations were, however, disrupted by both Inca and especially Spanish conquest. While the Incas did not share the racialized patriarchal gender formations and gendered and classed notions of private property that were brought by Europeans, they did mobilize a gendered hierarchy expressed in cosmological terms. When the Spaniards arrived in Peru, they struggled to make sense of local gender formations and often labelled rebellious Indigenous women as witches. Silverblatt's work captures how Indigenous women defended themselves against these forms of colonial abuse and delegitimation. European gender ideologies created divisions between public and private that could not easily be replicated in Indigenous societies. In Guatemala, what Gladys Tzul Tzul (2015) refers to as the Indigenous communal system based on collective land, work, *and* government relies on the full participation of women and men. There is then no divide between social reproduction and political life.

Breny Mendoza (2016: 118) asserts that such distinct approaches are not necessarily incompatible as they all "agree that the imposition of a European gender system had profound effects on relations between men and women in the colony, unleashing lethal forces against Native, enslaved, and poor mes-tizo women sufficient to be considered genocidal." There is no doubt that the Spanish colonizers, who raped Indigenous women and plundered Indigenous lands, brought with them European patriarchy, the Eurocentric gender binary, and the idea of compulsory heterosexuality. We can therefore understand con-temporary forms of misogyny, homophobia, and transphobia in Latin America as colonial legacies. There is also no evidence of homophobic pathology in precolonial Latin America but rather evidence of widespread acceptance of what we would consider diverse gender and sexual norms. According to Andil Gosine (2018), expressions of homosexuality or of alternative gender formations were constructed by the colonizers as evidence of the lack of the civilization or backwardness of the colonized. Gender and sexual liberation are therefore important components of decolonial struggle.

Camila Esguerra-Muelle (2019: 59) pushes this analysis even further. While she is supportive of Lugones' intersectional focus and concurs with her critique of Quijano, she is also concerned by Lugones' exclusion of people who are not cisgender, noting that "the term 'woman' with no other speci-fication is also cisggenderist and heterocentered, and even transphobic and lesbophobic, and that 'woman' is not only an ontic colonial locus but also a disputed terrain."

As Mendoza (2016: 17) writes, "European colonizers positioned men and women as antagonists" and so male violence against women is also a colonial legacy as the "bodies of women became the terrain on which indi-genous men negotiated survival under new colonial conditions." Colonialism is therefore destructive of communal solidarity and reciprocity between people of different genders. There is then a continuity between colonial vio-lence against women and today's forms of gender-based violence that express themselves in domestic abuse, sexual assault, and femicide. Sexual violence by European and Latin American men towards Latin American women as well as towards people with non-dominant genders and sexualities is an outcome of colonialism and it is a gender order in which Latin American cismen who themselves have been victims of racial domination or global capitalism are complicit.

GENDER AND SEXUAL FLUIDITY IN LATIN AMERICAN CULTURE: THE *MUXE* AND THE *MACHI*

While the Eurocentric gender binary is strongly reproduced in education, politics, legislation, and by both Catholic and Evangelical churches, it is con-stantly subverted in and by Indigenous cultures as well as by diverse forms of LGBTQ+ activism. In Mexico, binary gender is well established, and asserts itself through everyday forms of sexism, discrimination, and gender-based violence, the cultural concepts of marianismo and machismo, and through the meanings attached to iconic and contested Mexican women, including Malinche, the Virgin of Guadalupe, and La Llorona. In Juchitán in Oaxaca in Mexico, a community that displays high degrees of female empowerment and where the Zapotec language is widely spoken, there are three recognized genders, man, woman, and *muxe*. The *muxe* people (*lxs muxes*) live beyond binaries of sex and gender, adopting and mixing both masculine and feminine characteristics in their identities and self-presentation. They cannot be mapped onto the man/woman, homosexual/heterosexual binaries that characterize Eurocentric thought. *Muxe* marry or have sexual relationships with men, women, and other *muxe*. Some *muxe* assert that it is better to reject labels altogether rather than getting bogged down with confining terms like male, female, cisgender, transgender, gay, straight, bi, and so on. While they do not live completely free of discrimination, many local people speak warmly of their *muxe* friends, neighbours, colleagues, or family members, or admire their aesthetic or artisanal skills (Stephen, 2002).

While there are clearly precolonial as well as decolonial origins in these non-binary forms of gender expression, and many people have rightly identified *muxe* as belonging to older folkloric cultural practices, *muxe* people are not of course isolated from globalizing cultural phenomena around the rights of transgender and non-binary people and other forms of gender non-conformity. The wearing of traditional dress by *muxe* is, for example, a more recent adoption (McGee, 2018). So *muxe* identities are not static and are reconfigured in dialogue with cultural struggles over gender identities and for sexual citizenship taking place elsewhere and sometimes have contradictory outcomes. As Marcus McGee (2018) writes, drawing on his ethnographic research with the *muxe* in Juchitán, cosmopolitan struggles for sexual citizenship are often problematically entangled with colonial understandings of modernity and civilization which might curtail the forms of liberation attainable. He notes how in Oaxaca, *muxe* are represented simultaneously as politically progressive and as authentic and nostalgic expressions of the precolonial past, meaning that *muxe* identities cannot easily be read through a register of sexual citizenship. *Muxe* themselves are sometimes prepared to draw on these understandings of Indigeneity to perform an exoticized and partially commodified representation of being *muxe* that is of interest to development practitioners, visitors, filmmakers, and journalists. But this participation enhances the visibility of non-Eurocentric gender forms and might to help protect *muxe* from persecution and hate crime.

Given the way that *muxe* people challenge the colonial gender order, they do, however, sometimes find themselves at risk. In 2019, 62-year-old *muxe* activist and founder of the organization las Auténticas Intrépidas Buscadoras del Peligro (the Authentic Intrepid Seekers of Danger), Oscar Cazorla, was found murdered in his home in Juchitán. A year later, nobody had been brought to justice for his death (Cultural Survival, 2020).

In Chile, Mapuche spiritual healers known as *machi* also negotiate gender and sexuality in dynamic and culturally specific ways. Mapuche people, who could not be subdued by the Spanish, are subject to widespread discrimination in Chilean society and have been engaged in a long-term struggle against brutal Chilean colonial incursions. According to Ana Mariella Bacigalupo (2007), *machi* can be male or female but they often adopt a wide range of gender expression in their clothing and other markers and during rituals they move fluidly between genders. They are also misunderstood, othered, stigmatized, and exoticized by Chileans of European descent in particularly harmful ways which disrupts and erases the complexity of *machi* identities as well as their own self-definition. The *machi* work with these contradictions, drawing on conventional Chilean and Catholic forms of gender expression when it is politically expedient to do so and thus they both reproduce and subvert the Eurocentric gender system and the discourses that seek to confine bodies and identities. The *machi* do not fit into Eurocentric forms of feminist theorizing; they are native healers, intellectuals and "politically conscious subjects who interact with and resist colonialism and have fluid boundaries in which to constantly recreate themselves" (Bacigalupo, 2003: 51).

FEMICIDE/FEMINICIDE AND FACTORY WORK

Gender-based violence, including rape and torture, central to the conquest, was a key component of the human rights atrocities committed in Latin America during the military governments and dictatorships of the 1970s and 80s. Today, women continue to be assaulted, disappeared, and murdered by family members and partners, by strangers, and by the state.

The city of Ciudad Juárez on the US–Mexican border is well known for its *maquiladora* or export-processing factories that provide low-paid employment for thousands of Mexicans, especially young women. The preference for young, mostly single and childless women has been attributed to prevailing cultural understandings about gender that assume that young women are more docile and obedient, more suited to repetitive factory work and less likely to unionize and that their incomes are supplementary to the household so they could be paid lower wages (Elson and Pearson 1981; Tiano 1990; McClenaghan 1997). Early explorations of the relationship between gender and the New International Division of Labour (NIDL) tended to focus on exploitation, seeing women primarily as victims of industrialization (Benería et al., 2000). While wages are generally low and working conditions often extremely poor, much of the early work, informed by political economy approaches, tended to construct women as victims of neoliberalism and globalization. Some scholarship, while not denying the exploitation involved, has instead identified the complex motivations, diverse experiences, and forms of female agency that underpin women's factory work (see for example Laurie et al., 1999; Tiano and Ladino, 1999; Benería et al., 2000). Susan Tiano and Carolina Ladino's (1999) work in Ciudad Juárez in the Mexican border region suggests that women do not experience negative impacts directly but rather these are mediated through gender identities that shift over time. Through their factory labour, women are exposed to alternative ideologies which challenge traditional understandings of Mexican womanhood. They reported being able to socialize more freely and choose mates without parental supervision. Even in the context of arduous work and low pay, their autonomy expanded, they gained access to greater sources of personal enjoyment than their mothers' generation and as a result their understandings of femininity, motherhood, and domesticity were reconstructed in positive ways.

In Ciudad Juárez, however, it became extremely dangerous to be a female factory worker. Ciudad Juárez, just across the border from El Paso in Texas, is a violent city in which hundreds of drug-related homicides take place every year. Since the early 1990s, hundreds of women, most of them factory workers, have been kidnapped, raped, and murdered and their bodies dumped in the desert on the edge of the city. Most of them are poor women who have migrated to the border city in search of work and who are discursively constructed as both cheap labour and loose women by capitalism and patriarchy (Salzinger, 2000). The killers act with impunity as they are rarely, if ever, brought to justice, in part because they are seen as women who by migrating and taking on paid work in the transnational labour market have disrupted the dominant gender order (Volk and Schlotterbeck, 2007).

Marcela Lagarde (2006) began to refer to the femicide as feminicide or *feminicidio*. As Rosa-Linda Fregoso and Cynthia Bejarano (2010) state, the term *feminicidio*/feminicide rather than *femicidio*/femicide emphasizes that the murders of women and girls are embedded in structural and systematic forms of power and involve both public and private forms of violence and indeed the blurring of the public and private. While feminicide often co-exists with high levels of homicide, women who are killed are often subject to "specifically gendered forms of violence" (p.7). The concept underscores how gender injustice is also constituted by inequalities of race and class and enables us to position feminicide (and the forms of impunity that accompany it) as a crime against humanity. Furthermore, the use of feminicide can be understood as a feminist decolonial discursive turn that is "designed to reverse the hierarchies of knowledge and challenge claims about unidirectional (North-to-South) flows of traveling theory" (p.5).

For Segato (2010, 2016), the murders in Ciudad Juárez show how patriarchy, coloniality, and capitalism are inseparable. Her analysis has put paid to the idea that the murders are unintelligible. The murders emerge in a conjuncture in which neoliberal multiculturalism is exhausted, as cultural rights are recognized but there is no redistribution of wealth, and capital is moving into what she calls an apocalyptic phase in which the illegal accumulation of capital becomes structural, colonial forms of extractivism are intensified, and people who stand in its path are eliminated. The crimes in Ciudad Juárez are not random acts committed by anti-social individuals, they are systematic and carried out by highly organized mafia networks who maintain a vow of silence. Segato (2016) asserts that the unimaginable forms of cruelty inflicted on women should be understood as acts of communication that take both vertical – they punish the woman for her transgressive moral behaviour – and horizontal forms – they speak to other men in order to secure belonging in a patriarchal system. She describes the murders in Ciudad Juárez as "crimes of high intensity colonial modern patriarchy" (p.96).

The situation is desperately brutal, but many artists, human rights activists, and family members are engaged in courageous and creative acts of rehumanization of the victims as part of the pursuit for justice. In response to the murders and the failures of the Mexican state to address the issue, the city filled with pink crosses and posters of missing women and women began to form human rights organizations such as Nuestras Hijas de Regreso a Casa (Bring our Daughters Back Home) to demand justice. Local photographer Mayra Martell has movingly documented the bedrooms, life goals, clothing, and other personal belongings of the victims (Triquell, 2018). In 2017, the weather-worn missing posters of the victims were collated in an exhibition in the Museum of Memory and Tolerance in Mexico City. Performance artists and muralists such as Cesario Tarín Valdes and Maclovio Fierro Macias have attempted through their art to remember the victims and repair the city's broken social fabric (Mínguez García and Zamarripa Nungaray, 2016), while blogger Lluvia del Rayo Rocha, writes their stories on her blog Los Rostros del Feminicidio, or Faces of Feminicide. Given that as Karina Bidaseca (2015: 225) writes, feminicides occur at the point where "colonialism, imperialism, nationalism and fundamentalism intersect and where globalized racialist and sexist capitalism is stitched together," we require approaches that bring together political economy, cultural studies, and feminist decolonial thought. Sadly, feminicide is not restricted to Ciudad Juárez but is a continental-wide phenomenon. In Mexico, seven women are killed every day, while Argentina experiences a femicide every 30 hours and Nicaragua every five days.

WOMEN'S MOVEMENTS AND FEMINIST STRUGGLES

While there is no doubt that Latin American women and people of non-dominant genders have endured subordination and oppressive gender relations in a range of ways, it is important to recognize that they have a long history of political resistance and feminist struggle. In the 20th century, Latin American women mobilized in diverse and courageous ways; to overthrow dictators or contest human rights abuses, to confront economic hardship caused by economic globalization, and to call for an end to domestic violence. They became part of local and transnational women's and feminist movements to fight for change on gender-specific issues such as reproductive rights and gender-based violence. The revolutionary movements which swept Latin America in the 20th century involved substantial involvement by women. Women in Cuba, El Salvador, and Nicaragua, many of them mothers, joined guerrilla armies and militias (see Randall, 1981; Kampwirth, 2002; Cupples, 2006). Military participation in revolutionary struggles did not, however, automatically led to greater gender equality for women, as male-dominated revolutionary movements and governments dismissed gender issues such as domestic and sexual violence and reproductive rights as bourgeois, imperialist, or culturally irrelevant (Molyneux, 1986; Babb, 2010). In revolutionary Nicaragua, in spite of discourses of the *hombre nuevo* (new man), machismo remained stubbornly persistent (Lancaster, 1992). The Zapatista rebellion in Chiapas has taken a different path and has put feminist demands at the heart of its political agenda from the start (Kampwirth, 2002; Chapter 7).

Women also mobilized in the face of human rights atrocities committed by dictatorships and military governments in the 1970s and 1980s. The military's rise to power was generally accompanied by a call for a return to traditional values which it felt were being eroded, which included the idealization of "family values" and motherhood (Fisher, 1993). In Argentina, the military was extolling such values, while family members were being kidnapped, tortured, and disappeared in a state-sponsored campaign of terror (Fisher, 1989). Many women across the Americas but especially in Argentina, Chile, El Salvador, and Guatemala, lost husbands, sons, daughters, and grandchildren to state terror and were forced to mobilize in response. Jennifer Schirmer (1989) has termed these human rights mobilizations motherist politics as they were dominated by women searching for their missing children. These women frequently met in police stations, prisons, hospitals, or morgues as they searched for any news of their loved ones. Initially, many of these women did not see themselves as political activists, they were merely attempting to fulfil their roles as good mothers and wives, but the magnitude of the terror politicized them. The most famous of these motherist groups is the Madres de la Plaza de Mayo of Argentina, who have relentlessly protested the disappearance of their children every Thursday in the Plaza de Mayo in central Buenos Aires. Similar groups emerged in Chile (Group of Relatives of the Detained-Disappeared), Guatemala (Mutual Support Group or GAM), and El-Salvador (Co-Madres). Some Chilean women made tapestries or *arpilleras* to depict the repression, as "ways of saying without using words" (Boyle, 1993: 168; see Figure 6.3). What is particularly interesting about groups such as the Madres is that they were able to successfully turn the dictatorships' discourses on motherhood and the family against them. Motherist groups appropriated the cultural symbols of the repressive governments and used them to make their own political demands. At a time when many people were being arrested and assassinated for

Figure 6.3 An *arpillera* that depicts the first years living in exile. Many Chileans went to Aotearoa New Zealand during the Pinochet dictatorship

Source: Photo by Marcela Palomino-Schalscha and the Wellington Arpilleras Collective, reproduced with permission

involvement in political activities, these women presented themselves as good wives and mothers who were only trying to carry out their traditional roles in accordance with official ideologies. Since their creation, both the mothers and grandmothers of Argentina have fought tirelessly to bring perpetrators to justice, to locate the remains of those disappeared and to find their grandchildren who were delivered in captivity and then adopted by military families after their adult daughters were killed. Their work has also diversified into other areas and the Madres have created a bookstore, a university, a library, a cultural centre, a newspaper, a radio station, and a subsidized housing project. Their involvement in these political causes can be understood as an attempt to keep the revolutionary political aims that their children died for alive.

The harsh economic conditions generated by neoliberal economic policies had particularly gendered impacts. The philosophy behind IMF conditionality was that the cost of adjustment was to be borne by individuals rather than the state. Clearly, the increases in the costs of basic foods and services impacted most severely on the marginalized urban poor. Because women are generally responsible for reproductive activities, such as putting food on the table and caring for children and the elderly, they tended to be more severely impacted by these policies. As Amy Lind (2005) writes, women became the absorbers of the crisis, as governments assumed that women would take on additional unpaid caring or community roles, a policy approach that is "symbolically and materially violent" for low-income women and leads to the "institutionalization of poverty and survival" (p.17). Many women joined the labour market for the first time, taking on jobs in the informal sector, in domestic service, and in assembly factories. In Chile, Peru, and Uruguay, some women organized themselves into providing food for their families communally, creating community soup kitchens, the so-called *comedores populares* or *ollas comunes*.

One of the interesting gender dimensions of activities such as communal kitchens is that as women found collective solutions to reproductive tasks that had become too onerous as a result of the economic situation, in the process they also began to address more strategic gender issues. For many women, participation in these community-oriented activities proved to be an important source of self-education and empowerment (Chuchryk, 1991), dismantling any kind of binary separation between practical (providing food for familes) and strategic (pursuing feminist aims) gender issues (see Molyneux, 1986).

In more recent years, Latin American feminism has experienced a generational shift and is increasingly transnational, intersectional, and decolonial in its focus. Women and people of other genders are angry, tired of socioeconomic inequalities, corruption, impunity, femicide, homophobia, and transphobia. Some of the mobilizations against gender-based violence have been very large, they have often included diverse forms of artistic and digital activism and performance, and they make extensive use of social media to connect struggles across the continent. They are also much more focused on gender inclusivity and on dismantling binary gender (see Box 6.2). Black, Indigenous, and poor urban women have been decisively at the forefront of recent struggles, connecting gender-based violence and patriarchy to capitalist accumulation, extractivism, and racism.

In 2015, thousands of women in Argentina took to the streets to protest at the murder of 14-year-old Chiara Paez, who was beaten to death by her boyfriend after he found out she was pregnant. The protestors coined the slogans *Nos Están Matando* – They are killing us – and *Ni Una Menos* – Not One Less – meaning that not a single other woman should lose her life in this way. Both slogans became viral hashtags and they continue to circulate throughout the continent. *Ni Una Menos* builds on the struggles of the Madres de la Plaza de Mayo. One of their prime tactics has been the women's strike or *paro de mujeres* and these took place in Argentina and throughout Latin America in 2016, 2017, and 2018. These strikes involve bringing the private, the domestic, and the reproductive into the public sphere to denounce femicide, disrupt capitalism, and reveal extractivist logics with prefigurative actions (Quiroga, 2019; Motta, 2019). As Natalia Quiroga (2018: 565), writes they "show that, just as we impose limits on the different forms of violence suffered by the female body, so can we impose limits on the policies that devastate the conditions for reproduction." While Latin American women's movements have long fought for reproductive autonomy, today's movements are focused on the gendered and racialized dynamics of extractivism and dispossession and involve multiple and proliferating articulations. These elements become discursively linked through the concept of *cuerpo-territorio* or body-territory. As Sara Motta (2019) writes body-territory "is mostly centred around reproductive and bodily autonomy, as well as the right to defend indigenous and other territories from the expropriation of extractive capitalist and/or state forces" (p.21) and therefore is contributing "towards a re-imagining of revolutionary politics" (p.15).

More recently, Chilean women were able to further strengthen and extend this feminist energy. In 2019, as we've noted, much of Latin America was rebelling. In Chile, gender-based violence was articulated to inequality, state-led violence, and impunity. On 25 November (the International Day of Violence against Women) and in the middle of the protests, a feminist group called Las Tesis created a choreographed performance entitled Un violador en tu camino or A rapist in your path that they performed in Santiago (see Box 6.1 and Figure 6.4). The performance went viral and was subsequently performed by feminists all around Latin America and beyond, including Turkey, India, France, Spain, and the UK. It was also performed in New York outside the court where Harvey Weinstein was being tried. It's been performed by younger women, older women, disabled women, and has been creatively adapted to

Figure 6.4 Graffiti in Santiago, Chile: "the rapist is you"

Source: Carlos Figueroa Rojas, CC BY-SA 4.0, via Wikimedia Commons

different contexts with different kinds of grassroots organizing. Paula Serafini (2020) believes that A rapist in your path has prefigurative, participatory and horizontal qualities that are transnational in scope. She asserts that the reason why it is so effective is that "it marks a point of no return" because it reveals that it is "the system that is killing us" (p.294); consequently, people are no longer willing to be silenced and perpetrators can no longer expect impunity.

BOX 6.1 A RAPIST IN YOUR PATH

A rapist in your path: Lyrics	Un violador en tu camino: Letra
Patriarchy is a judge	El patriarcado es un juez
that judges us for being born,	que nos juzga por nacer,
our punishment	y nuestro castigo
is the violence you don't see.	es la violencia que no ves.
Patriarchy is a judge	El patriarcado es un juez
that judges us for being born,	que nos juzga por nacer,
our punishment	y nuestro castigo
is the violence you now see.	es la violencia que ya ves.
It's femicide,	Es feminicidio,
Impunity for my killer.	Impunidad para mi asesino.
It's disappearance.	Es la desaparición.
It's rape.	Es la violación.
And it's not my fault, nor where I was,	Y la culpa no era mía, ni dónde
nor what I wore.	estaba, ni cómo vestía.
And it's not my fault, nor where I was,	Y la culpa no era mía, ni dónde
nor what I wore.	estaba, ni cómo vestía.
And it's not my fault, nor where I was,	Y la culpa no era mía, ni dónde
nor what I wore.	estaba, ni cómo vestía.
And it's not my fault, nor where I was,	Y la culpa no era mía, ni dónde
nor what I wore.	estaba, ni cómo vestía.
The rapist was you	El violador eras tú,
The rapist is you.	El violador eres tú.
It's policemen,	Son los pacos,
Judges,	Los jueces,
The state,	El Estado,
The president.	El presidente.
The oppressive state is a rapist man.	El Estado opresor es un macho violador.
The oppressive state is a rapist man.	El Estado opresor es un macho violador.
The oppressive state is a rapist man.	El Estado opresor es un macho violador.
The oppressive state is a rapist man.	El Estado opresor es un macho violador.

The rapist was you	El violador eras tú.
The rapist is you.	El violador eres tú.
Sleep tight, innocent girl	Duerme tranquila, niña inocente
don't worry about the criminal,	sin preocuparte del bandolero,
your policeman lover is taking care	que por tu sueño dulce y
	sonriente
of your sweet dreams.	vela tu amante carabinero.
The rapist is you.	El violador eres tú.
The rapist is you.	El violador eres tú.
The rapist is you.	El violador eres tú.
The rapist is you.	El violador eres tú.

BOX 6.2 LATINX: MAKING GENDER INCLUSIVE LANGUAGE

In the past few years, various categories have been created to refer to Latin Americans in the United States, while people of Latin American descent in the United States have resisted various attempts to name them collectively, asserting their identities as Mexican, Puerto Rican, Cuban, or as Black or Indigenous. As Graciela Mochkofsky (2020) writes, the term Hispanic was adopted in the census of 1980 and became popularized by Spanish-language television channel Univision. But she notes that many objected to the connections to Spain and colonial oppression and so the term Latino began to circulate instead and was included as a census category in the 2000 census. The Spanish language has the gender binary embedded in its grammar and orthography. Every single Spanish noun has one of two genders – la ciudad (the city) is feminine while el país (the country) is masculine. Usually nouns that end in a are feminine (la mesa, the table) and those that end in o are masculine (el libro, the book) although there are exceptions, such as la mano (the hand) or el día (the day). Many scholars and activists have tried to find ways to make the language more inclusive, fluid, and gender neutral by replacing the gendered "o" and "a," with an "e," an "x" or a "@." For example, "todos" or "todas" meaning everyone become "todxs," "todes" or tod@s, or the terms Latin, Latino/a or Hispanic become Latinx, or sometimes LatinX. Latinx has been used since 2004, when many people were already using the term Latina/o (that doesn't dismantle the gender binary), but its use became much more widespread in 2015 (Marquéz, 2018). It is both a contested and contentious term. For some it is an important way of recognizing trans, non-binary, and queer gender formations and constitutes a political stance that promotes inclusivity, while for others it is seen as a fad that distorts the Spanish language, cannot be easily pronounced, and takes political correctness too far (Torres, 2018). In the United States, Latinidad was already a contested term, embraced by some and rejected by others, but the term Latinx seems to attract strong opinions, both in the United States and in the rest of (Latin) America. It is important to recognize that the use of the Spanish language in América/the Americas is a colonial legacy, but it is one that has resulted in many different ways of speaking Spanish. The language has inevitably been appropriated and transformed through use and it now belongs as much to Afrodescendant, Indigenous, and mestizo Latin Americans, North and South of the Río Grande, as it does to Spanish-speaking Europeans. The many creative regional differences in the use of Spanish have enriched the language. Nonetheless, a Madrid-based institution, the Real Academia Española (RAE), seeks to engage in a colonizing policing of the rules of Spanish, determining these not just for Spain, but for the whole of the Spanish-speaking world, putting "pressure Spanish speakers worldwide to adopt a Euro-centered linguistic identity" (Zentella, 2017:27). Ana Celia Zentella (2017: 24) describes their "policing and insistence on lexical and grammatical purity" as symbolic violence. The RAE has rejected the term Latinx as well as all alternative uses of the "x" and the "e," stating that the "o" already captures a mixed gender group.

But Latinx continues to be used and it clearly fulfils many functions. It emphasizes the presence of Latinx within US territory and especially "*as constitutive of the United States of America*" (emphasis in original), including those that were there before the United States took a third of Mexico's territory in the 1846–1848 war and those that have arrived since, those who are documented and those who are not (DeGuzmán, 2017: 215). Not only does it include those who are excluded by the gender binary and by the geopolitics of immigration, but it also "holds space for defiance and reminds the reader to critically consider how audiences interpret public markers of race, class and gender subject formation" (Valentina González, 2019: 1). It draws our attention to the coloniality of gender and helps us to acknowledge how Indigenous gender systems were quite different from those imposed by Europeans during the conquest (Blackwell et al., 2017). Latinx should not then be seen not a monolithic or fixed identity, but one that expands to include Indigenous- and Afro-Latinx diasporas as well as a range of genders, and one that moves in and out of the United States and Latin America. Claudia Milian (2019: 6) usefully urges us not to consider Latinx as something that we are for or against, but rather as "a point of orientation that allows to start charting the realm of LatinX inquiry." It is not perfect, but it is a useful mobilizing tool which complicates questions of nationality and belonging and has both queering and decolonizing ambitions.

Sources: DeGuzmán, 2017; Zentella, 2017; Blackwell et al., 2017; de Onís, 2017; Torres, 2018; Marquéz, 2018; Milian, 2019; Valentina González, 2019

LGTBQ+[1] STRUGGLES AND MOVEMENTS

Homophobia and transphobia are part of the colonial matrix of power and remain endemic and culturally entrenched social problems in Latin America. The strength of Eurocentric and colonial constructions of gender and sexuality are thus that homophobia and transphobia have been mobilized and produced by both right-wing dictatorships and left-wing revolutionary movements. A gay organization, the Frente de Liberación Homosexual (Homosexual Liberation Front) in Argentina was repressed by the military dictatorship that came to power in 1976 along with the other political groups that the regime deemed subversive. In the 1960s, the Cuban government even went as far as to incarcerate gay men in labour camps. Fears of social and familial rejection work to keep LGBTQ+ people in the closet and deter them from political mobilization against heteronormativity and in defence of their citizenship rights.

Today, LGBTQ+ organizing is challenging discriminatory attitudes and leading to changes in policy in some countries (see Box 6.3). But progress is slow. For example, equal marriage legislation has been passed in only six countries: Argentina (2010), Brazil (2013), Uruguay (2013), Mexico (2015), Colombia (2016), Ecuador (2019), and Costa Rica (2020) (see Kennon, 2020). Legislatures in other countries are continuing to assert that marriage can only be between a man and a woman. For example, Honduras and the Dominican Republic passed laws banning equal marriages and adoptions in 2005 and 2009 respectively. LGTBQ+ scholarship and activism draw on and contribute to growing dialogues between queer theory and decolonial scholarship that recognizes the close relationship between heteronormativity and coloniality. Activism by people with non-dominant genders and sexualities is therefore contributing to processes of decolonization.

BOX 6.3 COMING OUT IN LATIN AMERICA

1998 Ecuador – New constitution includes protections against discrimination based on sexual orientation

1999 Chile – Decriminalization of same-sex intercourse

2000 Brazil – The state legislature of Rio de Janeiro bans discrimination based on sexual orientation in public and private establishments

2003 Mexico – Federal anti-discrimination (including sexual orientation) law passed

2004 Brazil – Government introduces *Brasil sem Homofobia*, a programme to change social attitudes towards gays and lesbians

2004 Peru – Repeal of a law that banned members of the armed forces from having homosexual relations

2006 Mexico – Government passes a cohabitation law, which grants
cohabiting same-sex couples the same rights as cohabiting heterosexual
couples

2006 and 2007 Mexico – Coahuila and Mexico City legally recognize same-
sex civil unions

2007 Argentina – Buenos Aires allows same-sex unions

2007 Brazil – Rio Grande do Sul allows same-sex unions

2007 Uruguay – All cohabiting couples given access to health benefits,
inheritance, parenting rights, and pension rights, after five years of
cohabitation regardless of sexual orientation

2007 Colombia – Cohabiting same-sex couples given same rights as married
couples after two years of co-habitation

2008 Mexico – Trans persons permitted to change their legal gender and
name on official documents

2008 Nicaragua – Decriminalization of same-sex relationships and repeal of
anti-sodomy law

2008 Cuba – Government introduces free sex change operations

2008 Brazil – Three million people attend the LGBTQ+ parade in São Paulo,
the largest pride march ever held

2008 Brazil – The First National Conference of Gays, Lesbians, Bisexuals,
Transvestites, and Transsexuals is held in Brasilia

2008 Panama – The government repeals a law that criminalizes same sex
intercourse

2009 Bolivia – New constitution bans discrimination on the basis of sexual
orientation

2009 Chile – Social activists organize a mass wedding for sexual minorities in
front of the Metropolitan Cathedral

2009 Colombia – Same-gender couples granted same pension and property
rights as heterosexual couples

2009 Mexico – Legislature of Mexico City approves marriage and adoption
rights for same-sex couples

2009 Uruguay – Government approves a bill to end the restriction of adoption
to married couples

2010 Argentina – Legalization of equal marriage, the first Latin American
country and eighth in the world to do so

2011 Brazil – Supreme Court grants equal legal rights to same-sex civil unions
as those enjoyed by married heterosexuals, including retirement benefits,
joint tax declarations, inheritance rights, and child adoption

2013 Brazil and Uruguay legalize equal marriage

2016 Colombia legalizes equal marriage

2018 Brazil – Supreme Court allows trans people to change their names and
gender on official documents without undergoing surgery

2019 Brazil – Supreme Court rules that that transphobia and homophobia are
    criminal offences
2019 Ecuador legalizes equal marriage
2020 Costa Rica legalizes equal marriage

Sources: *The Economist*, 2015; Corrales, 2010; Corrales and Pecheny, 2010;
COHA, 2011; Kennon, 2020

LGBTQ+ activists have made important inroads in deeply homophobic legislatures by framing their search for equality as a human rights
issue and successfully connecting it to other human rights related struggles
(Encarnación, 2011). The fact that the struggle for human rights in countries such as Argentina has been so powerfully fought and felt and so costly
to so many people probably makes more people receptive to gay marriage on
the ground of human rights. It is clear that while legislative change comes
in the wake of cultural change and political demands, legislative changes also
foment cultural change and can make important contributions to social tolerance. A survey conducted in Mexico after the first "same sex" marriages were
celebrated showed people becoming much more accepting, with the numbers
of those opposed to such marriages falling substantially (Reid-Smith, 2020).
Increasingly, Latin American cities such as Buenos Aires and Mexico City
emphasize their gay friendliness to tourists and visitors (see Figure 6.5).

It remains, however, extraordinarily dangerous to be gay and transgender
in Latin America. According to Transgender Europe, which organizes the
annual Trans Day of Remembrance to denounce the murders of trans- and
gender diverse people, 82% of all of these murders globally took place in
Latin America, and 43% of them in Brazil (TGEU, 2020). Brazil has passed
a series of progressive laws that uphold gay and trans citizenship rights and
São Paulo is home to the largest gay pride march in the world, but homophobia and transphobia are widespread and frequently result in physical assault
and murder as well as discrimination. Indeed, President Jair Bolsonaro once
described himself as a proud homophobe. In 2020, at least 175 trans women
were murdered and 68% of these were Black and 72% of them were sex
workers (see Observatório de Mortes Violentas de LGBTI+ no Brasil, 2021;
Sudré, 2021). There is a close relationship between transphobia, racism, and
poverty.

While Brazil has the largest number of murders in absolute terms, in
per capita terms the most dangerous place on Earth to be a trans person
is Honduras (Orellana, 2021). Honduran trans people have no legal recognition and are subject to widespread violence and discrimination that has
increased exponentially since the 2009 coup, as has violence against environmental defenders. In 2019, the Inter-American Commission on Human

Figure 6.5 Pride march in Mexico City in 2018

Source: ProtoplasmaKid, CC BY-SA 4.0, via Wikimedia Commons

Rights (IACHR) initiated legal proceedings in the Inter-American Court of Human Rights regarding the case of Vicky Hernández, a human rights defender and trans women, who was murdered by state law enforcement officers in June 2009 during the coup when a curfew was in force. Vicky worked for an organization Unidad Color Rosa Colectivo TTT, and six of its seven founders, all trans women, have been murdered. Not only does the IACHR believe that the state is responsible for her death, and that she was murdered because of her gender identity, but that they also acted with impunity given the failure to investigate the case adequately. The murder thus violates the American Convention on Human Rights. This case is important because the IACHR is recommending not only compensation and support for Vicky's family but also the granting of legal recognition to trans people, the implementation of measures to address the structural causes of violence against trans people in Honduras, and the urgent need to bring to an end the conditions of impunity in which this violence takes place (see ISHR, 2020; OAS, 2019). If the case succeeds, it will set a legal precedent for all trans people in Latin America (Orellana, 2021).

Sheila Rodríguez Madera (2020: 6) has coined the term necropower to describe the everyday forms of discrimination that shape the lives of trans people, the "acts that do not emanate "from above" through policies and laws, but rather are reproduced in everyday relations with others (e.g., friends, family members, neighbors, co-workers, doctors, teachers, sex clients) and devalue trans people, affecting their sense of worth, and positioning them as unwanted in society." She refers to these acts as necropower as they always involve the threat of elimination. Necropower leads, however, to what she calls necroresistance that doesn't necessarily involve highly visible or spectacular forms of resistance but also involve everyday acts of defiance in the community or in public space that construct a sense of self-worth.

While it is crucial to document and condemn violence against LGTBQ+ people, we should also focus, as Florence Babb (2019) writes, on the ways in which these populations assert their right to both sexual pleasure and livelihoods, including sex work, using diverse forms of activism, art, and scholarship. For example, Operación Queer in Nicaragua is an "intellectually and culturally sophisticated" collective of artists, academics, and activists that "takes up questions of the body and identity and of forms of exclusion relating not only to gender and sexuality but to social class, ethnicity, age, ability, and aesthetics" (p.316). The collective Mujeres Creando in Bolivia, founded by María Galindo and Julieta Paredes, uses street performance, film and video, and graffiti art to critique capitalism as well as the patriarchy, racism, and homophobia of the traditional left in Bolivia, to express eroticism, and to visibilize women who are both Indigenous and lesbian (see Muñoz, 2015) or as María Galindo (2006) puts it "indias, putas y lesbianas" "Indians, prostitutes and lesbians."[2] Mujeres Creando advances a decolonizing agenda and rejects liberal notions of gender equity (see Galindo, 2006).

SUMMARY

- Sexism, racism, homophobia, and transphobia are all still alive and well in Latin America. These oppressive forces should be understood as a development issue because they prevent people from reaching their full potential by denying them access to opportunities that are available to people who are differently gendered or who express a different sexual orientation. They should also be understood as colonial legacies.
- The feminicide in Ciudad Juárez can be better understood when we combine political economy analyses with decolonial feminist ones. The dehumanizing actions of the murderers is responded to with humanizing actions of family members and activists.
- Gender-based violence and state-led violence against environmental defenders are part of the same conjuncture.
- Latin America has vibrant and sophisticated feminist and LGBTQ+ movements that are making their presence felt in scholarship, in policy-making arenas, in the street, and in the home. These movements are now much more intersectional in their focus.

- This activism makes a difference and that is why it is so fiercely opposed by the colonial and patriarchal power bloc.
- Gender and sexual liberation are important components of decolonial struggle.

DISCUSSION QUESTIONS

1 Gender is a key component of Quijano's colonial matrix of power. What is Lugones' critique of Quijano?
2 How does coloniality exacerbate gender-based disadvantage?
3 How does the idea that binary gender and patriarchy were colonial imports shape the way you feel about constructions and expressions of gender in your own life and cultural context?
4 What can the *muxe* and *machi* teach us about cultural constructions of gender?
5 Why do we need to understand the intersections between sexism and misogyny, political impunity, and the neoliberal economic model in order to explain the violence in Ciudad Juárez? What can we achieve when we combine political economy analyses with decolonial feminist insights?
6 What strategies and tactics are used by feminist and LGBTQ+ human rights organizations to fight against oppressions?
7 What are the politics that underpin the use of the term Latinx?

NOTES

1 LGBT (lesbian, gay, bisexual, trans) and LGBTQ+ (lesbian, gay, bisexual, trans, queer, with the + referring to genders and sexualities not captured by the other labels) are among a number of umbrella terms that are used to capture gender and sexual diversity and to signal inclusion. There are others in circulation in Latin America and elsewhere including LGBTI+ (lesbian, gay, bisexual, trans, intersex +) and LGBTTI (lesbian, gay, bisexual, travesti, transsexual, intersex +) and LGBTQIAA+ (lesbian, gay, bisexual, trans, queer, intersex, asexual, aromantic +). As umbrella terms they are all flawed and unwieldy, but they do at least attempt to signal inclusion. Latin American organizations fighting for the rights of people with non-dominant genders and sexualities use variations of this acronym or other terms altogether. As Florence Babb (2019) writes, some Latin Americans reject the label queer as Eurocentric and homogenizing, others appropriate it.
2 For examples of the performances of Mujeres Creando see http://mujerescreando.org/indias-putas-y-lesbianas-maria-galindo-y-la-desobediencia-feminista/

FURTHER READING

Routledge Handbook of Latin American Development Chapters 4 (Esguerra Muelle); 23 (Bradshaw et al.); 24 (Gideon and Alvarez Minte); 25 (Gutmann); 26 (Babb); 31 (Bickham Méndez); 47 (Quiroga).

Bacigalupo, A. M. (2007) *Shamans of the Foye Tree: Gender, Power, and Healing among Chilean Mapuche*. Austin: University of Texas Press.

A book about the *machi*, the Mapuche spiritual healers, and the ways in which they negotiate gender

Esquivel, L. (2005) *Malinche*. México: Suma de Letras.

A novel on the relationship between Malinche and Cortés and the conquest of the Aztec empire. English translation published in 2006 by Washington Square Press.

Fregoso, R. L and Bejarano, C. (2010) (eds) *Terrorizing Women: Feminicide in the Americas*. Durham, NC: Duke University Press.

An edited collection that explores the phenomenon of feminicide in Ciudad Juárez and other parts of Latin America and the kinds of citizenship practices and struggles for justice that it engenders.

Recommended films

On the Edge: The Femicide in Ciudad Juárez (2006), directed by Steev Hise

A documentary about the social, cultural and economic dimensions of gender-based violence and impunity that prevails in Ciudad Juárez.

Muxes –Mexico's Third Gender (2017), directed by Shaul Schwarz

A short film commissioned by *The Guardian* that deals sensitively with the *muxe* identity of Juchitán

Useful websites

Mujeres Creando (in Spanish)

The multimedia website of Bolivian feminist collective that covers their many campaigns and forms of direct action

http://mujerescreando.org/

Los Rostros del Feminicidio (in Spanish)

The blog of Lluvia del Rayo Rocha on the faces of feminicide in Ciudad Juárez

https://losrostrosdelfeminicidio.wordpress.com/

CHAPTER 7

Indigenous politics and movements

INTRODUCTION

The conquest of America by European powers involved genocide, dispossession, slavery, epistemicide, and ecocide and the Indigenous peoples were of course the first (Latin) Americans to experience the brutality of colonial rule and the first to enact decolonial resistance. Indigenous peoples have of course lived in what is now Latin America long before the conquest, so they have a very long history that we don't have time to cover here, and many aspects of Indigenous lives have nothing to do with colonialism. But colonialism was devastating as Indigenous peoples succumbed to coerced labour, European diseases, or malnutrition induced by land loss. Others died rebelling and resisting. Some colonizers were not in favour of outright extermination, but still believed in assimilation through evangelization, urbanization, and Spanish or Portuguese literacy. Either way a Eurocentric fantasy that distinct Indigenous cultures would not survive persisted. Even in the development and postdevelopment eras, extermination and dispossession have never ended and as we saw in Chapter 4, Indigenous environmental and human rights defenders continue to be criminalized and assassinated and Indigenous lands and resources are frequently targeted and contaminated by large agribusiness and extractivist operations. Being Indigenous in Latin America today means therefore continuing to negotiate coloniality.

But Indigenous peoples and cultures have survived; they are still here. As Rowe and Shelling (1991: 49) write, the conquest was catastrophic for Latin America's Indigenous civilizations, yet "neither the colonial nor the republican regime has been able to expunge the memory of an Andean, Aztec and Mayan civilization" and these cultures have persisted. Not only have they survived they are reshaping political cultures and development trajectories in all kinds of ways. Since the 20th century, the Eurocentric concept of "human rights" has frequently been mobilized by Indigenous people across the continent as

DOI: 10.4324/9781003110453-7

a means to respond to the colonial atrocities committed against them. Arturo Arias (2016: 327–328) captures the challenge for us as decolonial students and researchers:

> In the Latin American context, it is impossible to separate indigenous decolonial maneuvers from both the violence and the violation of human rights suffered by these populations in most countries with sizable native communities. We are still witnessing these deplorable conditions in nations otherwise heterogeneous among themselves, such as Guatemala, Peru, and Chile, to just cite a few, where violence, including rape, femicide, assassinations, or torture, still takes place on a daily basis against an array of indigenous groups, including Mayas, Quechuas, or Mapuches, to name but the larger and better known ethnicities. Indeed, most indigenous groups in the Americas have feared for their lives since at least 1521, the emblematic year of the fall of Tenochtitlan, the Mexica capital. (…) Since that fateful date, most indigeneities have lived in their own lands under military occupation, enduring a series of racialized regimes with no regard for the law or human rights. One of the inalienable ethical tasks for western scholars today should, in consequence, include coming to terms with the knowledge of indigenous racialized abjection and its grotesque implications to the fullest extent possible, if we are to reconfigure significant meaning of the phrase "human rights."

Indigenous peoples are now a formidable political force. All over the continent, they are marching, occupying, negotiating, writing, performing, and mobilizing to defend life and territory from destructive forms of development and capitalist incursions. They draw on their own ancestral knowledges to do so, and as noted in previous chapters, have gained constitutional recognition and asserted the right to exist on their own terms. Their resistance is a key source of hope to people the world over who are displaced by capitalist greed or by modernizing forms of development or who are threatened by or concerned about imminent climate collapse. Indigenous political action and intellectual production are forcing us to confront the failure of Eurocentric physical and social science and of existing forms of politics in tackling injustice. Taking Indigenous knowledges and struggles seriously is central to a decolonial approach.

Determining who is and isn't Indigenous is not straightforward. At the time of the conquest, there were hundreds of Indigenous groups and thousands of Indigenous languages. Some of these groups were nomadic hunter-gatherers; others had advanced civilizations and knowledge systems. Today, around 50 million Latin Americans identify as Indigenous; there many different tribes or nations all over the continent (see Figure 7.1). Some 90% of Latin America's Indigenous peoples live in just five countries: Bolivia, Ecuador, Guatemala,

Figure 7.1 Indigenous peoples in Latin America

Source: ECLAC, 2014, available at www.cepal.org/sites/default/files/styles/
infographic_aspect_switcher/public/infographic/images/indigenas_ingles.jpg

Mexico, and Peru (Yashar, 2005). Bolivia has a majority Indigenous population, in Guatemala Indigenous peoples account for at least 40%. Peru and
Mexico have Indigenous populations in excess of 10 million people. There
are also large Indigenous populations in Brazil, Chile, Colombia, Panama, and
Nicaragua. Some speak Spanish, Portuguese, or English, instead of or as well
as an Indigenous language and more than a thousand Indigenous languages
are spoken across the continent. Several Indigenous languages, including
Quechua, Guaraní, Aymara, Kekchí, and Nahua, have more than a million
speakers (for more detail see Kaufmann, 1994a, 1994b). Between eight and
nine million people across four countries, Guatemala, Mexico, Honduras,
and Belize, speak a Mayan language. But many Indigenous languages are
endangered, threatened by Spanish dominance. The Rama language spoken
in Nicaragua has only around 800 native speakers remaining.

Indigenous peoples live in forests and in cities; some are subsistence farmers, some are university academics, lawyers, and development practitioners. Indigenous identity is shaped by several factors, including language, community residence, diet, dress, customs, economic and cultural practices, and importantly is based on *self-identification*.

INDIGENOUS KNOWLEDGE SYSTEMS

As Quijano (2005: 58) writes, Indigeneity can only be understood in relation to what he calls the "coloniality of the system of power". If resources had not been and were not still concentrated largely in the hands of European descended peoples, the term Indigenous would have a very different set of meanings. Indigenous knowledge systems were and are quite different from European ones. In western epistemology and cosmology, religion is separated from science. In Mayan culture, whose territory stretches from central Mexico into the Yucatán Peninsula into the highlands of Guatemala and parts of Honduras and El Salvador, spirituality and science are seen as perfectly compatible and this compatibility can be observed in the ruins of the Mayan cities at Chichén Itzá, Tikal, and Copán. The ancient Mayans had developed mathematical and astronomical knowledges and they operationalized detailed and complex understandings of time. The grand pyramid at Chichén Itzá, which was possibly built in the 4th or 5th century CE (see Figure 7.2), is both a solar calendar – it has 365 steps equivalent to the days in a solar year – and a temple to Kukulkan, a feathered serpent god. Twice a year, on the spring and autumn equinox, special light effects show the serpent descending its stairway. The pyramid also produces a special acoustic effect where a single handclap from the plaza below the pyramid produces two quetzal chirps and an echo (Declercq et al., 2004; García-Salgado, 2010) As Declerq et al. (2004) write, the Dresden Codex shows the serpent as being connected to the quetzal, a brightly coloured long-tailed bird, and in Mayan culture an echo symbolizes a spirit. Mayan agriculture also has deep spiritual dimensions and Mayans understand themselves to be people of corn. Corn is both a sacred crop as well as a staple food in Mexico and Central America today.

As we noted in Chapter 2, the colonizers worked hard to destroy Indigenous knowledges and intellectuality but some of it from that period has survived in both oral and textual form. An important Mayan example is the *Popol Vuh* (The Council Book) an ancient K'iche text that contains the Mayan creation story and important cosmological and genealogical narratives. Today many different translations of it exist (see, for example, Recinos, 1950). Originally written in hieroglyphic form prior to the conquest, it was rewritten in alphabetic form in the middle of the 16th century and this version contains a preamble that wasn't in the original version. This

Figure 7.2 The grand pyramid at Chichén Itzá

Photo by author

was an extremely dangerous and subversive action at the time because Mayan intellectuals were often persecuted and killed and Mayan texts were destroyed (Henne, 2020). Nathan Henne (2020) calls the *Popol Vuh* one of the oldest surviving pieces of decolonial writing in the Americas. He notes how difficult it is to read it "on its own terms," as our reading practices are so thoroughly contaminated by Eurocentric thinking and the *Popol Vuh* had to be written in such a way to evade the wrath of the colonizers. Henne asserts that it requires "an alert and active decolonial reader in order to decode how the text undercuts and overturns colonial logic" (p.4). The European colonizers when confronted with advanced and complex civilizations attempted to understand and document them, even as they worked to destroy them. Diego de Landa (1978 [1937] [1552]) was responsible for the cruel destruction of much Mayan culture but paradoxically also managed to decipher Mayan hieroglyphics and document these intellectual endeavours.

The colonial period as discussed was characterized by Indigenous peoples' rebellions and uprisings (see Chapter 2) but also by everyday forms of accommodation, resistance, and transcultural exchange. Eurocentric knowledges were modified in the process and creative modes of border thinking were generated. Indeed, colonial culture was controlled through interpretation and creative hybridization (Rowe and Shelling, 1991; de Certeau, 1984). While Indigenous peoples were not able to escape the

Figure 7.3 Mexicans celebrate the Day of the Dead

Source: Photo by author

imposition of colonial cultural forms such as Catholicism, they blended them with their own traditions in ways that made them more palatable. Latin America is replete with examples of Indigenized and Africanized Catholicism, including the Day of the Dead (see Figure 7.3), the Virgin of Guadalupe, and Santa Muerte (see Chapter 10) in Mexico, Sihkru Tara in Nicaragua, Corpus Cristi in Bolivia, Candomblé in Brazil, Winti in Suriname, and Santería in Cuba. Indeed, at times, the Church has been a key source of Indigenous oppression but at others has acted as a key defender of Indigenous rights (Martí i Puig, 2010). For example, the Moravian Church in Nicaragua disapproved of Miskito spiritual practices but during the civil war in the 1980s began to defend the Miskito people from state-led aggression and violence (Hale, 1994; García, 1996; Hawley, 1997) and today there is a growing dialogue between Miskito and Christian cosmovisions. Elsewhere, liberation theology has been central to making sense of poverty and colonial violence by grassroots Ecclesial Base Communities (CEBs) in Brazil, Chile, El Salvador, Guatemala, and Mexico (Levine, 1988). *El Güegüence* is a play from the colonial era that is still performed every year in Nicaragua. It celebrates the creativity of Indigenous and mestizo resistance to colonial domination. It shows how through mimicry and pretence of complicity, Indigenous and mestizo actors were able to dupe and trick the Spanish colonizers, mock their authority, and subvert their aims (Cuadra, 1997, see Figure 7.4). Cultural exchange is a two-way process, European cultures have also been modified as a result of colonialism.

Figure 7.4 Characters from Nicaraguan anti-colonial play *El Güegüence*

Source: Photos by author

THE IDEOLOGY OF MESTIZAJE

Given these processes and the ethnic composition of the Latin American population, the post-independence period was a time in which the struggle for identity became paramount. Although Latin America was built on widespread mixing between Indigenous, African, and European-descended peoples, resilient racial hierarchies were established in which skin colour and phenotype were decisive. Indigenous and Black populations continued to be viewed by Europeans as inferior peoples, whose capacity for civilization was in doubt. The visible valorization of whiteness in all kinds of public spaces has been to the detriment of Indigenous peoples, while the myth of racial democracy (see Chapter 8) perpetuated in some countries has made it difficult to effectively respond to racism. For the Eurocentric elites of the continent, influenced by theories of scientific racism, there was a sense that Latin America was populated by the wrong kind of people and that social and economic progress would be hindered by their large Indigenous and Black populations.

The genocide and marginalization of indigenous peoples did not end after independence, with some national governments believing that Indigenous peoples should be eliminated or that they would at the very least disappear through miscegenation. Others implemented policies such as *indigenismo* which aimed to assimilate Indians into mainstream mestizo society through mechanisms such as Catholicism and education in the Spanish language. Mestizaje became a state ideology, in the sense that it was only by "conforming to a homogeneous mestizo cultural ideal" that one could enjoy the fruits of citizenship (Hale, 2004: 16). But mestizaje also became a creative response to dominant ideas of racial degeneration. Mexican José Vasconcelos posited that Mexicans were a cosmic race (*la raza cósmica*) because they were a mixture of European and native American elements (see Manrique, 2016; Chapter 8). Recognizing the Indigenous heritage of all Mexicans did not, however, eliminate racism towards Mexico's Indigenous peoples. Mestizaje mobilizes an individual model of citizenship which could not admit the culturally specific collective rights of Indigenous groups (Hale, 2004). As Daniel Runnels (2019: 140) puts it, mestizaje when posited as an "ideal racial configuration implies that indigeneity impedes the entrance into modernity, indigeneity being acceptable only as a distant trace in a newly miscegenated racial type." In the 20th century, some Latin American governments adopted a corporatist model into which Indigenous peoples were enrolled as peasants or campesinos and encouraged to abandon Indigenous identities in favour of class-based ones (see Yashar, 2005). Assimilationist ideas circulated alongside more contradictory discourses which saw Indigenous peoples as a pure racial form in need of protection. Policies such as *indigenismo* met with varying degrees of resistance, including movements like *indianismo* organized to resist the imposition of integration. Five hundred years after the conquest, Indigenous people were still being eliminated by state-led forces and lighter-skinned political elites as Guatemala's recent history reveals.

COUNTER-INSURGENCY AND GENOCIDE IN GUATEMALA

Geographical remoteness, political resistance, and cultural resilience enabled the Maya people of Guatemala to survive and adapt to colonization. Although Indigenous peoples constitute at least 40% of the Guatemalan population and Guatemala did not undergo the kind of whitening policies or exaltation of mestizaje that took place elsewhere, ladino (mestizo) culture is dominant in Guatemala and Indigenous Guatemalans are confronted with endemic racism and marginalization. Unequal land tenure means that landlessness and hunger have been persistent problems since the colonial period and have been the basis of Mayan resistance. The lack of democracy (abruptly ended by the CIA in 1954) and the appalling living conditions of Indigenous peoples and low-income ladinos gave rise to diverse forms of political struggle. While some Guatemalans formed a guerrilla army, the URNG (Guatemalan National Revolutionary Unity), and were mobilizing in the mountains, many Indigenous plantation labourers had created the Committee for Peasant Unity (CUC) to defend their interests. It was not long after the creation of CUC that the indiscriminate killing of poor Indigenous Guatemalans began, supported by large landowners, government-backed death squads, and US military aid. The Guatemalan civil war lasted for 36 years (1960–1996) and claimed 200,000 lives, most of them Mayans, and displaced a million people (CEH, 1999). The army killings were brutal and indiscriminate, those suspected of being enemies of the state were frequently tortured and mutilated, often in horrific army-led massacres. In 1980, a group of K'iche and Ixil protestors who had occupied the Spanish embassy were burned to death in a police raid (Menchú and Burgos-Debray, 1984). Under the government of General José Efraín Ríos Montt and his scorched earth policy, the Guatemalan army massacred thousands of Indigenous inhabitants and wiped more than 400 villages completely off the map. Several courageous human rights groups were formed by relatives of the victims to try and end the repression and bring the perpetrators to justice, including the Mutual Support Group (GAM), the National Widows' Coordinating Committee (CONAVIGUA), the Council of Ethnic Communities "Runujel Junám" (CERJ), and the Communities of Populations in Resistance (CPRs). In 1990, the residents of Santiago Atitlán, tired of army massacres and persistent intimidation, chased the army out of town using only stones and machetes, an event that later resulted in the town being declared a demilitarized zone. In 1992, Rigoberta Menchú, a 33-year-old K'iche woman whose father was killed in the raid on the Spanish Embassy, won the Nobel Peace Prize, an event that just a decade earlier seemed unthinkable.

The war continued until 1996 when peace accords were signed. The subsequent truth commission in Guatemala established unequivocally that the state committed acts of genocide against its own people and that most of the victims were Indigenous Mayans. The peace accords did not, however, bring an end to political violence and impunity. In the years following the accords, the rule of law continued to be absent and widespread criminal violence and political corruption persisted. Powerful grassroots social movements and institutional forces have combined to try and tackle the widespread organized crime, corruption, and impunity afflicting Guatemala. Bringing the perpetrators to justice has been extremely difficult. In 2012, however, Ríos Montt went on trial for genocide and crimes against humanity (see Figures 7.5 and 7.6). The case was based on the massacre of 1,771 Ixil people in the northwestern highlands in 1982 and 1983, a military strategy that also depended on the rape and sexual assault of Ixil women and the destruction of corn harvests. As Tzul Tzul (2018) writes, it was primarily Ixil women who courageously testified in the trial about the massacres in their communities and the brutal forms of sexual violence they had experienced; their testimonies revealed "the connections between territorial dispossession, sexual violence, and forms of resistance and of organizing for life" (p.405). These same Ixil women who according to Tzul Tzul possess an astonishing "will to live" (Castro Buzon, 2020) have also resisted

Figure 7.5 Efrain Ríos Montt testifies during his trial for genocide

Source: Elena Hermosa/Trocaire, Creative Commons Attribution 2.0, via Wikimedia Commons

the latest phase of violent dispossession from mining companies. Indeed, it was when they were "digging underground, looking for the bones of those killed during the genocide, when they came across mining company workers looking under the same earth for minerals and water sources" (Tzul Tzul, 2018: 404). The Guatemalan judicial system in which he was tried is a Eurocentric and Hispanic one that does not easily accommodate Mayan culture or customary law, although interpreters were provided so that the witnesses were able to testify in the Ixil language (Patterson, 2016). But as Elisabeth Patterson (2016: 241) writes, western legal concepts do not translate easily into Indigenous languages. For example, when one witness was asked "Will you state the truth?" he replied: "I will say what I have seen." Furthermore, as Arias (2016) writes, what the Ixil women really wanted was not Eurocentric justice, in which the general was sent to jail, but an apology, so that the unforgivable could be forgiven. But "even this minimal simulacrum of justice was too much to ask of the perpetrators of these genocidal practices" as it "would have conferred legitimation upon" the women, a move that 500 years after the conquest is still unthinkable within Guatemala's racialized and genocidal social order (p.329). Ríos Montt was convicted on

10 May 2013. The conviction was later annulled based on alleged procedural irregularities and Ríos Montt died in 2018 before the retrial could be completed. It was, however, an historic and unprecedented event (for detail on the trial, see Open Society Justice Initiative, 2013) that has multiple and ongoing political reverberations.

After the conviction of Ríos Montt, Guatemalans repeatedly turned out on the streets in their thousands to demand an end to impunity and corruption and the upholding of international law. These mobilizations strengthened the work of the UN-backed International Commission against Impunity in Guatemala (CICIG) that was instrumental in putting the former president and accomplice of the 1980s massacres, Otto Pérez Molina, in jail in 2015 for accepting bribes in return for discounted import tariffs, in the so-called La Línea case (see Ángel, 2016; Nelson, 2019; O'Boyle, 2019). When the CICIG began to turn its investigative attention to others including the then new president, Jimmy Morales, and in particular into his campaign financing, Morales declared Iván Velázquez persona non grata and an elite movement emerged to accuse the CICIG of foreign intervention in Guatemala's affairs. In 2018, Morales threw CICIG out of the country with congressional support. The struggle for justice in Guatemala suffers repeated setbacks, but the Mayan people and their ladino allies continue to demand justice.

DEVELOPMENT AND NEOLIBERALISM

Postwar development qua modernization and neoliberalism have had appalling consequences for the cultural and territorial integrity of Indigenous peoples. Neoliberal economic policies made many things worse. The social programmes that the corporatist model had implemented were dismantled and export-oriented growth led to the expansion of cattle-ranching, mining, logging, and oil exploration in and around Indigenous communities, displacing Indigenous families or polluting their land (see Chapter 4) and forcing them to mobilize to defend their lands and their livelihoods.

This mobilization became increasingly transnational in scope, as Indigenous peoples began to engage with a range of national and international NGOs, government agencies, and multilateral organizations such as the World Bank. Such transnational linkages enabled them to move "from political obscurity to political centrality" (Andolina et al., 2009: 1–2). In the process, they rearticulated dominant discourses of Indigeneity to gain new modes of cultural and legal recognition and to create institutions through which Indigenous identities and development concerns can be more effectively managed. Such policies have major implications for development thought, practice, and policy. Networks have become denser and more extensive in recent years, in part as a result of the development of new media technologies

Figure 7.6 Maria Soto and other Ixil women after former Guatemalan dictator Ríos Montt was found guilty of genocide

Source: Elena Hermosa/Trocaire, Creative Commons Attribution 2.0, via Wikimedia Commons

such as the Internet, which have enabled Indigenous communities to communicate and collaborate with other Indigenous peoples and with national and international NGOs, a kind of "scaling up" (Yashar, 2005) or "jumping scale" (Andolina et al., 2009).

Indigenous groups in Latin America have been able to take advantage of multilateral initiatives to advance political and territorial claims. In the 1920s, Indigenous peoples were denied access to the League of Nations but now they are highly active within the UN and other bodies. A number of important processes have been established by the UN and other multilateral organizations to promote and protect Indigenous rights. These include the creation of the Working Group on Indigenous Populations (WGIP) in 1982, the passing of C169 by the ILO in 1989, the World Bank's Operational Directive to promote Indigenous development in 1991, the establishment of the United Nations Permanent Forum on Indigenous Issues in 2000, and the approval of the UN Declaration on the Rights of Indigenous Peoples in 2007. In addition, many celebrities have endorsed Indigenous causes and got them into mainstream global media spaces. Sting, for example, has worked to support campaigns by the Kayapó in Brazil against deforestation and environmentally destructive dams. All of these initiatives, designed to

end discrimination against Indigenous peoples, have strengthened the negotiating power that these groups possess when dealing with their own national
governments. International collaborations have also been fundamental in legal
claims to land titling and the mapping projects that underpin them (see, for
example, Wainwright and Bryan, 2009). Transnational linkages can, however, be a double-edged sword, providing new room for manoeuvre and
sources of financial support and political solidarity, but they also sometimes
reproduce essentialized understandings of Indigenous peoples. As Suzana
Sawyer (1997: 71) writes, Indigenous peoples in Ecuador had to couch their
struggles in terms of first world environmentalism to gain leverage, mimicking
western apocalyptic understandings about the destruction of the planet, and
describing themselves as defenders of "the last frontier of uncontaminated
jungle remaining" and as managers of "the lungs of the world and the patri-
mony of all living species on the planet." When it comes to gaining land
titles, Indigenous people have had to reconfigure their lands as property and
map them by drawing on Eurocentric cartographic conventions, rather than
their own, in order for their claims to be legible in courts of law and to
non-Indigenous experts (Bryan, 2009; 2019). Contradictions abound: racist
governments that allow mining companies to dispossess Indigenous communities will often make use of exoticized images of Indigenous peoples in traditional dress in official tourist advertising. Paradoxically, for a time during the
neoliberal era in the 1990s and early 2000s, some Indigenous groups were able
to benefit somewhat from what Charles Hale (2005) refers to as neoliberal
multiculturalism.

NEOLIBERAL MULTICULTURALISM

Despite the hardships generated by neoliberalism, it also provided new opportunities for Indigenous action and neoliberal economic policies did allow for
a degree of cultural recognition. Hale (2004: 17) argues that some aspects
of neoliberal democratization proved to be compatible with Indigenous cultural rights and in particular, neoliberalism undermines mestizo nationalism,
including the "distinction between the forward-looking mestizo and the
backward Indian." In a series of articles (Hale, 2002, 2005, 2019), he notes
how neoliberal discourses of self-reliance had some appeal for Indigenous
peoples who were trying to evade state discipline and advance their own
agendas. At the same time, the proponents of neoliberal economic policies
became receptive to such demands for cultural rights. The World Bank, for
example, embraced the idea of an empowered Indigenous civil society able
to take responsibility for its economic and political future. This shift was not
motivated by a commitment to decoloniality. Hale (2018) believes that neoliberal states known for racist attitudes and actions against Indigenous peoples
supported collective rights because they could do so without redistributing

wealth and also because it facilitated new forms of state control. But some Indigenous peoples begin to work in and with the system to try to change it from within.

But neoliberal multiculturalism had significant limitations and indeed Hale (2004) observed how a discursive binary was drawn between the authorized Indian (*indio permitido*) who works within the constraints of neoliberal governance and gains a seat at the table, and the unruly, rebellious, and dysfunctional Indian (*indio insurrecto*) who challenges the premises of neoliberal multiculturalism and fights for territorial self-determination. Latin American governments began to use "cultural rights to divide and domesticate indigenous movements" (p.17). The dichotomy gave rise to divisions within Indigenous communities, between those compliant with the system and who benefit from it and those who continue to be excluded. There is often a disconnect between well-connected elite Indigenous leaders who work within a neoliberal script and more radical grassroots leaders who work outside of the state and are fighting for sovereignty and autonomy (see also Perreault, 2001; Postero, 2004). In other words, Indigenous peoples can gain access to spaces of power, if they do not take their political demands too far and the western liberal framework is left intact. This analysis explains the criminalization of Indigenous leaders by Pink Tide governments (see Chapter 4).

It appears likely the era of neoliberal multiculturalism is coming to an end and that we are entering a period of racial retrenchment in which dispossession is accelerating and racialized structural inequalities are worsening (Hale and Mullings, 2020). The violent murders of Berta Cáceres and other Indigenous leaders certainly suggest that that is the case, as do the anti-Indigenous pronouncements of Jair Bolsonaro.

THE ZAPATISTA REBELLION

There have been many Indigenous rebellions, but it is important to study some of them in depth, to understand their philosophy, their politics, and their tactics. The Zapatista rebellion that emerged in the mid-1990s is amongst the most significant political expressions of the anti-capitalist and decolonial movement in the world today. While it is rooted in the conjunctural politics of Chiapas in southern Mexico, it enjoys the solidarity of activists and intellectuals the world over. Spending time studying Zapatista philosophy and forms of struggle is a valuable component of a decolonial education.

The state of Chiapas is a region with a large Indigenous population, including Tzeltal, Tzotzil, Chol, Tjolobal, Zoque, Kanjobal, and Mame groups with a long history of rebellion dating back to the conquest. Many different Indigenous languages are spoken; Tzotzil and Tzeltal have several hundred thousand speakers. The conquest of Chiapas in the 1530s was extraordinarily violent and involved land theft and slavery and brought deadly pandemics such as measles, smallpox, and pneumonia. Some of the native peoples of Chiapas preferred to throw themselves from the top of the Sumidero Canyon than submit to Spanish invaders. In 1712, a Tzetzal army composed of several thousand rebels mounted an uprising, in response to the constant abuses of the Spaniards that included spiritual persecution, and many Spanish priests and officials fled or were killed (Klein, 1966). After independence, there was a strong resistance to being incorporated into the Mexican state. The Mexican Revolution between 1910 and 1920 called for land reform, and revolutionary leader Emiliano Zapata fought under the slogan "land and liberty." The revolution ultimately led to the constitutional protection of the *ejidos*, communal forms of land tenure, but land reform was messy and uneven and tensions over land tenure remained (Benjamin, 2000; Lewis, 2004). The PRI government that came to power in the wake of the revolution ruled through authoritarian and corporatist means and attempted to co-opt Indigenous groups. By the 1960s, the PRI could no longer contain the demands of the people in this manner and became increasingly repressive. The Mayan revival movement that emerged in the 1970s in this context of growing political violence emphasized the importance not only of agrarian reform and but also of Mayan histories and epistemologies, that asserts the history of "indigenous resistance to domination and exploitation" and "presents the Maya as protagonists, not passive victims in the past" (Benjamin, 2000: 423). An Indigenous congress held in 1974 captured:

> … the unhappy reality of Chiapas. Their greatest complaint was the insufficiency of good land. That was the primary cause of their hunger, misery, and exploitation. The agrarian reform process was decades in arrears due to corruption and illegal noncompliance. Ranchers employed gunmen to ensure that poor people did not cultivate part of their pasture lands. Land-poor Indian communities and ejidos (land-grant communities) still provided cheap, often child, labor to commercial plantations. Business in the highlands was controlled by Ladino and Indian caciques (bosses) who, allied with government agencies, bought their produce for little and sold them goods for a lot. Education and health care hardly existed, as high rates of illiteracy and infant mortality demonstrated. Where schools and clinics had been built, they were rarely staffed by teachers, nurses, and doctors or provided with books and medicines. Chronic alcoholism made every problem worse.
>
> (Benjamin, 2000: 426)

The Mayan cultural revitalization movement led to significant Mayan intellectual production and pedagogical activity, as well as political pressure. Liberation theologist and local bishop of San Cristóbal de las Casas, Samuel Ruiz, became a passionate critic of the abuses suffered by the Indigenous peoples of Chiapas.

In the 1970s, the pressure for land reform led the Mexican government to allow landless campesinos, most of whom were Indigenous, to settle in the Lacandón jungle. Different Indigenous and linguistic groups came together where they tried to make a living from cattle ranching, coffee production, and subsistence agriculture and in the process learned that "they were united by common grievances" (Kampwirth, 2002: 90). In the aftermath of the 1980s debt crisis and the implementation of structural adjustment policies, the prices of commodities grown in the area fell, and existing inequalities were exacerbated. The massacres in Guatemala in the 1980s also brought an influx of radicalized Guatemalan refugees to Chiapas. Economic misery persisted, in spite of Chiapas' substantial resource wealth, including corn, oil, gas, hydro-electric power, and a strong tourist industry fuelled in large part by the exist-ence of Mayan archaeological sites. Most of the Indigenous inhabitants have not benefited from that wealth, and instead have had to confront discrimin-ation, landlessness, and malnutrition.

The North American Free Trade Agreement (NAFTA) negotiations between Canada, Mexico, and the United States threatened further trade lib-eralization, including the importation of subsidized and genetically modified corn from the United States, more agribusiness and extractivism, and the dismantling of the constitutional protections afforded to the *ejidos* (Collier, 1994). For many chipanecos, NAFTA was the last straw and as a result many of the Indigenous peoples from the region formed the Zapatista Army of National Liberation (EZLN).

I was in Chiapas in July 1993 when the Mexican army discovered guer-rilla army training camps in the jungle. A few arrests were made but with only a few months to go until NATFA, the significance of this finding was played down. The government did not want Mexico to appear politically unstable to potential foreign investors. But on 1 January 1994, the day that NAFTA came into force, hundreds of armed Zapatistas wearing ski masks seized seven towns in Chiapas. Some had weapons; others had sticks carved into the shape of guns. It was a timely appearance as the inauguration of NAFTA meant global media attention was already focused on Mexico and the rebellion was able to draw attention to NAFTA's dark side. The Zapatistas came to the cities to announce that NAFTA would signal a death certificate for the Indigenous peoples of Mexico (Nations, 1994). They came to say "¡Ya basta! Enough is enough!"

After the uprising, the Zapatistas created their own autonomous spaces within Chiapas. They created Zapatista councils, known as *caracoles* within rebel territory, where health, education, land rights, and gender equality are all actively promoted and direct democracy is practised. Indeed, the Zapatistas, while drawing on well-established revolutionary traditions in Latin America, also put forward new modes of development which are shaping the texture of Indigenous political intervention throughout the continent. In Zapatista schools, for example, children are taught in their native languages including Tzeltal, Chol, and Zoque, they study maths, history, and organic agriculture, and they "learn to be critical of the way of life that is being imposed on them, and of the problems of the communities." (Davies, 2010). There is then a focus on autonomy, epistemic justice, and the defence of Indigenous ways of life. But the Zapatistas are not anti-modern; they make full use of the mediascape and modern media technologies.

Although they take their name from Emiliano Zapata, hero of the 1910–1920 Mexican Revolution, the Zapatistas differ greatly from other guerrilla movements in Latin America in that they do not seek to seize state power but rather attempt to enact alternative and autonomous ways of doing politics (Routledge, 1998). The Zapatistas and their former subleader, Subcomandante Marcos, who is not Indigenous but a mestizo intellectual from Mexico City, began to issue inspiring and charming declarations and communiqués, which were full of engaging political analysis, humour, and references to Indigenous cultures but also which refused any fixed definitions of Zapatista politics (see Subcomandante Marcos and Ponce de León, 2001). These communications circulated around the world on the internet (see Box 7.1) and were instrumental in garnering international solidarity for the Zapatista cause. Some of these were composed of political and literary dialogues between Marcos and a beetle called Durito (Subcomandante Marcos, 2005). Much to the chagrin of the Mexican government, Subcomandante Marcos very quickly became an international celebrity and media attention forced the government into negotiations, an unprecedented occurrence. As a weakly resourced army, the Zapatistas could easily have been annihilated by the Mexican army, but their superior discursive power turned this struggle into a war of words rather than a war of bullets (Subcomandante Marcos and Ponce de León, 2001). Global solidarity and internet activism mean that the Mexican army had to remain restrained in its use of repressive force (Froehling, 1999). At times, however, state-led violence against the Zapatistas has intensified. The Mexican state has clearly empowered paramilitary groups to launch attacks, including the devastating Acteal massacre of 1997, in which 45 Zapatistas were killed, and the attack on the community of La Realidad in 2014, in which a Zapatista teacher Jose Luis Solís López lost his life, and the autonomous school and health centre were destroyed.

BOX 7.1 FOURTH DECLARATION OF THE LACANDÓN JUNGLE

Brothers and Sisters: Many words walk in the world. Many worlds are made. Many worlds are made for us. There are words and worlds which are lies and injustices. There are words and worlds which are truths and truthful. We make true words. We have been made from true words. In the world of the powerful there is no space for anyone but themselves and their servants. In the world we want everyone fits. The world we want is a world in which many worlds fit. The Nation which we construct is one where all communities and languages fit, where all steps may walk, where all may have laughter, where all may live the dawn. We speak of unity even when we are silent. Softly and gently, we speak the words which find the unity which will embrace us in history and which will discard the abandonment which confronts and destroys one another. Our word, our song and our cry, is so that the most dead will no longer die. So that we may live fighting, we may live singing. Long live the word. Long live Enough is Enough! Long live the night which becomes a soldier in order not to die in oblivion. In order to live the word dies, its seed germinating forever in the womb of the earth. By being born and living we die. We will always live. Only those who give up their history are consigned to oblivion.

Source: EZLN, 1996

The creative imagination of the Zapatistas repeatedly confounds and outwits the repressive measures of the Mexican state, rendering them politically ineffectual. In a symbolic and politically effective gesture, in 2000 it was announced that the Zapatista Air Force had attacked the barracks of the Federal Army with hundreds of planes. The planes were, however, made of paper and each was loaded with a message or "discursive missile" to protest at the ongoing military incursions into Zapatista lands (Lane, 2003). During the Mexican electoral campaign of 2006, the Zapatistas launched the Other Campaign, a parallel tour that drew attention to the corruption and political bankruptcy of the official campaign. In 2014, Subcomandante Marcos "disappeared" in order to bring an end to the cult of personality that surrounded him (desinformémonos, 2014). In 2017, the Zapatistas presented a presidential candidate, an Indigenous woman, for the 2018 elections. María de Jesús Patricio, or Marichuy, used the campaign to hold mass rallies, not to try and win the election but to draw attention to the violence of capitalism, racism, and patriarchy in Mexico. In May 2021, a delegation of Zapatistas departed from the Isla de Mujeres in Mexico and set sale for Europe. They planned to arrive 500 years after Hernán Cortés arrived in Mexico to ransack and destroy. In a communiqué, they announced they were on the way to Spain to let the Spanish people know that they didn't conquer them and that their resistance and rebellion continues. They travelled not in search of "difference, superiority, confrontation, forgiveness or pity," but rather "in order to find what makes us equal" (EZLN, 2020).

The Zapatistas have developed a detailed and complex anti–capitalist and decolonial philosophy. Sergio Tischler (2019) details some of the contents of this philosophy, focusing on Zapatista understandings of the revolutionary subject, the Zapatista concept of time, and gender-based subversion as follows. First, the revolutionary subject is not a singular figure but is instead a world in which "many worlds fit" so must be understood in its multiplicity and multivocality. Through dialogue, it rejects the hierarchies of conventional political organizing and works instead for an autonomous politics outside of the state. Second, the Zapatistas who "created a rupture in the time of capitalist domination that seemed unbreakable" (p.255) understand three different kinds of time: exact time, just time, and necessary time. Exact time is linear and linked to capital, while just time is about "the reproduction of the community fabric and the relation with the land" and necessary time is "the time of the revolution understood in horizontal, anti-capitalist terms" (p.258). Finally, the anti-capitalist struggle is by necessity also an anti-patriarchal one as both patriarchy and capitalism are part of the global economic and cultural system that exploits and dehumanizes.

THE POLITICS OF *BUEN VIVIR*

In previous chapters, we've discussed how the Indigenous politics of *Buen Vivir* have been mobilized in constitutions and other formal spaces. We return briefly to this question here to think in a little more depth about some of its implications for decolonizing development. As we've outlined, *Buen vivir* (or sumak kawsay/suma qamaña) means good life or living well in a holistic manner; it means good life for Nature as well as for people and harmonious relations between the two. While it draws on Andean traditions, it is now used by Indigenous peoples and decolonial social movements all over the continent and has broad applicability to a range of modern-colonial and capitalist contexts. Gudynas (2011) notes how the concept is hard to translate as it addresses but also defies Eurocentric understandings of both well-being and development. Indeed, it dismantles the binary separation between people and Nature that is central to western thinking; "Nature becomes part of the social world (...) and the concept of citizenship is widened" to include animals, plants, ecosystems, and spirits who have both can also think, feel and exert agency (p.445). A form of development that makes someone more prosperous but pollutes a local river is not in keeping with the principles of *Buen Vivir*. Gudynas (2011) describes development as a zombie idea, as it is one that is declared obsolete, as it has clearly failed, but nonetheless, along with economic growth, continues to be proposed as the only way forward. He sees *Buen Vivir* as a way to get beyond Eurocentric thinking and develop decolonial alternatives that do not destroy or harm. It is an appealing concept as it contains a clear critique of capitalist consumerism and the forms of development that bring harm to people and environments and it involves the foregrounding of Indigenous ways of knowing. It involves the pursuit of spiritual and material well-being that is rooted in social and environmental justice and collective solidarity. But *Buen Vivir* is not anti-modern or anti-science. As Gudynas (2011: 446) writes, "Buen Vivir will not stop building bridges, and will not reject the use of Western physics and engineering to build them, but the ones that it will propose may well have different sizes and materials, will be placed in other locations, and certainly will serve local and regional needs and not the needs of global markets." While *Buen Vivir* mobilizes non-Eurocentric ways of thinking, and activates the kind of ontologies discussed in Chapter 4, there is some overlap with western notions of sustainable development and Global North initiatives around de-growth and commoning (Walsh, 2010). These overlaps enhance the possibility for North-South solidarity in the building of a different world.

As Chapter 4 made clear, the institutionalization of *Buen Vivir* has not prevented harmful extractivist forms of development. While this should not result in the dismissal of *Buen Vivir* as an aspiration, it is a concept that has attracted some criticism. By providing a critique of development, we need to ask, as Catherine Walsh (2010: 17) speculates, whether it is really possible for *Buen Vivir* to transcend "the modern-colonial imposition" especially when we consider that the "very idea of development itself is a concept and word that does not exist in the cosmovisions, conceptual categories, and languages of indigenous communities." Similarly, noting that the concept of *sumak kawsay* cannot be found in ethnohistorical or ethnographic records or in ancient Quechua or Aymara texts, Carmen Martínez Novo (2018: 390) questions its supposed Indigenous roots suggesting that it emerges instead from "the interface of development, indigenous intellectuals, environmentalism, and populist politics" and furthermore it has been promoted by MCD scholars who have also sometimes acted as advisors to the Pink Tide governments. It results in a phenomenon that she calls ventriloquism, where white-mestizo elites speak on behalf of Indigenous peoples. She believes that the scholarship that celebrates *Buen Vivir* "has been insufficiently self-critical and reflective of its own complicity with the state's repressive project vis-à-vis indigenous communities" (p.390) and that scholars have overstated the degree to which these governments departed from neoliberal and colonial models of development.

Indigenous cultures are, however, neither pure nor static, and Indigenous peoples creatively rework their own cultural references and borrow from other cultures. As Gudynas (2011: 443) writes, "*Buen Vivir* should not be understood as a return to a distant Andean past, pre-colonial times. It is not a static concept, but an idea that is continually being created." As Indigenous activists, intellectuals, and policymakers take up the concept, they are able to gain some control over the term and how it is articulated. The Ecuadorian government is now widely understood to have betrayed the principles of *Buen Vivir* by pursuing extractivist policies, despite being the government that has most incorporated the concept into policy. So maybe it doesn't matter so much whether the term is truly an ancestral one, but rather whether it can do decolonial work or the extent to which it can resist co-optation by neo-liberal or neocolonial forces. As de Sousa Santos (2020a:19) writes, the politics of *Buen Vivir/Sumak Kawsay* draws on established ancestral knowledges and practices and its institutionalization gave rise to a period of innovation in which Indigenous peoples "struggled for the recognition of their rights, respect for their ways of life, and the dignity of their existence as survivors of the great modern colonial genocide, perpetuated today by the new colonialism and the racism that for decades has been practised by both right-wing and left-wing political parties" (my translation). *Buen Vivir* discourses have then played an important role in displacing the authority of neoliberal capitalism and contesting the colonial matrix of power.

BOX 7.2 THE CONFEDERATION OF INDIGENOUS NATIONALITIES OF ECUADOR

The colonial matrix of power has been particularly brutal in Ecuador where Indigenous peoples have had to fight to protect their ancestral lands from mining, logging, cattle ranching, African oil palm cultivation, and oil exploration. CONAIE, or the Confederation of Indigenous Nationalities of Ecuador, was formed in 1986, to protest at the national and multinational incursions into their ancestral lands and to convert Ecuador into a plurinational participatory democracy. It is now one of the most prominent and successful Indigenous organizations in Latin America (Zamosc 2004; Yashar, 2005). CONAIE brought together the fight for land with the fight for cultural recognition and the demands for a plurinational state, forming what Tom Perreault (2001) calls "an identity/territory nexus." For Deborah Yashar (2005), CONAIE has successfully articulated cultural demands such as the right to be Indigenous and practise Indigenous customs with class-based ones such as access to land, water, fair prices, and credit. Since the early 1990s, they have organized large high-profile uprisings and demonstrations, blocking roads, and occupying government buildings, which have forced the government into negotiations and into both adopting and abandoning particular policies. CONAIE led important uprisings in 1990, 1994, 1997, 2000, 2002, and 2005. In 1996, CONAIE created a political party, Pachakutik, and gained seats in congress. Increasingly Indigenous demands began to resonate with non-Indigenous populations, and it became harder for aspiring politicians to ignore Indigenous voters. CONAIE's activism put concepts such as plurinationality and territoriality into mainstream political discourse, concepts which were later enshrined in the 2008 constitution. As Sarah Radcliffe and Isabella Radhuber (2020: 6) state, CONAIE's vision of plurinationalism is an expansive one that "emerges out of colonial-modern spatial-geographical exclusion of racialized subordinated nationalities, and a proposal to reorganize state-space-power for plural marginalized subjects." It is a vision that is accompanied by a cogent critique of the conquest and its multiple legacies of dispossession. CONAIE's support was instrumental in the coming to power of Rafael Correa, Ecuador's first Pink Tide government, but they rapidly became a vocal critic, maintaining that he had failed to deliver on electoral promises to Indigenous peoples. Under Correa, CONAIE leaders were brutally repressed and criminalized and Indigenous gains, such as those in intercultural and bilingual education, were undermined. In 2012, CONAIE organized a march to Quito to protest at Correa's neoliberal and extractivist approach to land and water. The march attracted thousands of demonstrators and support from other Indigenous organizations. In 2015, CONAIE, along with other Indigenous organizations, led a massive protest against the government of Rafael Correa in opposition to proposed constitutional amendments that would allow him to serve a fourth term. CONAIE is a heterogeneous organization, often beset with contradictions. This is because it is not a hierarchical organization with a single leadership and it does not count on a disciplined following (Bonilla and Mancero, 2020). Nonetheless, CONAIE's capacity to mobilize has not diminished. In October 2019, CONAIE led a successful Indigenous uprising against austerity measures imposed by the government of Lenín Moreno that the government was then forced to abandon. In the 2021 elections, Pachakutik gained 27 seats in the National Assembly, its best electoral performance ever.

Sources: Perreault, 2001; Zamosc, 2004; Yashar, 2005; Bonilla and Mancero, 2020; Radcliffe and Radhuber, 2020

INDIGENOUS MEDIA

As indicated, the Zapatistas have made full use of the global mediascape and especially the internet to spread their message and build solidarity, demonstrating that that there is "no necessary contradiction between techno-logical modernization and grassroots mobilization" (Yúdice, 2003: 106, see also Halleck, 1994). Indigenous media practices constitute a form of border thinking that defy conventional binaries of tradition and modernity. While western media has often been viewed with suspicion by Indigenous peoples, as they have for such a long time been subject to the colonizing gaze of first world photographers and filmmakers who were there to document a culture assumed to be on the brink of extinction (Ginsburg, 2002), in more recent decades it has become clear that Indigenous media, particularly videomaking has becomes "a crucial technology for the reinvention of indigenous cultures, directed against the long-standing discrimination of indigenous people, languages, medicinal practices, as well as their social and economic relations" (Schiwy, 2003: 116). Western media technologies when controlled by Indigenous media makers themselves can be put to alternative uses and they are proving to be a useful tool for telling stories using Indigenous peoples' own words and images and for protecting and revitalizing Indigenous languages and cultural traditions. As a result, many Indigenous groups throughout Latin America have become active producers of their own media and deal with themes that are often marginalized in mainstream media sites, such as land rights, traditional medicine, and bilingual education. For example, the Caribbean Coast of Nicaragua has a vibrant participatory mediascape that broadcasts in Miskitu, Creole English, and Spanish. There are many community radio stations that broadcast in Miskitu and Creole as well as Spanish and that routinely include the perspectives of Miskito listeners via phone in and studio interviews. Taking advantage of technological developments that have facilitated the democratization of media access, the region also has several community television channels, including Canal 5 and TV7, that broadcast news and current affairs in Miskitu (Glynn and Cupples, 2011). Indigenous media is threatening to the mestizo social order and the authoritarian government led by Daniel Ortega and if these organizations express dissenting political views, they face the withdrawal of state advertising and death threats, or their reporters are assaulted or the programmes are forced off the air through political pressure (Cupples and Glynn, 2018).

The Rama-Kriol community of Bangkukuk Taik in the South Caribbean that was facing imminent destruction because of a proposed interoceanic canal with Chinese investment has used participatory video to document the absence of consultation and to articulate how the canal would destroy their culture and livelihoods. Two videos that circulated on YouTube and Vimeo assert "an alternative epistemology, rooted in a rich history and a self-sufficient economy beyond the state, an anti-capitalist rejection of money and a sense of responsibility instead to land, water, flora and fauna, in which the elders transmit cultural practices and the Rama language to the children, and in which time is understood to be nonlinear, as posterity and future well-being for Rama depends on conserving the efforts of ancestors" (Cupples and Glynn, 2018: 44–45). De Sousa Santos (2004, 2014) has developed the concept of sociology of absences to capture how colonial forces attempt to restrict what can be considered to exist, in order to privilege Eurocentric understandings of time, productivity, and racial classification. Drawing on de Sousa Santos, we suggested that people of Bangkukuk Taik use video to make alternative knowledges, temporalities, and economic systems visible and that their mediated activism enables them to construct a counter-hegemonic ecology that defends territorial rights, Indigenous cultural identities, and the life of humans and non-humans (see Cupples and Glynn, 2018).

SUMMARY

- Indigenous peoples, their cultures and their knowledges are still with us in spite of the brutality of colonialism, the ideology of mestizaje, and the persistence of Eurocentric fantasies of elimination.
- Indigenous knowledges survived the colonial era in part through creative and subversive forms of hybridization.
- In 20th-century Guatemala, many thousands of Indigenous Mayans were massacred by their own government backed by US military aid.
- While neoliberalism has been largely harmful for Indigenous people, it allowed some space for the securing of Indigenous cultural rights. There is, however, an ongoing disciplining of counter-discourses.
- The Zapatista rebellion is an inspirational form of Indigenous politics that has captured the global imagination.
- The mobilization of discourses of *Buen Vivir* has had quite contradictory outcomes. But they have played a role in disrupting the authority of neo-liberal capitalism.
- Many Indigenous peoples are making their own media to fight for self-determination, protect their languages and cultures, and to put Indigenous issues such as land rights onto the political agenda.

DISCUSSION QUESTIONS

1   How did Guatemala's Indigenous peoples get entangled in the counter-insurgency policies of the Guatemalan government in the 1980s?
2   What does Charles Hale mean by neoliberal multiculturalism? What are its consequences for Indigenous movements and struggles?
3   How are indigenous movements becoming transnationalized? What are the advantages and disadvantages of transnationalization for Indigenous struggles?
4   What are the key elements of Zapatista thought and philosophy and how do they challenge dominant notions of power, development, and governance?
5   Discuss how Indigenous peoples are using media to fight against racism and discrimination?

FURTHER READING

Routledge Handbook of Latin American Development Chapters 3 (Postero); 6 (Hale); 19 (di Giminiani); 21 (Tischler); 22 (Bryan)
EZLN (1994–2005) The Six Declarations of the Lacandon Jungle

All six are available in English here:

https://radiozapatista.org/?page_id=20278&lang=en

Postrero, N. G. and L. Zamosc, L. (eds) (2004) *The Struggle for Indigenous Rights in Latin America*. Eastborne: Sussex Academic Press, pp. 189–216.

An edited collection examining Indigenous movements in eight countries in Latin America, with an emphasis on struggles for rights and democratization and negotiations of neoliberalism.

Yashar, D. J. (2005) *Contesting Citizenship in Latin America: The Rise of Indigenous Movements and the Postliberal Challenge*. Cambridge: Cambridge University Press.

A book which takes a comparative approach to Indigenous mobilizations and attempts to illuminate the apparent differences in political effectivity.

Warren, K. B. (1998) *Indigenous Movements and their Critics: Pan-Maya Activism in Guatemala*. Princeton: Princeton University Press.

An ethnographic account of cultural activism in Guatemala that centres Indigenous voices.

Ranta, E. (2020) *Vivir Bien as an Alternative to Neoliberal Globalization: Can Indigenous Terminologies Decolonize the State?* London: Routledge.

A useful assessment of the potential and downsides of the concept of *Buen Vivir*

ECLAC (2014) *Guaranteeing Indigenous People's Rights in Latin America: Progress in the Past Decade and Remaining Challenges*. Santiago: United Nations.

A detailed report that contains an overview of the state of Indigenous rights across the continent, including territorial rights, health, education, and the role of censuses.

Recommended films

1994 (2019), directed by Diego Enrique Osorno

A 4-part miniseries about the events of 1994 when presidential candidate and PRI reformer Luis Donaldo Colosio was assassinated. Episode 2 ("Revolution") on the Zapatista rebellion and its consequences for Mexican politics is particularly recommended.

When the Mountains Tremble (1983); *Granito: How to Nail a Dictator* (2011); and *500 Years: Life in Resistance* (2017), directed by Pamela Yates

Pamela Yates has directed three films about the human rights atrocities in Guatemala and the struggle for political redress. All of these films are highly recommended. The filmmakers have produced an excellent detailed guide to accompany the films that provide a number of questions and ideas to enhance your learning (available here http://500years.skylight.is/en/film-guide-for-educators/).

A Little Bit of So Much Truth (2006), directed by Jill Friedberg

A highly recommended feature length documentary about the struggle in Oaxaca to overthrow the governor that involves ancestral knowledges and creative use of local media.

A Place Called Chiapas (1998), directed by Nettie Wild

A documentary focused on the early days of the Zapatista rebellion. It features interviews with Subcomandante Marcos and other Zapatistas.

Serpiente Emplumada (2018), directed by Ricki Lopez Bruni

The endeavour to capture the quetzal on camera results in a journey in which the filmmaker learns about Indigenous cosmovisions, environmental destruction, and political violence and how scientific knowledges and Indigenous knowledges can be brought into dialogue in a non-hierarchical way.

Useful websites

The Archive of the Indigenous Languages of Latin America (AILLA)
Based at the University of Texas, AILLA is a digital language archive that contains texts, recording and media in and about the Indigenous languages spoken in Latin America.
www.ailla.utexas.org/

Cultural Survival
The website of Cultural Survival, an Indigenous-led NGO, that supports the struggles of Indigenous communities for self-determination.
www.culturalsurvival.org/

Enlace Zapatista
A multilingual digital repository for Zapatista news, communiqués and publications
https://enlacezapatista.ezln.org.mx/

CHAPTER 8

Afrodescendant politics and movements

INTRODUCTION

The conquest of Latin America led to the transatlantic slave trade and was built on the coerced labour of millions of Africans. From the 16th century until the abolition of slavery in the 1800s, a staggering 12 million Africans were kidnapped and transported to the Americas, five million of these to Brazil alone. Those that survived the journey were converted into a commodity to be bought and sold. Once in the Americas, they endured violence and intolerable working conditions and many more died as a result. Others survived and they enacted many different forms of resistance. Of course, Black cultures in Latin America did not begin with slavery. Those enslaved had their own histories, knowledges, cultures and spiritualities that they brought with them to the Americas, and which continue to circulate. This chapter introduces Black and Afrodescendant politics and cultures in Latin America, focusing on the legacies of the transatlantic slave trade, the form of erasure and discrimination that Afro populations confront, and the forms of anti–racist resistance and intellectual production that challenge this state of affairs.

BLACK MOBILITIES

There are many ways that slaves secured their freedom before slavery ended. As Sidney Chalhoub (2018) writes, in Brazil in 1872, 16 years before slavery was abolished, three out of every four people of African descent were free or freed. All over the Americas, slaves often escaped too to form their own maroon communities.

In Brazil, slaves escaped to the interior to form *macombos* or *quilombos* (from the Angolan word *kilombo*) where they grew their own food and governed themselves. There are thousands of *quilombo* territories in Brazil

DOI: 10.4324/9781003110453-8

today inhabited by descendants of the *quilombolas*. One colonial era *quilombo*, Palmares, had a population of more than 10,000 maroons. It functioned like an African state within Brazil. It traded extensively with the surrounding areas, practised syncretic religions, and was able to resist several Dutch and Portuguese military campaigns for most of the 17th century (Anderson, 1996). Zumbi, a Brazilian of Congolese descent and the last leader and king of Palmares, who was murdered and decapitated by the Portuguese in 1695, is an important Afro-Brazilian icon for the resistance he mounted against slavery and colonial rule. The 1978 manifesto of the Unified Black Movement Against Racism and Racial Discrimination recognizes the role of Zumbi and quilombist consciousness in ambitions to achieve racial equality (do Nascimento, 1980). In November 1995, 300 years after Zumbi's death, the Black movement in Brazil held a march, the Marcha Zumbi de Palmares contra o Racismo, pela Cidadania e a Vida (the Zumbi de Palmares March against Racism, for Citizenship and Life) to "protest against the subhuman conditions in which the black people of Brazil lived" (Rocha, 2020: 166). The rights to land of the *quilombolas* were recognized in the 1988 Constitution and an organization CONAQ was formed to support the process, but titling is proceeding at a very slow pace and only a small proportion have received titles (de Sousa Santos, 2018b; Phillips, 2018).

Figure 8.1 The 20 November celebration of Black Consciousness Day in Quilombo dos Palmares

Source: Ministério da Cultura/Janine Moraes, CC BY 2.0 via Wikimedia Commons

In early 17th-century Colombia, escaped slaves from Cartagena de Indias created their own free city-republics or *palenques* that could not be reconquered by the colonizers. Some of these, including Palenque de San Basílio, maintained their political autonomy for centuries, but haven't, as Bernd Reiter (2015) argues, received the same scholarly or political attention as the free city-republics of Europe, a failure that is driven by anti-Black racism within Latin America and by the assumptions made by Eurocentric thinkers such as Max Weber. He writes:

> Palenques are the true birthplaces of democracy in the Americas and also the places where a strong, active, and equal citizenship was created and continues to be practiced. For its part, Palenque de San Basílio created a participatory and egalitarian political system, distributing responsibilities by age groups through their kuagro system.
>
> (Reiter, 2015: 335)

He goes on to note that the post-independence Colombian state rather than learning from and incorporating these models of participatory democracy worked to marginalize and impoverish the *palenques* through persecution. Today Palenque de San Basílio is often treated by the state as a site of cultural heritage and important tourist attraction, rather than as a site of creative political autonomy. But its *kuagros* still exist and perform a range of important social, cultural, economic, and political functions based on the ancestral knowledges of the *palenqueros*. The *palenqueros* also make use of more recent constitutional and legislative gains such as the 1991 Constitution and Law 70 of 1993 which enshrines cultural recognition and the right to collective land titles for all of Colombia's Afrodescendant populations.

Not all Black people that came to Latin America did so as slaves. In the 19th and 20th centuries, there was substantial migration of Black people from Anglophone Caribbean islands such as Jamaica and from the US South, who migrated to Central America and worked on construction projects, including the railways and the Panama Canal, and in logging and large agri-business (Harpelle, 2001; Pineda, 2006). The many Afro-Antilleans who came to Limón, Costa Rica, including activists and intellectuals such as Marcus Garvey (see Box 8.1), created a vibrant and thriving Black culture (Palmer, 2004 [1993]).

BOX 8.1 THE LIFE OF MARCUS GARVEY, 1887–1940

Marcus Garvey was born in Jamaica in 1887 and went on to become an important Black leader, intellectual, and journalist in his home nation and globally. He travelled widely around the world, spending much time in Latin

America as well in the UK and the United States. He believed that Black people in different parts of the world were united by common forms of suffering and discrimination. While in Central America, he experienced first-hand the horrendous working conditions experienced by Black workers on banana plantations in Costa Rica and in the building of the canal in Panama and he dedicated himself to improving working conditions through organization. He went on to found the United Negro Improvement Association (UNIA), an organization dedicated to Black improvement through education, hard work, and economic self-determination, gaining a presence in many parts of the world. He tried to establish a Black-owned and operated shipping company, The Black Star Line, a venture that failed for a variety of reasons, including mismanagement and deliberate sabotage by the US government, and led to Garvey's incarceration in the United States after he was unjustly convicted of mail fraud associated with the company. While he was seen as a dangerous subversive by many governments, his support for racial separatism and the Back-to-Africa movement were quite controversial, given that such beliefs that were seen by some, especially in the United States, as being dangerously close to the views of white supremacists. He often clashed with other African American intellectuals, such as W.E.B. Du Bois, over such matters. Garvey used the expression "emancipate ourselves from mental slavery because whilst others might free the body, none but ourselves can free the mind" that would later be popularized in the lyrics of Bob Marley's "Redemption Song." There are many other reggae artists that also pay tribute to Garvey in their music. I first experienced the enthusiasm for Garveyism and the strength of his legacy in Puerto Limón, Costa Rica, where the UNIA branch, referred to as the Black Star Line, on Calle 5 is a restaurant and thriving hub of cultural and intellectual exchange for Afro-Costa Ricans. The building is currently being rebuilt after being seriously damaged in a devastating fire in 2016. Garvey died from a stroke in London in 1940; many years later his remains were repatriated to Jamaica.

Sources: Garvey and Garvey, 2013[1923]; Harpelle, 2003; Palmer, 2004[1993]); Lawler, 2005

WHITENING, RACIAL DEMOCRACY, AND ANTI-BLACKNESS

After independence, white-mestizo and *criollo* political elites and intellectuals were very influenced by (now discredited) theories of scientific racism that had taken hold in Enlightenment Europe that connected race to intelligence and posited the superiority of whiteness. Latin American intellectuals such as Carlos Octavio Bunge, José Ingenieros, Alcides Arguedas, Euclides da Cunha, and Domingo Faustino Sarmiento were all concerned about the implications of racial diversity and viewed Indigenous and especially Black populations as obstacles to national development. For example, Bunge, an Argentinian positivist of German descent, believed that for Latin America to evolve like Europe or the United States, it would be necessary to "eugenically correct the inadequate assimilations" and he welcomed diseases such as smallpox and tuberculosis that were decimating Indigenous and Black populations in Argentina (Miranda and Vallejo, 2006: 77, my translation). His compatriot, Domingo Faustino Sarmiento, an important writer and president of Argentina, promoted democratization and public education but also held racist views, believing that immigration from Europe would exert a positive civilizing pressure on Argentina. He believed that "this alone would be enough to cure in ten years, at most, all the homeland's wounds" (Sarmiento, 2004[1845]: 248). He set out how this would be achieved. This aspiration meant that many post-independence governments embarked on official policies of whitening (*blanqueamiento/branqueamento*), encouraging immigration from Europe to lighten the population. In early 20th century Brazil, for example, policymakers encouraged European immigration while banning African and Asian immigration. Latin American elites had absorbed the tenets of Eurocentric race science.

This wasn't the only approach in circulation at the time; by the early 20th century some intellectuals such as Mexican José Vasconcelos or Colombians José María Samper or Luis López de Mesa embraced and celebrated mixing as a positive and democratizing process and suggested that mestizos were physically, socially, and culturally superior to Europeans (Wade, 2017). But as Peter Wade (2017) writes, even in these formulations, racial hierarchies and white supremacy remained. Mexican José Vasconcelos, who argued for the superiority of what he called the "cosmic race" made up of European, native American, Black, and Asian elements over the soulless Anglo-Saxons, could not move beyond racial stereotypes. In his essay, he stated that the "infinite quietude" provided by the native American "is stirred with the drop put in our blood by the Black, eager for sensual joy, intoxicated with dances and unbridled lust" (Vasconcelos, 1997[1925]: 22). Vasconcelos was a racist thinker and would go on to support fascism and anti-Semitism during the Second World War.

In 1933, white Brazilian intellectual Gilberto Freyre asserted that Brazil could never become white or European, but instead was or could become a racial democracy, in which people of different races peacefully co-existed and in which Brazil avoided the racial violence and segregation that characterized the United States (Andrews, 1996). This vision, which did involve the valorization of Black and African culture and an attempt to accept that all Brazilians were shaped by the nation's multicultural mix, was very appealing. But the discourse of racial democracy rather than being seen as an aspiration that needed to be built was mobilized in such a way that racism was rendered irrelevant. It was strengthened by comparisons with the US racial order that was based on the black/white binary and the "one-drop rule" as in Latin America race relations were based on a colour continuum whereby the lighter you were, the greater your social status (Wade, 2010, although see Skidmore, 1993 and Andrews, 1996 who call into question the idea of a bipolar US and multiracial Latin America). This is, of course, a contradictory phenomenon, as it is possible to transcend one's racial status and become whiter by moving to the city, getting an education, and adopting particular forms of dress, speech and occupation (Appelbaum et al., 2003), but it does of course still reproduce the perceived superiority of whiteness. It encourages Blacks not to identify with their Black ancestry in order to gain some degree of social mobility.

Racial democracy became a pervasive foundational notion and led to a mainstream denial of the existence of racism that made it extremely difficult to denounce or even talk about ongoing forms of racial discrimination, even though lighter-skinned Brazilians continued to enjoy far more socio-economic advantages. Even some Black intellectuals, including W.E.B. Du Bois, initially believed that racial democracy had been a success although realized later they were mistaken (Andrews, 1996; Araujo, 2015). For a long time, in spite of the fact that Afro-Brazilians were more likely to be unemployed, have insecure housing, suffer from educational disadvantage, or be a victim of police brutality, criticizing the concept of racial democracy was seen as an unpatriotic "act of subversion" (Andrews, 1996: 491). More recently, a leading Afro-Cuban commentator, Roberto Zurbano Torres (2013), caused a scandal in his home country when he published an article in the *New York Times* on the persistence of racial discrimination in Cuba (see also Bush et al., 2018).

AFRO ERASURE

The consequences of the racial thinking of the 19th and 20th centuries are still with us, but they affect Indigenous and Black populations in different ways and racial formations differ from one part of the continent to another. In the Mexican context, Christina Sue (2013) shows how the influential modes of thought outlined above created three key ideological pillars; first, the valorization of the mestizo; second, the idea that racism does not exist, and finally the idea of nonblackness, the negation of African heritage. These are elements that are certainly present elsewhere in the continent. The presence of the first and the second make the third, the problematic Afro erasure, harder to address.

While Africans, along with native Americans and Europeans, are foundational in the creation of what is now Latin America, African heritages occupy the most difficult and least visible place. According to developing ideologies of mestizaje, Indigenous peoples were understood "as ancestral contributors to the new, hybrid mestizo nation and culture, even if they are seen as marginal and traditional in the present" (Hooker, 2005: 301). All over the continent, we see a denial of African ancestry as well as the failure to recognize the contribution made by Afro-Latin Americans to each nation's development (Ramírez, 2007). Many white-mestizo Latin Americans simply do not imagine that Mexicans or Nicaraguans can also be Black (Sue, 2013; Vílchez, 2017). So although Indigenous peoples are subjected to multiple forms of racism, discrimination, and exclusion, Afro-Latin Americas are subject to even more discrimination. As Juliet Hooker (2005) argues, Indigenous groups have had much more success in claiming collective rights than Afro-Latin Americans in part because Indigenous peoples are seen to possess a distinctive cultural identity and culture which has made states more receptive to Indigenous claims, a situation that leads to what she refers to as Indigenous inclusion/ Black exclusion. Afro-Latin Americans, as descendants of slaves, are seen as always diasporic, which means their claims to land and other collective rights are seen as less legitimate than Indigenous ones. Scholars have documented how Afrodescendant Nicaraguans and Hondurans, the Black Creoles and Garífuna, have often strategically adopted an Indigenous identity to facilitate political redress (Gordon, 1998; Hooker, 2005: Anderson, 2009). Latin American regions with large Black populations are frequently Othered and stigmatized as dysfunctional sites that do properly belong to the imagined mestizo nation-state. Stigma towards the Caribbean Coasts of Costa Rica and Nicaragua, whose Black inhabitants speak a Creole English rather than or as well as Spanish, is activated by their distinct colonial histories. In Nicaragua, the long-term historical relationship with the Anglophone world "provokes suspicion and serves to place Costeño society as a suspect internal order" (Pineda, 2006:3), while the Afro-Costa Rican inhabitants of Puerto Limón are seen by white-mestizo Costa Ricans as "not really *ticos*" which places them "outside the term Costa Ricans use to refer to their national identity" (McCoy-Torres, 2016: 2). Bocas del Toro in Panama, home to a large Afro population, attracts the same kind of neglect and stigmatization and is seen as a "forgotten, unwanted, and unsafe place believed to be nothing more than a wild jungle" (Guerrón Montero, 2014: 427).

Afro erasure is accompanied by systematic everyday and institutional racism that is manifested in a range of ways – in levels of poverty, education, the labour market, and the judicial system. Afro-Colombians talk about being denied entry to nightclubs or being mistaken for domestic servants. Black Creole Nicaraguans and Costa Ricans talk about being mistaken for drug dealers. Afro-Brazilians and Afro-Colombians, like African Americans, must confront horrifying forms of highly racialized police brutality and are much more likely to be incarcerated (González, 2020; Muñoz, 2020; Noriega, 2020). Racialization (and resistance to racism) takes different forms in different parts of the continent, and it is important to do geographically specific work while noting the widespread applicability of Afro erasure, discourses of racial democracy, and ideologies of mestizaje (see Box 8.2)

BOX 8.2 BLACKNESS AND WHITENESS IN ARGENTINA

While almost all Latin American countries engage in some form of Afro erasure, it is probably at its most extreme in Argentina. There is a tendency for Argentines to describe themselves as people that descended "from ships" but they mean ships that came from Europe full of Italians and Spaniards, not slave ships. But some 200,000 African slaves were brought to the River Plate region (Argentina and Uruguay) during the colonial era and a 1778 census recorded the African population as reaching 37% (Goñi, 2021b). In the 19th century, white European immigration was encouraged, while Afro-Argentines were often drafted in the military which increased their death rate relative to whites and mestizos (Andrews, 1980; Anderson and Gomes, 2021). But most people of African descent were eliminated through discursive and ideological means. In other words, their descendants still live in Argentina today, but it is as if they don't. Even though Argentinean national identity is built on tango (along with football), the Afro origins of tango are often denied. Yet most of the early tango practitioners and performers, including dancers, musicians, and composers, were of African descent (Karush, 2012). Even Jorge Luis Borges believed that Argentina could only draw on European traditions and dismissed the significance of Black and Indigenous contributions to Argentinean culture and identity (Goñi, 2021a). Argentinian school children do not learn about their own African histories and ancestries. So Blackness gets invisibilized and many Black people fail to inhabit Black identities. As Patricia Gomes notes "many Afro-descendant families emphasized their whiteness to save themselves. They ripped up old photos and denied the existence of a black relative" (in Goñi, 2021b). In the census, many Black Argentines don't acknowledge their Black ancestry.

Today, there are many Afro-Argentines, who along with the many Afro-Latin American and African immigrants to Argentina, are organizing, making political demands, and engaging in diverse acts of cultural and political visibilization. Judith Anderson and Patricia Gomes (2021) have documented this organizing and its multiple achievements in detail. During the Perón era, racism was rife, but a more inclusive sense of Argentinean nationhood opened, where Black and Indigenous people from the interior who were migrating to the cities "were widely celebrated and promoted by Peronists as exemplifying an authentic national culture" (p.8, drawing on Chamosa, 2010). In the 1980s, as the dictatorship ended, Black activists formed the Comité Argentino Latinoamericano contra el Apartheid (The Argentine Committee against Apartheid) to fight for Black liberation in Argentina but also for the end of apartheid in South Africa, and began to organize many conferences and cultural events, designed to learn about Black history, present proposals, and instil Black pride. Afro activism has been characterized more by spaces of dialogue, intellectual production, and the maintenance of cultural practices rather than by protest and has led to important forms of institutional and legislative change. In 1995, the government created the National Institute Against Discrimination, Xenophobia, and Racism (INADI) which was a first step in recognizing the existence of Black populations and the need to tackle racism. In 2005, the state produced a National Plan against Discrimination in which it recognized for the first time the "historical presence of black communities in Argentina" and their contribution to the nation and the formation of national identity (p.21). The 2010 census specifically allowed for Afrodescendant identification and in 2015, an anti-discrimination law was passed. But anti-Black and anti-Indigenous sentiment remain strong. In June 2021, during the state visit of Spanish prime minister, Pedro Sánchez, to Argentina, the Argentinian president Alberto Fernández offended many Latin Americans when he reproduced the white supremacist image of Argentines as Europeans and therefore as superior to other Latin Americans as follows:

> The Mexicans came from the Indians, the Brazilians came from the jungle, but we Argentines came from the ships. And they were ships that came from Europe.
>
> (Reuters, 2021)

Sources: Andrews, 1980; Karush, 2012; Anderson and Gomes, 2021; Goñi, 2021a, 2021b

ANTI-RACIST ACTIVISM AND SCHOLARSHIP

In response to systemic Afro erasure, Black and Afro-descendant anti-racist activism, scholarship, and media production are making a tangible difference, especially in terms of visibilizing Black struggles and achievements and claiming collective rights. All Latin American countries have active Afrodescendant NGOs, social movements, and cultural groups. The 2001 3rd World Conference against Racism, Racial Discrimination, Xenophobia, and Related Intolerances held in Durban in South Africa was seen by many as a turning point in getting Black issues onto the political agenda. At the current time, many activists, intellectuals, and community leaders are making use of the UN Decade for People of African Descent (2015–2024) to lobby their national governments for change.

Anti-racist activism and scholarship began during the colonial era and have intensified in recent years. Bearing in mind that coloniality tends to deny and discredit Afrodescendant intellectuality, it is important to read and engage with Afro-Latin American thought and the influential contestation of racism developed by intellectuals such as Manuel Querino, Abdias do Nascimento, Beatriz Nascimento, and Marcus Garvey (see Box 8.1). As Niyi Afolabi (2013) writes, Manuel Querino (1851–1923), who was raised by a teacher after his parents died of cholera when he was only four, was a particularly inspirational and meticulous Afro-Brazilian cultural historian whose work has reversed "the historical dislocation and disempowerment of Afro-Brazilians by using the power of research and dissemination to confront the white racist establishment of 19th century Brazil" (p.260) and reveal the Black contribution to Brazilian history and development (see also Gates Jr., 2011). More recently writer, politician, and artist, Abdias do Nascimento (1914–2011) authored a stringent criticism of the concept of racial democracy, produced important theatrical works, and introduced affirmative action proposals to the Brazilian legislature (see do Nascimento, 1989). He starts an essays on the Quilombist model as a motivating idea and source of energy in Afro-Brazilian politics as follows:

> I want to begin this text emphasizing the urgent need of the Brazilian Black people to win back their memory, which has been systematically assaulted by Brazilian Western-inspired structures of domination for almost 500 years. A similar process holds true with the history of Africans on the Continent and their descendants scattered through all the Americas.
>
> The memory of Afro-Brazilians, very much to the contrary of what is said by conventional historians of limited vision and superficial understanding, does not begin with the slave traffic or the dawn of chattel slavery of Africans in the fifteenth century. In my country, the ruling class always, and particularly after the so-called abolition of slavery (1888), has developed and refined myriad techniques of preventing Black Brazilians from being able to identify and actively assume their ethnic, historical and cultural roots, thus cutting them off from the trunk of their African family tree. (…) Never in our educational system was there taught a discipline revealing any appreciation or respect for the cultures, arts, languages, political or economic systems, or religions of Africa. And physical contact of Afro-Brazilians with their brothers in the continent and the diaspora was always prevented or made difficult, among other methods, by the denial of economic means permitting Black people to move and travel outside the country. But none of these hindrances had the power of obliterating completely, from our spirit and memory, the living presence of Mother Africa. And even in the existential hell we are subjected to now, this rejection of Africa on the part of the dominant classes has functioned as a notably positive factor, helping to maintain the Black nation as a community above and beyond difficulties in time and space.
>
> (do Nascimento, 1980: 141–142)

There is also an important body of Black feminist scholarship written by Afro-Latin American intellectuals that is focused on the intersection between race, class, and gender and on the specific experiences of Black women. Black Brazilian feminist scholar, poet, and activist Beatriz Nascimento, sadly murdered in 1995 at the age of 52, produced an important body of work on gender, race, and the politics of the *quilombo*. As scholars and translators of Nascimento emphasize, her work is important and has been very influential in Brazil, but epistemic racism and sexism mean that her work has been undervalued and virtually unknown in the Anglophone academy (Smith et al., 2021). Nascimento (2021b[1988]) captured how Black Brazilian thought is made obsolete through a form of Eurocentric erasure, and how the *quilombo* is not just a physical space created and inhabited in the past by escaped slaves, it remains a space of ongoing political potential in the world today, given the ways in which their traditions of popular resistance that they established continue to inspire Black artistic and intellectual production. As Nascimento (2021a [1985]: 304) notes, *quilombo* which emerged "as a reaction to actually existing colonialism" returned in the 1970s "as a form of reaction to cultural colonialism" that inspired notions of freedom and justice among Brazilians of African descent. Nascimento wrote about the epistemic erasure inflicted upon intellectuals such as herself:

> Recently, when I returned again to my studies, I found myself on the familiar soil of an obsolete territory. Obsolete not because this territory has ceased to exist, or has been surpassed—in truth it is continuously in flux—but because it has been reduced to a status of minority, with all that implies: the slight, the inferior, the preliminary, the impotent and the infantile. This territory is both the path already taken, and the one that lies ahead.
>
> (Nascimento, 2021b [1988]): 305)

As Christen Anne Smith (2016) writes, Nascimento along with other Afro-Latin American women writers, who write in Spanish or Portuguese, are erased by coloniality, patriarchy, and language, and thus exist in a space of limbo. She includes here the work of Epsy Campbell (Costa Rica), Andreia Beatriz dos Santos (Brazil), Francia Márquez (Colombia), and Victoria Santa Cruz (Peru). We might add here the work of Lélia Gonzalez (1982, 1998), Luiza Bairros (2008), and Sueli Carneiro (2011) who have articulated the liberation of white women to the oppression of Black women, particularly given the ways in which race and class intersect in the Black body, such as the Black domestic servant who makes it possible for the middle-class white woman to join the professional labour market. They all draw our attention to the Eurocentrism of much feminist thought that is not capable of understanding how gender subordination is shaped by race and racism. Gonzalez (1988) developed a transnational analysis of gender and race and invented the concept of *Amefricanidade* to refer to people of African descent living in America (where America is North, Central and South America) (in Williams, 2018). Before she died of cancer in 2016, Luiza Bairros was a member of President Dilma Rousseff's cabinet as Minister for Racial Equality. Another important Black Brazilian politician and intellectual who set out to address Brazil's colonial condition and the ways that racism, misogyny, homophobia, poverty, and violence all intersect was Marielle Franco (see Box 8.3).

BOX 8.3 MARIELLE FRANCO: EU SOU PORQUE NÓS SOMOS

Marielle Franco was a Black Brazilian feminist, left-wing politician, and human rights and LGBTQ+ activist (Figure 8.3). She grew up in a favela, so her early life was characterized by racialized inequalities and persistent police violence, and she wrote a Master's thesis on the militarization of the policing of the favelas. She went on to become a city councillor in Rio de Janeiro, where she had to confront both sexism and homophobia that often took the form of interruptions which she mobilized to draw attention to injustice (see Silva and Won Lee, 2020). Her campaign slogan was "Eu sou porque nós somos" "I am because we are" and her election brought hope to the city's marginalized populations (Santos de Araújo, 2018). She was a fierce critic of the racist violence that afflicted the city, especially the police brutality targeted at Afro-Brazilians in the favelas. As Gabriela Silva Loureiro (2020:51) writes, Marielle, who drew on her identity as a Black woman from the favelas "is situated within a tradition of Brazilian Black women who used their intersectional identities to assert a different way of doing politics." In March 2018, the day after she tweeted her condemnation of the police murder of Matheus Melo as he left church, she too was gunned to death, along with her driver. Her death sent shock waves through the favelas as well as through Black communities across Latin America. Silva Loureiro believes that Franco's anti-racist and feminist epistemology constituted a threat to the dominant racial order and her assassination is "part of the necropolitical governance of Rio, where violence is the last resort technique for enforcing racial oppression" (p.53). Franco's life experience, political activism, and her embodied access to Black ancestral knowledges gave her a profound insight into how colonial oppression results in the dehumanization and premature death of Black people from the favelas who are seen as disposable. Not only was did she grow up in the favela, but as she noted she used "this place of favelada to make claims and to make politics in a different way" (Franco, 2016, cited in Silva Loureiro, 2020: 54). Franco also made connections between the military dictatorship and the current rise of authoritarianism in Brazil and the role of the United States in reproducing racialized inequalities in Latin America

(Santos de Araújo, 2018). It is clear that Franco's struggle was an epistemic one as well as a broader political one that involved a decolonial locus of enunciation. Like Berta Cáceres discussed in Chapter 4, Marielle's legacy lives on and continues to inspire anti-racist activism in other Black women and her murder directly encouraged unprecedented numbers of Black women all over the country to run for office. As Santos de Araújo (2018: 207) writes:

Marielle lived and fought as a black lesbian woman, a *favelada,* an activist, an intellectual, a politician, a partner, a friend, a mother, and a daughter. She was at the intersections of many struggles and she was a warrior. Let's not forget that.

Sources: Santos de Araújo, 2018; Silva Loureiro, 2020

While many Black and Afrodescendant groups are gaining new forms of political visibility, some of these movements are also profoundly contradictory (de Santana Pinho, 2010). In Bahia, often referred to the most African part of Brazil, the inhabitants are reinventing Blackness in creative and self-affirming ways and by so doing are attracting African American tourists to the region. The tension between the need to reject essentialist understandings of

Figure 8.2 Marielle Franco in 2016

Source: Mídia Ninja, CC by-SA 2.0, via Flickr

race, while adopting strategically essentialist modes of valorizing blackness as a means to contest racism is, however, highly complex. She argues that the reinscription of Africa on the body in a way which restores dignity also paradoxically tends to fix the Black body and convert it into a site of commercial exploitation (see Box 8.4).

BOX 8.4 BAHIA AND THE AFRICAN TRADITION

Salvador in the state of Bahia in Brazil is one of the most African cities in Latin America. Here Blackness is both visible and celebrated in African-derived religions such as Candomblé, which combines elements of Angolan Catholicism with Nigerian Yoruba worship of Orishas. While Candomblé was an important means of slave survival and resistance, it remains a significant religion in Brazil today with thousands of practitioners. Yoruba and other African traditions are also visible in the celebration of carnival where groups such as Afoxé and Blocos Afro perform samba-reggae and other kinds of music. They sometimes have Yoruba names derived from Candomblé such as Ilê Aiyê and they are a key source of both Black pride and the conservation of African heritage. Such groups have strong activist and grassroots dimensions. They have demanded rights and opened up new spaces for anti-racist political action.

For much of the 20th century, Afro-Bahian cultural practices, including Candomblé, capoeira, and the blocos were subjected to repression and vilification by white elites and the police. In the 1970s, while Brazil was still a dictatorship, these movements, inspired in part by movements for Black liberation elsewhere, began to flourish. Famous blocos such as Ilê Aiyê and Olodum were formed at this time. They played a key role in the re-Africanization of carnival and constituted an important platform from which to challenge racial discrimination and discourses of racial democracy in Brazil. Both groups engage in pedagogical work to transform Brazil's education system and centre people of African descent and Afro-Brazilian history in school curricula. They have also become highly profitable enterprises that engage with big business and mainstream electoral politics which subjects the Black body to complex forms of commodification. While commodification is not automatically negative as it often enables subordinated populations to gain some control of their own representations and generate dignified livelihoods, some scholars have expressed concerns that the success of such movements reproduces essentialized understanding of Blackness that dilutes the radical edge of these movements and makes the anti-racist rearticulation of hegemonic power harder to achieve.

Sources: de Santana Pinho, 2010; Gates Jr., 2011; Dixon, 2016

Figure 8.3 A poster at URACCAN, the local university in Bilwi Nicaragua, promoting the use of the Creole language

Source: Photo by author

Despite many setbacks and persistent challenges, Afrodescendant activism and intellectual production are shaping policy, securing rights, and changing attitudes. There are some very prominent and important organizations including the PCN in Colombia (created in 1993) and ONECA/CABO in Central America (created in 1995) that have been instrumental in bringing about some of these changes. In Colombia, Ecuador, and Nicaragua, Afrodescendants have formal constitutional recognition and in Brazil, Colombia, and Nicaragua their rights to land are enshrined in legislation, although such rights are often weakly respected and slowly enacted. In some countries, racism is deemed a crime or anti-discrimination plans of action have been created that have led to affirmative action initiatives. In others, day or weeks have been set aside to celebrate Afrodescendant culture. Others are discussing the question of reparations for slavery or making Afrodescendant media and film (see Bush et al., 2018) and defending and promoting the Creole language in a way that confronts Spanish dominance (see Figure 8.3). In many places, there is a struggle for data justice that is making a difference.

CENSUS ACTIVISM AND DATA JUSTICE

Even though racial ideologies such as those outlined above discourage Black identification and result in invisibilization, the Black-identified population in Latin America is significant. Today there are at least 150 million people of African descent in Latin America. They are composed of many different communities. After Nigeria, Brazil is the country with the largest number of people of African descent anywhere in the world. Afro-Latin Americans go by different names: Afrodescendant, Black (negros/pardos), Creole, Garífuna, Palenquero, and Raizal or as Afro-Brazilians, Afro-Cubans, Afro-Panamanians, Afro-Ecuadorians, and so on. As you can see from Table 8.1, Brazil, Cuba, Haiti, and the Dominican Republic have majority Black populations, but even a country such a Mexico with a very small percentage of Afro-identified people is home to more than a million Afro-Mexicans. In most of these cases, the actual number of Afrodescendants is much larger, as many Afrodescendants cannot admit their Black ancestry (see for example González-Rivera, 2020). For example, while only 10% of the Colombian populations identifies as Afro-Colombian in the census, it is estimated that 25% is of African descent (Afro-descendant Women Human Rights Defenders Project, 2012).

A particularly interesting mode of continental Black activism is that developing around the census that can be understood as a struggle for data justice (Andrews, 2018). Racial counting and classification were widespread

Table 8.1 Afrodescendant populations in Latin America

Country	Total Afro population, 000s	Percentage
Haiti	9,801	95
Dominican Republic	7,556	80
Brazil	96,795	50.7
Cuba	3,905	39.4
Colombia	4,311	10.4
Panama	313	9.2
Nicaragua	590	9
Costa Rica	334	7.8
Uruguay	255	7.8
Ecuador	1,042	7.2
Venezuela	952	3.5
Peru	846	3
Mexico	1,382	1.2
Argentina	149	0.4
Bolivia	35	0.3
Chile	16	0.1
Paraguay	8	0.1

Source: From Arteta, 2017

during the colonial era, because it enabled the colonizers to establish hierarchies through which power could be exerted and racist practices legitimized. Many of these categories related to degrees of mixing between Blacks, white, mulattos, Indians, and so on or to skin colour – people could be the colour of wheat, coffee, or cinnamon – rather than to African or Indigenous ancestry. As a result, this practice was understood to be racist and was abandoned after independence. This move fitted comfortably with the assumptions of the time that whitening would eventually lead to the elimination or the assimilation of the Black population. We don't really know the actual size of the Black population because until recently national censuses did not ask questions about racial identification (Hooker, 2005). Increasingly, failure to count was seen by activists as a kind of statistical racism that perpetuated Afro erasure and made it difficult to formulate public policy. It is easier to claim that there are no Black people that need special policies as a result of colonial histories and persistent discrimination if you don't know how many there are.

Some Black organizations began to lobby for their inclusion in the national census, while some such as Kamba Cuá in Paraguay and Lumbanga in Chile also tried to conduct their own censuses (Antón et al., 2009). The Kamba survey suffered from a weak methodology but did reveal that there were many Afro-Paraguayans that lacked political recognition (Wilson, 2021). Between 1990 and 2018 most Latin American countries added this census category. There is still substantial underreporting – the way in which the question is asked is important and, as we've noted, some Afro-Latin Americans do not recognize their own Black ancestry and Black histories are often excluded from the school curriculum. Consequently, activists in many countries have tried to raise awareness of the importance of self-identification. Afro-Colombians and Black Creole Nicaraguans have produced campaign videos aimed to get people to tick the right box on the census (see Conferencia Nacional de Organizaciones Afrocolombianas, 2015; Cumbre Internacional Mujeres Afro, 2017).

AFRO-COLOMBIANS AND RAIZALES: RACISM IN COLOMBIA

In Chapter 3, we outlined the deadly armed conflict in Colombia in which many Colombians were killed and even more were displaced (Box 3.1). In the 1990s and 2000s, the Afro-Colombian populations on the Pacific Coast were systematically and violently attacked, killed, and displaced by paramilitary and guerrilla forces. As Richard André (2011) notes, few people, even within Colombia, paid attention. This is in spite of the fact that Afro-Colombians had gained recognition in the 1991 Constitution and in 1993 Colombia passed Law 70, the Law of Black Communities, an important albeit often limited legal mechanism for the securing of the land rights of Afro-Colombians.

This section focuses on some of the consequences of the Colombian conflict for Afro-Colombian people. It draws on the work of Afro-Colombian scholar Aurora Vergara-Figueroa (2018) who has done long-term research in the Chocó where more than 80% of the population of this region identifies as Black or Afrodescendant. It is an area that is highly stigmatized by mestizo elites in Bogotá and Medellín. She writes:

> Indeed, this region has been represented as a space of darkness where it is impossible to live, and as a space of extreme poverty and ignorance, which is unable to administrate its own resources. On the other hand, this region is seen as having potential for territorial expansion, as an area open for exploration, and continuing colonialization.
>
> (Vergara-Figueroa, 2018: xx)

On the morning of 2 May 2002, 119 people were massacred in the Catholic church of Bellavista in the municipality of Bojayá in the state of Chocó. Some 1,744 families had to abandon the territory to escape the atrocity of this event and became what is known as Internally Displaced Peoples (IDPs). Many returned later and are still trying to rebuild their lives in the aftermath of this trauma. Vergara-Figueroa understands these forms of violence as underpinned by what she calls "historical-emptied-spaces" and "the routinization of erasure" (p.xxi). She sees the massacre in decolonial terms as part of a larger history of deracination and systematic killing of Afrodescendant people in Colombia. Drawing on Aníbal Quijano's coloniality of power, she stresses the importance of looking at the long history of dispossession before the massacre, how the massacre exacerbates racial and regional inequality, and prevents the community members securing political redress. By deracination she means "a set of economic, social, political, cultural and ideological process, which involves the violent dispersing of the inhabitants of a territory. It undertakes the effacement of the population and the appropriation of their lands" (p.xxxi). She is motivated by the need to move away from dominant social science frameworks and debates about forced migration which end up reproducing exploitation by not exploring the historical roots of processes such as land dispossession. In other words, the massacre was not an isolated event and should not be treated as such. And furthermore, "using the term 'forced migrant' for people who have been abused and dispossessed, in essence legitimizes a process of deadly land dispossession and economic exploitation. It makes us accomplices to it" (p.7).

Vergara-Figueroa describes how "the territories of Chocó and its rivers have been kept as laboratories of death" stressing that struggles to survive "are shaped by the legacies of slavery and the colonial foundation of the power, which reconfigure in every historical period to dispossess the population that resides in the land where they have worked, fought, and produced. It is a history of a racialized and gendered labor exploitation" (p.28). Her decolonial historical and geographical analysis of racism and dispossession stretches over four centuries and doesn't just restrict itself to the period of the Colombian civil war. The region has been subject to two distinct worldviews, one based on ancestrality and the collectivity, and another that sees it as a site for top-down development, extractivism, and exploitation. Her long-term ethnographic work with survivors has enabled her to centre the voices of the people of Bellavista, and capture the horror that they lived through and the forms of resistance that they enacted. This approach has facilitated what she calls "an Afrodiasporic feminist critique of the epistemologies of dislocation" (p.84). In their accounts, you can see the forms of racialized dehumanization and historical erasure to which they were subject.

> I was eighteen years old in 2002. In the news there was a famous story about a woman who gave birth in the Church because she was too nervous, and who died instantly with the baby after the explosion. They were my woman and my child. How can one possibly recover from that? I am about to tell that I could be considered a displaced, a deracinated, an uprooted all of that, and none of the above at the same time. That is me. What can we use that for if at the end nobody likes us, or cares about us? My family is pretty much falling apart. We, my mom and I, made it alive, but we all have our own traumas. If someone is to tell my story is to get people to know that I am a human being, I also have feelings. I really don't know why this happened to us. Is it because we are Blacks? I don't know.
>
> (pp.64–65)

> We have been resisting and demonstrating our presence in the Old Bellavista celebrating regular masses and meetings to demand material reparations of the damage that has been done to us. We have placed marks in the land, and the owners of the cows took them off. We are now planning on building a community museum or memorial in the reconstructed Church to tell the world our story. If we don't have a presence in our land what history are we going to tell? What is going to be written about us? [Who] are the massacred Bellavista people the world will see? In the memory of our *muertos* (dead) we find the strength to survive.
>
> (p.68)

The survivors have used song as a means of resistance and memory formation, in order to resist historical erasure and official indifference (see Box 8.5). They changed the lyrics of traditional songs known as *alabaos* as a form of mourning. Here is just one of them – you can read the others in Vergara-Figueroa's book.

BOX 8.5 JORGE LUIS MASO PALACIOS: AN *ALABAO* FROM BOJAYÁ

He was humanitarian in the river of Bojayá
And with all his patience
And he was going to visit us

That Father Jorge Luis Maso was going to Quibdó
And arriving at the Mercedes a panga killed him

The panga that killed him
Belonged to the paramilitaries

He received a heavy blow
That threw him into the Atrato River

He was going up with Iñigo along the Atrato
They went in search of the stores
To buy cheaper

Goodbye Father Jorge Luis
You left for no more

The Pogue Community
Will always remember you.

Source: Vergara-Figueroa, 2018

Coloniality, slavery, dispossession, racism, extractivism, and counter-insurgency have all combined in the Colombian Chocó and have resulted in unimaginable suffering and premature death inflicted on Black lives. Vergara-Figueroa's analysis shows how cultural expression is used by the community members to resignify the meanings of armed conflict, to denounce the complicity and indifference of politicians, to overcome trauma and build dignity and psychological strength, to reclaim their own (Black) bodies, and counter dominant colonial narratives that construct Chocó as a backward and empty space that can be violently occupied and used for extractivist activities such as gold-mining and palm oil cultivation.

There is another Black population living in Colombia, also exposed to racism and discrimination, but positioned differently within the colonial matrix of power. The Raizal population lives on an archipelago in the Caribbean Sea made up of the islands of San Andrés, Providence, and Santa Catalina, along with a number of cays and banks. These islands are just off the Caribbean Coast of Nicaragua but belong administratively to Colombia.

The islands were first inhabited in the 17th century by English Puritans and their slaves. During the colonial era, England and Spain as well as pirates such as Henry Morgan fought for control of the islands. There were many slave rebellions and some slave-owning families, such as the Livingstons and the Archbolds, began to free their slaves. Philip Livingston went on to become an abolitionist and founded the first Baptist Church in San Andrés, which became the most important religion on the islands. After the abolition of slavery in 1853, most freed slaves received plots of land and began to trade and intermarry heavily with the rest of the Anglophone Caribbean and the region known as the Mosquitia, with whom they had a shared language and culture. They began to call themselves Raizales but their cultural affinity was with the Black Creole people of Bluefields and Corn Island in Nicaragua, Puerto Limón in Costa Rica, Bocas del Toro in Panama, and the Cayman Islands (see Figure 8.4). For many decades, people moved freely around this region by boat to trade, play sport, visit family, and create new families. Colombia was far away and exerted little influence.

In the 20th century, however, Bogotá stepped up its geopolitical ambitions towards the archipelago and a process of Colombianization began. This involved the attempt to replace Baptism with Catholicism and English with Spanish. In 1928, Nicaragua and Colombia signed the Esguerra-Bárcenas treaty in which Colombia recognized Nicaragua's sovereignty over the Mosquito Coast and the Corn Islands, while Nicaragua recognized Colombia's sovereignty over the islands of San Andrés, Providence, and Santa Catalina. This ruling made little practical difference for the local people and for most of the 20th century, people travelled from San Andrés and Providence to Bluefields and Port Limón and back without needing passports or other official documentation. In the second part of the 20th century, San Andrés was declared a free port and an airport was built, so Colombian traders and tourists started to arrive in ever larger numbers. Later they were joined by Colombians, including Afro-Colombians, displaced by armed conflict. Many mainlanders stayed to set up businesses or marry islanders. Large tourist hotels filled the coastline. San Andrés began to suffer from overpopulation, biodiversity loss, and conflicts over land, water, sanitation, garbage (especially plastic bottles), and fishing resources. The local Raizal population also felt that their language and culture were starting to be overwhelmed by Spanish and mainland cultural practices. It also became harder to travel between the islands and Nicaragua, the direct flights came to an end and the frequent excursions by boat became much less frequent, cutting family members off each other and putting an end to the trading exchanges. In 2001, Nicaragua resumed its sovereignty claim on the islands, via the International Court of Justice in the Hague, claiming that the 1928 Treaty had been signed by a dictatorship so wasn't valid. In 2012, the Court ruled that the islands were in fact Colombian, but that the sea (apart from the 12 miles surrounding the islands) belonged to Nicaragua. Many Raizal people see the ongoing process of Colombianization (see Figure 8.5) along with the

Figure 8.4 San Andrés is part of Colombia but is Raizal ancestral territory

Source: Photo by author

Figure 8.5 The people of San Andrés celebrate 20 July, Colombian Independence Day

Source: Photo by author

Hague ruling as a form of racial violence and are convinced that the Colombia state seeks to eliminate them as a people. As a result, some Raizal people are less enthusiastic than other Afro-Colombians about the inclusion of racial identity categories in the census, as it provokes fear that the data could be used to further drive elimination. One Raizal leader told us:

> We are one people. The Mosquitia is one territory. Bogotá and Managua *decidieron dividir nuestro territorio* [decided to divide our territory]. This relationship is one of blood. I never said I wanted to be Nicaraguan. I never said I wanted to be Colombian either. I am a Raizal from San Andrés. All that is ours, *todo lo nuestro*, is foreign to the Colombian state. We've tried to be Colombian in our own way but they don't let us do so.

In response to this state of affairs, there is an important struggle underway for Raizal emancipation and for some that includes the recovery of the Creole nation where Black Creole and Raizal people can continue to interact. The discourse of Colombianization is a powerful motivator of autonomy and identity reconstruction, but this perspective can sometimes act as a barrier to solidarity with Afro-Colombians. It can also pose a difficulty for the many islanders who have one Raizal parent and one parent from the Colombian mainland.

BLACK AND INDIGENOUS

Finally, even though we have dealt with Indigenous and Black Latin Americans in separate chapters, in order to deal with histories and struggles in more depth, it is important to recognize there is substantial solidarity between them. Black and Indigenous populations are seen as distinct and mutually exclusive categories, in part, because of the ways in which Blackness is associated with diaspora and uprootedness and Indigeneity is associated with deep cultural roots (Hale, 2004; Wade, 2006; Anderson, 2009). Yet in reality, Latin America is also populated by an Afro-Indigenous population. The Garífuna people, who originate from the island of Saint Vincent and now live in Belize, Honduras, and Nicaragua, have African as well as Arawak and Kalinago origins. On the Caribbean Coast of Nicaragua, there are both Miskitos and Creoles, with the former adopting an Indigenous identity and the latter a Black one. In practice, however, the two groups are thoroughly mixed and the boundaries between them are highly fluid. Many Costeños have a Creole parent and a Miskito parent (or a Chinese parent). Whether one identifies as Miskito or Creole depends in part on the community in which one lives and the discursive positioning with which one more closely identifies (Dennis, 2004). While Indigenous peoples can strategically draw on discourses of territoriality and understandings of themselves as pre-modern in order to gain cultural rights, Blacks are able to identify with potentially empowering and resistant modes of style and consumption. In Honduras, the Garífuna people are practising what Mark Anderson (2009) calls a "black indigeneity" or an "indigenous blackness" in which the racial distinctions between Black and Indigenous are refused or at least creatively negotiated. These developments are important not just for those who identify as Afro-indigenous but also for all racialized Latin Americans as they potentially constitute a means to overcome the discursive and material divides separating racialized populations and with them the ability to work together to forge an anti-racist politics.

SUMMARY

- After independence, many Latin America countries tried to whiten their populations through European immigration. These policies were often accompanied by discourses of racial democracy that led to a denial of racism.

- Afro-Latin Americans are subject to structural and everyday forms of racism and discrimination. As a result, many Afrodescendants do not identify as Black.
- Hegemonic discourses of mestizaje work to erase Afro histories and identities; Black struggles and histories are not taught in schools; and Afrodescendant contributions to development, art, literature, music, media, sports, and science are overlooked. Invisibilization is one of the key barriers to dismantling institutional and everyday racism.
- The attempt to dismantle racism while expressing pride in Blackness sometimes has contradictory outcomes.
- Yet Afrodescendant cultures, languages, and spiritualities remain hugely influential, and the continent is characterized by vibrant Afrodescendant social movements that are changing attitudes and policies in tangible ways
- But racist and colonial logics remain active. Black community leaders and activists have been intimidated and murdered for attempting to defend their communities, livelihoods, and ways of life.

DISCUSSION QUESTIONS

1   Why is it harder for Black people to claim territorial rights than Indigenous people?
2   How have Latin American intellectuals thought about race?
3   How and why did whitening policies come to be established after independence?
4   Who was Marcus Garvey and why does he inspire many Afro-Central Americans?
5   Why are census data on Black populations so unreliable? Why do they matter? Why do some Black populations want to be counted and others not?
6   Why should our anti–racist politics be intersectional?
7   In the case of the San Andrés archipelago, can we consider Colombia to be an internal colonizer?

FURTHER READING

Routledge Handbook of Latin American Development Chapter 22 (Bush et al.)
de Santana Pinho, P. (2010) *Mama Africa: Reinventing Blackness in Bahia.* Durham, NC: Duke University Press.
A text that explores the influence of Africa on Black identities in Bahia, the role of the blocos afros and their contradictory cultural and political outcomes in the context of international tourism and other processes.
Hooker, J. (2005) *Race and the Politics of Solidarity.* Austin: University of Texas Press.

This book explores the politics of multiculturalism, struggles for racial justice, and the need to make whiteness visible.

Wade, P. (2010) *Race and Ethnicity in Latin America*, 2nd ed. London: Pluto Press.

A book that explores the complex and shifting ways that the concept of race is understood and mobilized.

Recommended films

The Black Creoles (2011), directed by María José Álvarez and Martha Clarissa Hernández

A film that deals with Black identities and memories on the Caribbean Coast of Nicaragua and Costa Rica.

Black in Latin America (2011), directed by Ricardo Pollack and Ilana Trachtman

A four-part PBS documentary series hosted by Harvard professor Henry Louis Gates Jr to explore Black history and culture in Brazil, Cuba, the Dominican Republic, Haiti, Mexico, and Peru.

El Testigo (2018), directed by Kate Horne

A documentary about Colombian photographer Jesus Abad Colorado, captured twice by guerrillas, who photographed some of the most moving images of the armed conflict. It contains a section on the resistance of the survivors of the Bojayá massacre.

Marcus Garvey: Look for me in the Whirlwind (2001), directed by Stanley Nelson

A PBS documentary that tells the story of the life of Marcus Garvey and the impact his thinking had for the African diaspora.

Tango Negro: The African Roots of Tango (2013), directed by Dom Pedro

An Angolan filmmaker travels to Buenos Aires to document tango's African origins.

They are We (2014), directed by Emma Christopher

A film about Afro-Cuban culture and how music reunited a people separated by the transatlantic slave trade.

Useful websites

Creole Connections

An interactive video platform that maps the cultural, geopolitical, and familial connections between Nicaragua, Colombia, and the San Andrés archipelago

www.creoleconnections.cahss.ed.ac.uk/interactive/#WELCOME

Slave voyages

An interactive website that documents the transatlantic slave trade, with databases, maps, timelines and data visualizations of the movements of slave ships and information about people, places and ships

https://slavevoyages.org/

Florescer por Marielle

A website dedicated to the life and legacy of Marielle Franco (in Portuguese)

https://florescerpormarielle.psol50.org.br/nao-serei-interrompida/

PERLA (The Project on Ethnicity and Race in Latin America)

A continental project that is particularly concerned with racial identification, data, and censuses in Latin America. It is particularly focused on Brazil, Colombia, Mexico, and Peru.

https://perla.princeton.edu/about/

Renacientes (in Spanish)

The website of the Proceso de Comunidades Negras (PCN) in Colombia, with published reports and information about the cultural and political processes, achievements and human rights abuses that affect the lives of Afro-Colombians

https://renacientes.net/

CHAPTER 9

Disastrous development

INTRODUCTION

Geological and climatological conditions combine with practices of dispossession to make Latin America susceptible to disaster. Half of the continent lies on a very active earthquake fault line, known as the Pacific Ring of Fire and it experiences hundreds of earthquakes every year. Some of these are devastating and cause massive loss of life, human injury, and damage to infrastructure. In 2010, an earthquake in Haiti killed 230,000 people and left half a million people homeless. In addition, much of Central America and the Caribbean is in a hurricane belt and usually experiences a number of tropical storms and hurricanes every year and many of these trigger landslides. Like earthquakes, some of these storms can cause catastrophic loss of life and other forms of devastation. The most devastating hurricane to hit Central America was Hurricane Mitch in 1998, which killed more than 11,000 people in Honduras and Nicaragua (see Box 9.1). The Pacific coast of the continent is prone to the El Niño Southern Oscillation (ENSO). Every five to seven years, a combination of higher-than-normal surface temperatures in the Pacific Ocean with high air surface pressure produces extreme weather, including droughts and floods and often impacts negatively on farming and fishing. Latin America also has a number of active volcanoes. In 1985, 25,000 people were killed by lahars created by the eruption of the Nevado del Ruiz Volcano in Colombia. Melting glaciers are also serious hazards. People who live in Peru's Cordillera Blanca have repeatedly faced glacier disasters, such as glacial lake outburst floods, glacier avalanches, or glacier landslides created by thinning or fracturing ice (Carey, 2008). The 1970 Yungay earthquake in Peru triggered a glacier avalanche which buried tens of thousands of people. Table 9.1 is a selective list of the some of the major disasters to occur Latin America in the last few decades. It is important to recognize that in addition to the major events which attract (for a short period) international media attention, there

DOI: 10.4324/9781003110453-9

Table 9.1 Selected list of large disasters in Latin America since 1960

Date	Place	Disaster event	Fatalities
2020	Central America and Caribbean	Hurricanes Eta and Iota	300
2019	Brazil	Floods and mudslide (dam collapse)	300
2018	Guatemala	Eruption of Fuego Volcano	200–2000
2017	Puerto Rico and Caribbean	Hurricane Maria	64–4600
2017	Mexico	Earthquake, magnitude 7.1	369
2016	Ecuador	Earthquake, magnitude 7.8	663
2015	Chile	Earthquake, magnitude 8.3	7
2014	Chile	Earthquake, magnitude 8.2	7
2011	Brazil	Floods	800
2010	Haiti	Earthquake, magnitude 7.0	230,000
2010	Chile	Earthquake, magnitude 8.8	400–500
2010	Colombia	Flood	418
2009	El Salvador	Hurricane Ida	275
2007	Peru	Earthquake, magniture 8.0	514
2007	Nicaragua	Hurricane Felix	188
2005	Central America	Hurricane Stan	1600
2001	El Salvador	Earthquake, magnitude 7.7	800–1000
2001	Peru	Earthquake, magnitude 8.4	77
1999	Venezuela	Floods	30,000
1999	Mexico	Floods	636
1999	Colombia	Earthquake, magnitude 6.8	1000
1998	Central America	Hurricane Mitch	11,000
1997–1998	Andean region	El Niño	600
1997	Venezuela	Earthquake, magnitude 6.8	81
1997	Peru	Landslide	300
1994	Colombia	Earthquake, magnitude 6.8	1000
1992	Nicaragua	Earthquake, magnitude 7.7, and tsunami	116
1991–1993	Whole continent	Cholera epidemic	9000
1991	Chile	Landslide	141
1988	Central America	Hurricane Joan/Juana	200
1988	Mexico	Hurricane Gilbert	433
1987	Ecuador	Earthquake, magnitude 6.9	1000
1987	Colombia	Landslide	640
1986	El Salvador	Earthquake, magnitude 5.5	1000
1985	Colombia	Eruption of Nevado del Ruiz Volcano	23,000
1985	Mexico	Earthquake, magnitude 8.1	5000–20,000
1982	Guatemala	Floods	620
1982	Mexico	Eruption of El Chinchón Volcano	2000

Table 9.1 Cont.

Date	Place	Disaster event	Fatalities
1976	Guatemala	Earthquake, magnitude 7.5	23,000
1974	Honduras	Hurricane Fifi	6000
1972	Nicaragua	Earthquake, magnitude 6.2	10,000
1970	Peru	Earthquake, magnitude 7.9, and avalanche	70,000
1961	Central America	Hurricane Hattie	275
1960	Chile	Earthquake, magnitude 9.5	2000–6000

Sources: US Geological Survey; National Oceanic and Atmospheric Information; EM-DAT

are also many small-scale or slow onset hazards such as drought, desertification, and air pollution that are often catastrophic in terms of their impacts on people and livelihoods.

BOX 9.1 GENDER AND DISASTER

One of the most tragic events to hit Central America at the end of the 20th century was Hurricane Mitch. Ranked category five on the Saffir-Simpson scale, Mitch produced torrential rain for five days, washing away homes, livestock, roads, and bridges and causing the deaths of at least 11,000 people. In Nicaragua, 3,000 people were killed when a massive landslide swept down the sides of the Casita Volcano, burying several communities that lay in its path. I interviewed a number of single women who had been made homeless by the hurricane and joined a self-help housing project led by a local NGO. They were all low-income single mothers who worked in the informal sector, but they responded to the disaster in quite different ways.

While attention to gender is a crucial part of improving disaster response, relief, and research, there is a danger of oversimplifying how gender shapes responses to disaster or is responsible for generating certain kinds of vulnerabilities or strengths. Women, for example, are often seen in quite essentialist or deterministic ways as mothers or as more vulnerable than men. In the aftermath of disaster, gender relations and identities are often reworked, in ways that are empowering, disempowering, or contradictory. My research with survivors of Hurricane Mitch showed how participants' involvement in the disaster process impacted on their subjectivities and led to the destabilization or reproduction of certain gender identities. It revealed how the experience of the disaster is shaped not only by pre-disaster vulnerabilities and forms of resilience but also by the discursive positioning facilitated by the disaster itself and how these intersect with previous experiences as workers, farmers, political activists, single mothers, or victims of domestic violence. It is then not possible to say that women, because they are women, will behave or respond to a disaster in a particular way. Even women with similar superficial background characteristics will not react to the same event in the same manner.

At the time of the hurricane, Marcia Picado and Ramona Dávila were both low-income single mothers in their mid-thirties who had secured new homes as part of a post-disaster housing project. Marcia's condition as a *damnificada* (disaster victim) accorded her the discursive capability to incorporate suffering into her gender identity, to see herself not as a survivor, but as a long-suffering woman, or *mujer sufrida*.

For Ramona, acquiring a house in the project brought housing stability to her life for the first time. As a child and an adult, she had often been made homeless. During Hurricane Mitch, the river flooded and filled their house with water in the middle of the night. Some of the houses in this community were made of brick (and hence were sturdier), but Ramona's house was made of wood and could not withstand the strong currents. All of her children managed to get out safely with their father except for her fifth child, Orlando, whom she had informally adopted from a local woman just before the hurricane. Ramona stayed behind to look for him. At this point, the electricity failed and the scene was plunged into darkness. The neighbours outside called on her to get out, as the entire house was starting to move and was about to be swept away. Suddenly, Ramona remembered that there was a pile of mango wood in the corner of the house, which she described to me as a "message from God." She scrambled around until she felt the boy's legs sticking out of the top, his face buried in the firewood and the water. She grabbed him by the legs, pulled him out, and extricated them both from the house with seconds to spare before it disappeared down the river. Orlando was badly injured but survived. After Mitch, Ramona was homeless once again, now with five children to support. Her partner of 16 years, who was an alcoholic and had been violent, left them at this stage and never returned. Her new home with its legal title and the end of the domestic violence she suffered prior to the hurricane, as well as having rescued Orlando a second time, appear to have brought about a more positive sense of self as well as spiritual revitalization.

Sources: Cupples, 2004, 2007; see also Bradshaw, 2001

While there are many disaster events that involve earthquakes, hurricanes, landslides, and floods and many of these have led to substantial loss of life and livelihood destruction, it is important, however, not to simply blame nature for such losses. *There is no such thing as a natural disaster!* (O'Keefe et al.,

1976; Smith, 2006). Large earthquakes and powerful hurricanes happen in the Global North, but the degree of suffering is much greater in the Global South (UNDP, 2019) and the poorer populations in Latin America are far more at risk than the affluent ones. This is because they are more likely to live in hazardous locations – such as on the slopes of volcanoes or hillsides – and in flimsy homes without foundations. They are also less likely to have savings or insurance that would help them to recover. In 1976, people began to refer to the deadly earthquake that hit Guatemala as a classquake because of the way it targeted those who were poor and already vulnerable; the 90,000 people made homeless lived in the poor neighbourhoods, while the homes of the rich withstood the shaking (Susman et al., 1983; Smith, 2006). As we shall see, Latin America's hazard susceptibility is exacerbated by development models in place and these development models frequently turn hazards into disasters. We must therefore consider the role played by colonialism, imperialism, structural adjustment, and uneven globalization in the making of disasters.

WAYS OF KNOWING DISASTER

Scholars such as Terry Cannon (1994) and Ken Hewitt (1995) produced a stringent critique of mainstream hazard and disaster management, planning, and research for their failure to pay attention to the social, political and cultural factors that lead to losses. The field has been dominated by a belief that disasters are created by extreme geophysical or climatological processes and can be solved by a technocratic approach and the implementation of engineering knowledge. Indeed, the emphasis on the impact of nature can cause potentially dangerous interventions and argues that for a natural phenomenon to become a disaster, it has to affect vulnerable people (Cannon, 1994). In other words, what is in place when the earthquake or hurricane occurs, including the levels of vulnerability and resilience of the affected population and the preparedness and willingness of institutions to respond, will strongly influence the extent of the disaster that follows.

Disasters bring the failures of development into sharp relief. The 1985 Mexico City earthquake produced loss of life and homelessness on a large scale. But the earthquake also revealed the extent of government corruption in Mexico and how politicians had been complicit in the disregard of seismic building codes by construction companies during the 1980s and these were among the buildings that had failed with tragic consequences. Adherence to building codes matters and housing people in poorly constructed buildings that are likely to kill them in an earthquake is criminal, but we need to push our analysis much further. As Erin Durban–Albrecht (2017) writes in relation to the earthquake in Haiti, Haitians died in such large numbers because of the way Haiti has been denied its own existence and its own history and has been repeatedly punished by colonial and imperial powers (see Chapter 2).

She calls for "an antiracist historic frame" that "helps us to understand that it was these Euro-American imperialist legacies rather than 'poverty' or 'lack of building codes' that determined the extent of bodily damage in the earthquake" (p.201). As Karen Salt (2019) writes, it is clear that some of the development practitioners who came to "help" and some of the journalists who came to report on the quake made things worse, through the militarization of development assistance and through representational strategies that without context posit Haiti as poor, unstable, and dysfunctional and Haitians as subhuman. When it comes to disaster reporting, however, not all journalists rework colonizing tropes, there are also journalists who report differently and well, as Salt (2019) emphasizes.

These dynamics, awful though they are, mean that disasters are catalysts for political change. They mobilize survivors often in ways that could never have been anticipated, they reveal existing social, economic, and racial inequalities, and they often call the legitimacy of existing leaders into question (see essays in Buchenau and Johnson, 2009; Solnit, 2010). The 1972 Managua earthquake was a key contributing factor to the fall of the Somoza dictatorship in Nicaragua. The Somoza dictatorship embezzled most of the overseas aid that had been sent to help the disaster victims and as a result it turned many more people against him and increased the support for the revolution (see Chapter 5). Similarly, the devastated survivors of the 1985 earthquake in Mexico City began to engage in collective, spontaneous, and creative acts of solidarity and to organize themselves (Poniatowska, 1995). Mexico was due to host the 1986 World Cup and the PRI government was determined to prove to the rest of the world that it was able to host the tournament despite the disaster. But at least 250,000 people had been left homeless and so homeless survivors began to effectively mobilize around the games in pursuit of housing and other forms of political redress. This included the threat two months before the World Cup that 50,000 homeless survivors would occupy the Azteca Stadium, where some of the matches would be held, if their demands for housing were not met (Eckstein, 2001). The poor and homeless could not of course afford to attend the matches. Nonetheless, during the opening ceremony, 11,000 middle-class football fans booed at President Miguel de la Madrid (Walker, 2009), an indication of the profound conjunctural shift that had been brought about through popular mobilization after the earthquake. The earthquake also gave rise to an important labour movement led by seamstresses many of whom had died trapped in factories by their bosses (Poniatowska, 1995) and to Superbarrio, a masked activist that drew on the symbolism of Mexico's *lucha libre* (wrestling) tradition (see Figure 9.1). Superbarrio began to turn up whenever people were threatened with eviction, to defend and to testify, and with indignation and humour juxtaposed "the symbolic power of suits, ties, titles and other symbols of distinction that bureaucrats, politicians, and the upper class use, with the symbolic power of popular superheroes" (Cadena-Roa, 2002: 209). This mobilization would be one of a number of factors

Figure 9.1 Superbarrio appears in Mexico City

Source: Gobierno CDMX, CC0 1.0 released into public domain, via Wikimedia Commons

leading to the gradual political decline of the governing party, the PRI, over the coming years, which started with the 1968 massacre of protesting students and culminated with their dramatic removal from office in 2000 after 70 years in power.

On 19 September 2017, the 32nd anniversary of the 1985 quake and a day dedicated to both commemoration and earthquake preparedness drills, Mexico City and Puebla were struck by another deadly earthquake. As in 1985, the magnitude of the disaster exceeded the capacity of the state to respond, and so once again it was ordinary people who mounted the relief effort in their communities. Destructive earthquakes test the citizens and the state. In both quakes, that of 2017 and 1985, we saw "the best of human beings and the worst of governments" as the citizenry responded with "compassion, empathy, loyalty, brotherhood, friendship and solidarity" while the government responded with "ineffectiveness, lethargy and disorder" (Viveros-Wacher and Kraus-Weisman, 2018: 106, my translation). Indeed, the same capacity to organize, mobilize, feed, heal, denounce, and show solidarity that Mexicans repeatedly display in political action seems to be the best form of disaster preparedness and response.

In an article that connects disaster studies and urban studies, Jess Linz (2021) shows how although the 1985 earthquake in Mexico City produced democratizing forms of social mobilization and solidarity, it also laid the groundwork for subsequent gentrification and displacement in neighbourhoods such as Roma, Condesa, and Juárez, that became dominated by corrupt real estate speculation. Drawing on the work of Lauren Berlant, she describes the 2017 earthquake, as an affective impasse. The earthquake, which suspended capitalist business as usual, transformed the city into a relief operation and brought new smells of dust, gas, and death, provided a moment of political possibility, an opening, and a chance to intervene in and disrupt the process of gentrification that produces housing injustice, eviction, and displacement. The inevitability of gentrification was called into question as "the earthquake had shaken loose the narratives governing urban life under gentrification" and constituted "a reminder that these narratives were only ever provisional to begin with" (p.290). It was a process shaped by the cultural memory of the 1985 event. 1985 was doing "affective work" in 2017 "stirring city atmospheres" and making urban political transformation palpable (p.296).

The dominant hazards paradigm does not only fail because it doesn't take into account the political ecology of the disaster, it also does not and cannot accommodate Indigenous ways of knowing earthquakes and volcanoes. Entities that for Eurocentric geologists are merely "natural hazards" that can be managed through a technocratic and highly neoliberalized idea of "hazard/disaster management" are subject to quite different understandings within Indigenous cosmovisions. Volcanoes are also spiritual, sacred, and agentic beings that are important repositories of meaning (Anderson, 2011), while earthquakes are often understood as omens that need to be heeded (see Box 9.2). In precolonial times, the Pinchincha Volcano in Ecuador, as Tamara Estupiñán Vitero (1998) writes, was the object of a deep religious veneration and was used as a tomb where the remains of Incan leaders would be interred. After the conquest, Pinchincha began to be known through Catholic lenses and then scientific ones. Humboldt, Reiss, Stübel, Wolf, and Dressler were some of the first European scientists to study Pinchincha. Indigenous and scientific ways of knowing volcanoes co-exist. Eurocentrism retains the colonial fantasy that nature is something that can be controlled or dominated by man (the gendered pronoun here is deliberate) through monitoring or engineering, while Indigenous knowledges tend to contest the idea of human mastery of nature and work instead with a "relational ontology marked by contingency and connection" (Williams, 2008: 1128, see Chapter 4). In the colonial era, there was, however, some overlap between Indigenous spiritualities and Catholic religious understandings, as both admit the role and agency of supernatural beings, something that western science struggles to accommodate.

We also need to develop an understanding of the epistemological consequences of the different modes of resistance and political action that disasters engender. In an article on Hurricane Felix in Nicaragua and Hurricane Katrina in the United States, we emphasized that disaster events don't come equipped with any necessary and essential meanings, but instead unleash a hegemonic struggle over what and how the disaster is made to mean

(Cupples and Glynn, 2014a). In post-Katrina New Orleans and post-Felix Bilwi, this struggle was able to disrupt some of the ideologically conservative and neocolonial meanings that tend to accompany disasters and put an alternative set of discourses into circulation that contributed to the affirmation of alternative ways of being and knowing that negated coloniality and neoliberal rationalities. This kind of work can help us to understand the ways in which disasters perpetuate coloniality and facilitate particular forms of decolonial resistance that destabilize dominant ways of knowing disaster.

THE COLONIALITY OF DISASTER

My own intellectual engagements with the politics of disaster, including Hurricane Mitch in 1998, Hurricane Felix in 2007, and the eruption of the Fuego Volcano in 2018 (Cupples, 2004, 2007, 2012; Cupples and Glynn, 2014a), have underscored the importance not only of disrupting through cultural, feminist and ethnographic methods the dominant hazards paradigm that see disasters as one-off, inevitable and natural events that need to be managed, but also of understanding the ways in which disasters are underpinned by racist colonial logics that deem some people and places as less worthy of saving. Anthony Oliver-Smith's (1986, 1994) work attributes the devastation created by the 1970 Peruvian earthquake which buried the city of Yungay to a process that began 500 years earlier with the conquest, one that included the failure of the colonizers to learn from Indigenous peoples' sophisticated anti-seismic practices. People die because we don't implement the knowledges that we need to keep us safe, in the way that pre-conquest Indigenous peoples did, *and* because those in power today don't care enough about the lives of the poor and vulnerable and see them as disposable.

As global disaster losses have continued to increase in a conjuncture characterized often by revanchist and extractivist neoliberalism, many governments, researchers, and development practitioners have embraced the concept of resilience as a means to address growing vulnerability to socionatural hazards and disasters. Resilience does not have fixed meanings – and there is no doubt that many Latin American disaster survivors often display traits that we might call resilience and this resilience enables them to cope, help others, rebuild, and lobby governments or NGOs to respond. Resilience is, however, perhaps often too easily articulated to neoliberalism and to the withdrawal of social safety nets and in that sense, it can be a very dangerous concept. Within this discursive regime, a resilient community is one that displays self-reliance and doesn't need support. And so if people are not resilient enough to survive, cope, or rebuild, it is then easier to blame them, rather than analyse the conditions of structural violence in which they live and which exacerbate the effects of a hurricane or earthquake. A focus on the coloniality of disaster is a useful way to avoid such theoretical pitfalls. While the concept of natural disaster puts the blame on nature (it was simply a very large earthquake or hurricane), resilience puts the onus on individuals or communities to respond appropriately. We need to ask instead how colonial dispossession, racism, and state-led violence turn a hazard into a major disaster. This approach means making connections between disaster losses and land dispossession, structural racism, and imperialism (see Sealey-Huggins, 2018; Lloréns, 2018) and analysing the ways in which disasters are taken advantage of by capitalist actors to further dispossess or make extraordinary profits at the expense of the victims, a phenomenon that Naomi Klein (2007) calls disaster capitalism. This book has outlined many instances of state-led and state-endorsed violence in Latin America, where poor people or those racialized as Black or Indigenous, are displaced, criminalized, or exposed to premature death for defending their lands or their ways of life. So why would the same colonial structures of power seek to protect these lives when affected by an earthquake or a hurricane?

Indeed, we frequently see the denial of humanity on the part of the power bloc in disaster events. In 2007, a group of mainly Indigenous Miskito lobster divers and female traders, known as *piquineras*, were working out in the Miskito Keys in the Caribbean Sea when the Nicaraguan government was warned about the approach of Hurricane Felix. There was absolutely no government-led evacuation of the area, even though the Nicaraguan navy frequently patrolled those waters in search of drug traffickers. Indeed, the authorities stated that the people had refused rescue, something that is contested by the survivors. Hurricane Felix devastated the keys and those that survived say the sea fill with corpses (see Cupples, 2012). Drawing on Agamben's notion of the state of exception and Foucault's work on biopolitics, I noted:

> As Felix pounded its ferocity onto the Keys, the divers and *piquineras* who were abandoned out there and deprived of protective state biopower, found themselves trapped in between what Agamben (1998) refers to as zoe (humans as bare life) and bios (humans as valued citizens), a zone of indistinction between life and death.
>
> (Cupples, 2012: 22).

A few years later, I witnessed the same kind of colonial logics applied to the people of San Miguel Los Lotes in Guatemala. In June 2018, hundreds of people in this community died buried by a rapidly moving pyroclastic flow when the Fuego Volcano erupted (see Figure 9.2). Across the road was a luxury resort and golf club that managed to evacuate all of the foreign tourists and middle-class Guatemalans before it was buried in volcanic material. It is clear that CONRED, the state organization that charged with disseminating information about risk and knew that this eruption was serious, failed to communicate the warnings to the people of San Miguel Los Lotes. We can see further examples of colonial attitudes in the official disaster response. The Guatemalan government has refused to properly count the dead and it

Figures 9.2 San Miguel Los Lotes buried in volcanic material after the eruption of the Fuego Volcano

Source: Photo by author

privileged repairing the main road from Antigua to Guatemala City that was damaged by volcanic material rather than exhuming the remains of those buried by pyroclastic flows so that their families could give them a proper burial. Our ongoing work in Guatemala is exploring the modes of articulation between the volcanic disaster and the violence of the civil war, both of which resulted in a popular clamour for exhumation. As we noted above, disasters and the deep social inequalities that underpin them lead to resistance and struggles for justice and these often have decolonial motivations. Let's turn now to post-María Puerto Rico where some of the best recent scholarship on the coloniality of disaster has been done.

HURRICANE MARÍA IN PUERTO RICO

Unlike most of the rest of Latin America, Puerto Rico never became independent after the end of Spanish colonialism in 1898[1] but has been occupied by the United States for almost 125 years. In that period, Puerto Ricans have had to confront a range of colonizing interventions that have produced a high level of precarity and vulnerability on the island. Constitutional protection was not conferred until 1917 and was done so in part to suppress the demands for independence. Throughout the 20th century, independence movements were repressed, often brutally. As US citizens, Puerto Ricans have the right to live in the United States – they don't have to cross the desert illegally like Mexicans and Central Americans and many of them have done so. As we shall see, this migration is, however, often not a free choice and is not necessarily less painful. The Puerto Rican population that resides in the United States has been there for a long time (see Box 5.3), but it has grown in recent years as a result of austerity politics and the effects of Hurricane María.

The 20th-century US colonization of Puerto Rico involved opening up the island to US investors who would be attracted by low taxes and cheap labour (Operation Bootstrap) as well as the mass sterilization of Puerto Rican women driven by a discourse of overpopulation and a logic of elimination (Gutiérrez and Fuentes, 2010). A gag law passed by the US Congress in 1948 meant that Puerto Ricans were not allowed to write about independence, sing a pro-independence anthem, or display a Puerto Rican flag. Nine years later, the law was declared unconstitutional by the Supreme Court and repealed but the FBI continued to carry out surveillance of independence sympathizers. In 1952, Puerto Rico was declared a Commonwealth, a move that did enhance the rights of Puerto Ricans but also further opened up the island to US cap- italist penetration. Some US entrepreneurs made vast profits that were not reinvested in the island, or they left behind environmental pollution. Juan Declet-Barreto (2020: 252, my translation) describes this move as "a colonial arrangement left intact but disguised as democracy." Puerto Rico would go on to lose much of its comparative advantage in the 1980s when structural adjustment-mandated trade liberalization opened up other Latin American economies to foreign investment. As a result, Puerto Rico started to borrow heavily and by the early 2000s had taken on an unsustainable debt burden which resulted in harsh austerity measures being imposed. These measures included quite drastic cuts to the health and education budgets.

By 2015, Puerto Rico had an unpayable US$72 billion debt and the US congress passed PROMESA (Puerto Rico Oversight, Management, and Economic Stability Act), a law designed to restructure the debt but in a way which involved the removal of Puerto Rican autonomy. The law is overseen by an unelected body of federal officials who do not have to answer to the local government or to the residents and it enjoys full legal immunity; indeed locals call the committee "la junta" (Rebollo Gil, 2018). Because Puerto Rico is neither a US state nor an independent nation-state, it can't refinance or default. This law involved more austerity, more cuts to public spending, more unemployment, and therefore more precarity. The result of these interventions has been massive, forced migration to the mainland. Guillermo Rebollo Gil (2018) outlines how debt management in Puerto Rico is accompanied by a discourse which focuses on the inability of Puerto Rico to manage its own finances responsibly, it is a "self-effacing discourse" that "evades the fundamental issues of US colonialism and thus, of US responsibility for Puerto Rico's dire socio-economic predicament" (p.40).

This was Puerto Rico's colonial condition in 2017 when Hurricane María hit the island. María was a category 4 hurricane that destroyed homes, roads, bridges, water supply, electricity, internet, and cell phone service. Some people lost all their belongings. While the official death toll stood at 64, independent sources put the death toll at over 1,000, while a team of researchers in the United States and Puerto Rico documented 4,645 excess deaths in the three months since the hurricane made landfall, many of which can be attributed to the lack of access to health care (Kishore et al., 2018) or to the fact that the lack of running water resulted in many Puerto Ricans consuming water from toxic wells (Bonilla, 2020).

After more than a century of union with the United States and given the scale of the devastation that Maria left behind (see Figure 9.3), the US citizens living in Puerto Rico might have expected an efficient disaster response. But as Yarimar Bonilla (2020) notes, many people remained without electricity, running water, or internet for months and some people just headed to the airport to migrate to the United States, abandoning their cars in the car park with the keys left in the ignition for easy removal. Those that remained struggled without electricity and growing insecurity. In the weeks

Figure 9.3 Puerto Rico: the morning after Hurricane María

Source: Roosevelt Skerrit, released into the public domain, via Wikimedia Commons

after the hurricane, people waited for the state to clear up the mess and provide drinking water and when aid did arrive, it was slow, inadequate, and pitiful (Bonilla, 2020: 3). While they waited, they set up their own drinking water systems, formed community kitchens, and built their own mutual aid networks.

Puerto Rico's colonial condition was revealed to the world and to Puerto Ricans themselves. The disaster response "revealed, in an embodied way, the racialization of Puerto Ricans as colonial subjects" (Ficek, 2018: 102) forcing them "into an affective reckoning with the kinds of structural violence they had been enduring for decades" (Bonilla, 2020:1). For one Puerto Rican activist, "[v]ery little about the hurricane is natural. This is mass murder" (Llenín Figueroa, 2019:97) and the responsibility lies with the United States "that has built itself on the basis of our misery" (p.99).

As Puerto Rican scholars and activists emphasize, what is going on in Puerto Rico is not a recovery, it is a continuation of the colonial disaster, where powerful forces are seeking to return to the status quo of extraction and exploitation to allow colonial capitalism to continue (see Klein, 2018). By 2019, Puerto Ricans had brought down a corrupt governor in massive protests. They began not only to rebuild and help each other but also to address the root causes of the disaster through art, protest, and analysis. A collection of proliferating and articulating political movements see the "need for decolonization to serve as the centerpiece of a just recovery for Puerto Rico and the Caribbean as a whole" (Bonilla and LeBrón, 2019a: 16). Bonilla (2020:9) cites Puerto Rican scholar and filmmaker Mariolga Reyes Cruz (2018), who sees these struggles as "gestating sovereignties" of different kinds; "food sovereignty through back-to-the-land movements, territorial sovereignty through land occupations and community trusts, energy sovereignty (*soberanía energética*) through solar power initiatives, and even political sovereignty through a reimagining of the means and forms of collective action." These movements are based on horizontal modes of decision making, autonomous modes of organizing, and directed not at personal profit but at the common good (Reyes Cruz, 2018; see Bonilla and LeBrón (2019b) for examples of these social movements). This vision is, as she asserts, underpinned by decolonial ambitions.

> Now, we have another possible future to forge. It will require tons of courage, creativity, humility and love to weave the framework capable of strengthening our interdependencies, from diversity, sustainability and democratization. It will be an arduous road, full of contradictions, experiments, doubts and hopes. Hopefully in the process we can become more Aymara and less western so that "we can all go together, that no one is left behind, that everything reaches for everyone, and that no one lacks anything." There is not much time left.
>
> (Reyes Cruz, 2018, my translation)

COVID-19 IN LATIN AMERICA

This book was written during a global disaster that is still unfolding and that hit Latin America very badly. Like other disasters, it has both revealed and exacerbated coloniality and it intersects with ongoing struggles for democratization, dignity, and decolonization in geographically specific ways. From the Eurocentric world of public health and political disease management driven by datafication and biopolitical concerns, we repeatedly hear that we are living through an unprecedented situation and that we will soon return to (some kind of) normality. The Indigenous people of the Americas whose ancestors died in the many pandemics brought by the conquest would be the first to tell you that there is absolutely nothing unprecedented about the situation we are living through now, expect perhaps that it has restricted the freedoms and mobilities of affluent Europeans.

Along with evangelization and slavery, the European conquest brought disease, including influenza, smallpox, measles, and typhus. As George Lovell (1992, 2020) writes, smallpox wiped out the Indigenous population of Hispaniola and decimated the Aztec population of Teotihuacán. The Andes fared equally badly. As Lovell (2020) shows, the Indigenous peoples carefully documented these experiences. As the Mayan book of Chilam Balam of Chumayel tells us:

> There was then no sickness. At that time the course of humanity was orderly. The foreigners made it otherwise when they arrived here. They brought shameful things when they came.
>
> (*The Book of Chilam Balam of Chumayel* 1782,
> cited in Lovell, 2020)

A similar experience is also recorded in the Annals of the Cakchiquels:

> Great was the stench of the dead. After our fathers and grandfathers succumbed, half of the people fled to the fields. The dogs and the vultures devoured the bodies. The mortality was terrible.
>
> (Recinos and Goetz, 1953: 115–116, cited in Lovell, 2020)

We don't know for sure how many Indigenous people died in pandemics during the colonial era, as many were killed in conflicts with colonizers and as a result of appalling working conditions in mines or plantations. But one estimate is that disease and violence combined led to the deaths of about 50–60 million native Americans (see for example Koch et al., 2019).

While pandemics are nothing new, contemporary Latin Americans are particularly vulnerable to a global pandemic such as Covid-19. The continent is still trying to "contend with its epidemiological backlog, eradicating traditional diseases linked to poor living conditions and inadequate nutrition as well as making headway against the ills of modern society such as violence, AIDS, and heart disease" (Franko, 2019: 465). The failure of development means that in Latin America there is still hunger and malnutrition, as well as diabetes and obesity. Two US Americans of Mexican descent are part of a movement "utilizing ancestral knowledge to help communities of color respond to the public health crises and the decimation of our food system's foundation, brought upon us by the industrialized, Western diet" (Calvo and Rueda Esquibel, 2015: ii), encouraging Latin Americans in the United States to cook and eat like their grandmothers did. Their cookbook *Decolonize Your Diet* is based on the fact that ancestral Mexican and Central American foods – beans, squash, and corn of many colours and varieties, wild herbs, chiles, tomatoes, avocadoes, cacti – were enormously healthy. The colonizers brought diseases such as smallpox and measles, but they also brought wheat, sugar, beef, cooking oils, and cheese, the basis of many of today's so-called lifestyle diseases.

It is clear though that ill health and suffering are not randomly distributed. In Europe and the United States, as well as in Latin America, you are much more likely to get infected and die from Covid-19 if you are poor and/or Black. While Latin American countries are more used to dealing with disasters, including pandemics, than many parts of the Global North, the ability to contain the pandemic is limited by the extent of informality especially in the urban labour market, by overcrowded housing, by the absence of social safety nets, by inadequate and overstretched health systems, and by poor data gathering systems that compromise the ability to track and model the disease. As Stefania Milan et al. (2021: 26) write, with respect to the global datafication of the pandemic, the lack of reliable data within Latin America might well lead to the universalization of the problem and the import of models from other parts for the world that are not designed for Latin American realities, creating "an explosive cocktail that is likely to create more problems than it solves." While the data on deaths and infections that are published daily might be reliable enough for decision-making in many countries, they are certainly not reliable in others. While Figure 9.4 gives some indication of the impact, it is important to recognize that the data are not at all accurate. In Nicaragua, many deaths of people with Covid-19 symptoms are attributed to other causes such as pneumonia or are simply not counted at all (Cupples, 2020).

Latin American countries have had very varied experiences of Covid-19. Authoritarian populism on the right and the left has played a key role in allowing the virus to spread. Brazil under President Jair Bolsonaro stands out as having managed the pandemic atrociously, but Andrés Manuel López Obrador (AMLO), president of Mexico, Nicolás Maduro, president of Venezuela, and Daniel Ortega, president of Nicaragua have all underplayed the seriousness of the virus, displaying and encouraging social contact. Both Bolsonaro and AMLO contracted the disease and have presided over some of the worst death

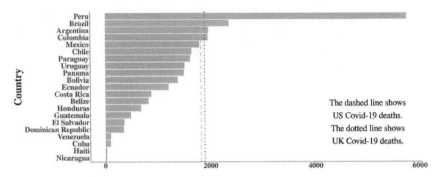

Figure 9.4 Cumulative Covid-19 deaths per million people, until 19 June 2021

Note: Limited testing, failure to register Covid-19 deaths, and poor data collection in some Latin America countries means that the number of deaths is not accurate

Source: Data from https://github.com/owid/covid-19-data/tree/master/public/data and compiled by Ruben Vine

tolls in the world. By June 2021, more than half a million Brazilians had died from the disease, the second highest number of deaths in the world after the United States. Indigenous groups in the Amazon believe they are facing extermination as coronavirus spreads through their communities (Philips, 2020). In April 2021, former president Luiz Inácio Lula da Silva, just after the Supreme Court overturned his conviction on corruption charges, described Brazilian deaths from Covid–19 as "the biggest genocide in our history" (Blasberg et al., 2021).

While the situation in Brazil is horrendous and has received much international media attention, in per capita terms Peru has had many more deaths than any other country in Latin America. In both countries, there is a real concern that the pandemic will be used to further intensify and accelerate extractivism (see Chapter 4). There is evidence that the Peruvian government is using the pandemic to expand extractivist operations and suspend and weaken the protections for environments and people living on targeted lands (Greenfield, 2021) and that in Brazil loggers are using the cover of the pandemic to accelerate deforestation (Democracia Abierta, 2020).

Ecuador has also faced a challenging situation, especially in the coastal city of Guayaquil. In 2020, the health system collapsed and people started dumping the corpses of coronavirus victims in the street. As Daniel Borja and Viviana Buitrón Cañadas (2020) write, people in Guayaquil were saying we can't go back to normality because "normality" was the problem (*ya no podemos volver a la normalidad porque la "normalidad" era el problema*). Guayaquil has a high degree of housing informality and precarity and is riven with inequalities. In some parts of the city, the inhabitants face extraordinary levels of overcrowding and unequal and inadequate access to basic services, while in others, such as Puerto Azul, they have luxury homes, a yacht club, and excellent services including private health care (Borja and Buitrón Cañadas, 2020)

The Nicaraguan government's response to the pandemic has been nothing short of absurd, especially given the extent of household overcrowding, urban precarity, and the weakness of the health system. Rather than implementing measures advised by the public health experts and the WHO, the government publicly rejected any kind of quarantine or lockdown, the borders remained open, as did businesses, markets, government departments, schools, and universities. Furthermore, the government continued to encourage and organize mass gatherings while failing to collect and publish accurate data on infections and deaths. One of the most ludicrous of such events was a march organized for 14 March 2020 under the banner of Love in the Time of Covid-19, featuring carnival-style floats, in which state workers in the capital were forced to participate. In 2018 and 2019, the Nicaraguan government severely repressed a series of popular uprisings – more than 300 people were killed and many more incarcerated – but it denied all wrongdoing and worked hard to feign normality. It appears that the management of the pandemic is consistent with and a continuation of that behaviour (see Cupples, 2020).

Cuba, on the other hand, appears to have done much better, in spite of the US economic blockade that it has endured for the past six decades. The Caribbean island nation is dealing with a severe economic crisis induced first by the re-intensification of the blockade during the Trump presidency and then by the pandemic which has led to the collapse of international tourism. Foodstuffs and ordinary drugs such as paracetamol are in short supply. But Cuba has good disaster management. As mentioned in Chapter 5, it has supported many international humanitarian healthcare efforts around the world, including in Western Africa during the Ebola crisis. Cuba's relative success can be attributed in part to its superior public health care system, one of the hallmarks of the revolution, but also because of it responded rapidly to the crisis, putting plans in place before anyone on the island tested positive. As Helen Yaffe (2020) writes, Cuba put in place strict border controls and airport testing; it immediately began to work on developing testing and vaccinations, and financial support was provided for workers. Furthermore, Cuba has developed two vaccinations of its own, Soberana 2 (Sovereign) and Abdala (named after a poem written by José Martí) that are proving to be highly effective in clinical trials (see Burki, 2021). These have taken place on the island but also in Iran, that has much higher rates of infection.

DECOLONIAL ANALYSIS

While the pandemic is still unfolding and might be with us for many years to come and any analysis might be premature, it can be analysed through a decolonial lens. The pandemic has revealed the coloniality of power on a global scale. There are extreme racial inequalities in the Global North and Global South in terms of who is more likely to get infected and to die from the disease. Indeed, the poor in Latin America first contracted the disease from the affluent who had returned from trips to Europe (Graham–Harrison, 2020). While Europeans started to received vaccines as soon as they were developed and approved, vaccination rates in Latin America have been desperately slow. As of June 2021, Haiti had still not received or administered a single vaccination (Wyss, 2021). It is clear that poverty and inequality exacerbate the effects of the pandemic and the pandemic exacerbates poverty and inequality. Meanwhile, in the absence of good government guidance and support for self-isolation, people are taking measures into their own hands, staying home where possible, wearing masks, meeting outdoors, closing communities to outsiders, creating mutual aid networks, and encouraging others to take care (see Figure 9.5).

Figure 9.5 A pharmacy in Rabinal, Guatemala raises awareness of Covid-19 precautions

Source: Photo by Alejandra Colom, reproduced with permission

According to Marcos Scauso et al. (2020), the othering tendencies of liberalism that attribute ideal subjectivities to some people while racializing and gendering others in ways that exclude have been exacerbated by the current crisis. Just like Hurricane María in Puerto Rico, Covid-19 has exposed the forms of social and material injustice and the forms of gendered, classed, and racialized othering linked to coloniality and neoliberalism that are concealed by discourses of progress and democracy. Indeed, in Puerto Rico, people have recognized that it is its colonial subordination to the United States that has left it unable to cope with large disasters, whether they are triggered by hurricanes or coronavirus. And the two disaster events intersect and interact. Many of those who migrated after María were health workers, which has further weakened the ability of the health system to cope with the pandemic. As a result, Puerto Ricans have started calling the virus #coloniavirus (Declet-Barreto, 2020). Nonetheless, disaster capitalism in San Juan and elsewhere in Puerto Rico continues apace and the ideal liberal subject – who can work at home, self-isolate, wash hands regularly, stockpile food, and socially distance – remains hegemonic. Only Latin Americans that enjoy middle-class privilege can occupy that mode of being. It is therefore accompanied by an epistemic violence where those that can't – especially the urban poor, who do not always have reliable access to running water, rely on informal sector work and live from hand to mouth – are constructed as irresponsible spreaders of the disease. It is also a colonizing discourse because it conceals the forms of predatory capitalism that include a disregard for non-human animals and non-human forms of life that have allowed the virus to flourish.

De Sousa Santos (2020b) has suggested the pandemic provides us with a cruel pedagogy. He notes that since the 1980s, when neoliberalism became the dominant form of capitalism, the world has been in a permanent crisis. Of course, the idea of permanent crisis is an oxymoron, as a crisis is conventionally understood as something exceptional and temporary or short-lived. When a crisis is temporary, it can be explained by the factors that contributed to it, but when a crisis is permanent it becomes something that explains everything else. We can see how the pandemic – like neoliberalism in general – is used as an excuse to make cuts to public budgets and bring about reductions in wages. That is happening in the Global North as well as in Latin America. But de Sousa Santos asserts that proponents of neoliberalism don't actually want to resolve the pandemic because it enables them to legitimize the scandalous distribution of wealth in the world and to boycott any necessary measures that would prevent ecological collapse. We should therefore stop thinking of the pandemic as a crisis that is in direct opposition to normality.

De Sousa Santos also recognizes that the pandemic has generated dramatic changes in our lives and has enabled some of us to stay at home, spend more time with children or families, and consume less. An alternative to hypercapitalism becomes visible. But at the same time, for those of us with privilege or relative security, it has underscored the fragility of human life.

Many people already lived with insecurity – especially in low-income and informal urban neighbourhoods and in communities that have substantial mineral resources that are appropriated by extractivist practices – but now even those in the Zone of Being (see Chapter 11) have started to feel insecure, as businesses collapse and jobs disappear.

De Sousa Santos suggests we look at the pandemic allegorically. It is invisible and yet all powerful, and in this sense, it is like the market that is omniprescent and involves mutations that can't be predicted. Both the virus and the market make lives fragile and can result in premature death. Indeed, since the 17th century, the world has been dominated by what he calls three unicorns – capitalism, coloniality, and patriarchy – and they produce a common sense that is contradictory. Capitalism affirms that all people are equal while patriarchy and coloniality assert that there are natural differences that mean that the equality of the superior human is not equal to the inferior human beings. The thing about these three unicorns is that they cannot dominate alone. They are only powerful because they function together. But they have consequences – and these consequences are extreme social inequality (fuelled by racism, misogyny, transphobia and so on) and imminent ecological collapse.

The pandemic for de Sousa Santos contains four important lessons. The first is that the climate crisis just as serious as Covid-19 but doesn't take up media attention and political debates in the way the pandemic does, even though the pandemic is a manifestation of ecological crisis. Second, the pandemic can kill anyone, but it does not kill indiscriminately. Third, as a development model, capitalism has no future. Austerity and neoliberalism have weakened our ability to confront the pandemic and it will be followed by more austerity and more neoliberalism and there will be more pandemics and even weaker ability to respond. Finally, it is clear that coloniality and patriarchy are alive and are being strengthened at the current time. In 2019, just before the pandemic, Latin America had exploded in mass protests as people were contesting the ways in which coloniality, capitalism, and patriarchy were producing human suffering and exclusion and planetary destruction. De Sousa Santos believes it is likely that when this phase of the pandemic comes to an end, governments will try to reduce people's expectations, the protests that we saw in 2019 will resume, and they will be repressed. As the time of writing, this is exactly what is happening in Colombia. In May and June 2021, after a national strike was called, many protestors were killed, injured, and disappeared by state forces, especially in Cali. As Betty Lozano Lerma (2021) writes, the government of Iván Duque

> is treating legitimate protest guaranteed by the constitution as a war. Duque has given the army command in what is already a self-coup, deploying disproportionate military force against young people who are only armed with their dreams and rocks.

In this conjuncture in which precarious life is further precarized, we need to bring about an epistemological, cultural, and ideological turn in support of policies that will allow human life to continue on the planet. If not, we are facing a situation where life continues but human life does not. As Ailton Krenak (2020: 23) puts it, the pandemic is forcing us to abandon our anthropocentrism, it makes clear that biodiversity doesn't need us and we will become one of the many species at risk of extinction. The virus is not infecting the birds or bears and "the bitter melon continues to grow at the side of the house" (p.23). De Sousa Santos is clear about what we need to do and that is:

> to create a new common sense, the simple and obvious idea that, especially in the last forty years, we have lived in quarantine, in the political, cultural and ideological quarantine of a capitalism closed in on itself, as well as in the quarantine of racial and sexual discrimination without which capitalism cannot survive. The quarantine caused by the pandemic is, after all, one quarantine within another. We will overcome the quarantine of capitalism when we are able to imagine the planet as our common home and nature as our original mother to whom we owe love and respect. It does not belong to us. We belong to her. When we overcome that quarantine, we will be freer from quarantines caused by pandemics.
>
> (p.28, my translation)

Segato's (2020: 16–17) reflections on possible postpandemic futures underscores the uncertainty we face and the recognition that these futures could be terrifying or so much better. As she writes, the pandemic might shatter the neoliberal allusion and emphasize to neoliberal actors that we need to put life before capital accumulation. Or we might end up in a dystopian world of surveillance and digital control that will destroy our humanity. Or authoritarian forces might use it to embrace social Darwinism in which the virus is allowed to eliminate the weakest in society. It might intensify fascist violence against anyone perceived to be spreading the virus, such as immigrants or marginalized urban youth. On the other hand, the virus might emphasize to us that this trajectory of suicidal planetary destruction cannot continue. It might "impose a feminine perspective on the world: retie the knots of communal life based on reciprocity and mutual aid" that is quite different from the "patriarchal, bureaucratic, distant and colonial State" that currently governs us and in which we no longer have any trust. Krenak (2020) draws our attention to how the Eurocentric world refuses to abandon a linear concept of time; in spite of the pandemic, the same people who ridicule him "for acting in communion with what they call nature" (p.27) are living in suspension, postponing activities and projects for when normality returns. In his view, these are the people who are living in the past, as that world might not come back. Like the people of Guayaquil, he hopes that normality does not return.

When we look at the situation in pandemic Latin America, there is clearly much suffering. But the pandemic is also allowing alternative decolonial and democratic imaginaries to come to the fore. People are talking more about the forms of structural violence that puts marginalized populations at greater risk and simultaneously devalues their knowledges. In many cases, the drive for self-government and autonomy that we see for example in Zapatista communities is intensified. People are developing new forms of horizontal solidarity. Let me end with an example from Chile, drawing on the work of César Guzmán-Concha (2020).

In October and November 2019, Chile and especially the capital Santiago was in full blown insurrection. Chileans were tired of rampant inequalities that manifest themselves in elite privilege, poverty, unemployment or underemployment, corporate abuses, insecure housing, gender-based violence, and racism especially towards Indigenous Mapuche. The hiking of metro fares drew multiple accumulated grievances to the surface and resulted in widespread massive demonstrations that lasted for several weeks, despite militarized repression on behalf of the state. Guzman-Concha describes this as a contestation over who had the right to occupy the city, to be in the streets. From 18 October onwards, the protests spread all over city very quickly, often resulting in clashes with the police. People started to demand a new constitution and began organizing in their neighbourhoods to debate what it should include. The protests led to an agreement that there would be a new constitution to replace the one written during the dictatorship that would start with a referendum to be held in April 2020.

By March, however, the pandemic had hit Chile, brought in by wealthy Chileans who had travelled abroad. The government declared a quarantine and postponed the constitutional referendum to 25 October. The virus spread from upper middle-class areas to working class parts of the city and the social and economic costs have been extraordinary. More and more of the population was plunged into severe income vulnerability. But the networks formed during the protests transformed into solidarity networks that provided social and economic support to communities; organizing sanitation brigades to clean communal areas, delivering clothing and medicines, and running soup kitchens (*ollas comunes*) to feed the vulnerable, resurrecting a response to economic crisis that started in the 1930s and was common again during the dictatorship in the 1980s. Once again, and as in Puerto Rico, the people stepped in where the state failed, further eroding the legitimacy of the political class. While the *ollas comunes* got food to hungry people, people were also discussing the question of food sovereignty and the right to healthy food and constitutional referendum. There is a tangible process of democratization and political engagement. In the October 2020 referendum, 78% of Chileans voted to scrap the Pinochet-era constitution and in May 2021, when Chileans voted for the delegates of the new constitutent assembly, the political elite was overwhelmingly rejected in favour of independent and Indigenous delegates. The government's right-wing coalition did not secure enough delegates to block the inclusion of progressive articles (see Bartlett, 2021). In the midst of the global pandemic, Chile achieved something rather extraordinary.

There is then some hope amidst a fairly grim and still unfolding situation. The pandemic has been devastating but as with other disaster events it reveals compassion and solidarity and unleashes struggles for rights, democracy, and dignity. These struggles are accompanied by an interrogation of the levels of structural violence that afflict the continent, and especially Indigenous and Afrodescendant populations and the rural and urban poor. It's clear, however, that more extractivism, more neoliberalism, more coloniality are not the answer. It is also clear that while some of the Eurocentric ways of knowing and responding to the disasters are desperately needed, including in the scientific development of effective vaccines, they are also limited in other ways as they rely on economic growth to get us out of the crisis. It is important that we pay attention to Indigenous and non-Eurocentric ways of knowing disaster and disease (see Box 9.2). Unfortunately, the voices of capitalist developmentalism remain hegemonic. These are the voices that assert that the pandemic is unprecedented, has nothing to do with predatory capitalism, and that we will soon return to normality – a normality that values economic growth over wealth distribution or ecological protection.

BOX 9.2 COVID-19 FROM A MAPUCHE PERSPECTIVE

Covid-19 is a global phenomenon that is being managed by drawing on Eurocentric science. It is constructed as something biological that affects human beings as individuals, that is disconnected from the broader trajectories of capitalism, and that requires a public health (biopolitical) response. But many Indigenous peoples do not understand disease in that way. For example, according to Gustavo Esteva (2019: 503), "the Rarámuri and the Yucatán Mayas, do not have words for health and disease. A perturbation in communal relationships may manifest itself in the form of an earthquake or the pneumonia of don Rafael. If don Rafael gets pneumonia, the community will not look for the condition of that individual patient. They will ask why the community got 'sick' in don Rafael." Similarly, as Mapuche intellectual José Quidel Lincoleo (2020) writes, according to *Mapuche rakizum* (Mapuche thought), we as a species are entangled with a multitude of other lives and so the virus is not an isolated phenomenon but rather results from a series of interconnected planetary events in which human and nonhuman species are implicated. While western politicians and public health experts constantly refer to the pandemic as something unprecedented, Mapuche oral culture has repeatedly circulated fears of catastrophic events that would involve mass sickness. Mapuche and especially the *machi* (see Chapter 6) are also attuned to the *webe* or the omens brought by birds, winds or solar or tectonic phenomena that draw attention to the disruptions to come. They knew the pandemic was coming because it was predicted in the solar eclipse and in Chile's recent devastating earthquake. The eclipse and the quake both foreshadow a cycle in which *küme mogen* (*Buen Vivir*) will be undermined. The flowering of the *caña coligüe* (Chilean bamboo) also indicated bad times ahead. So the arrival of the pandemic is therefore something that has been discussed and was anticipated. Furthermore, such events happen because of transgressions. In Chile and elsewhere in Latin America, forests have been cleared and rivers, seas and lakes have been polluted, which has made the *geh*, the spirit owners, very sad and angry and therefore those responsible need to be sanctioned. Those responsible are usually not Indigenous people; they are often people who work for transnational corporations who do not see the cultural and spiritual meanings attached to these entities but regard them instead as sites of extractivist profit from which the original inhabitants can be eliminated. They displace people and other species to create mines, plantations and hydroelectric dams that leave behind a trail of contamination and destruction. These extractivist practices are connected to the current predicament. The pandemic has emerged because of the disrespect some people display towards other living beings, both biological and spiritual ones. Furthermore, the Mapuche are horrified at the current discourse that asserts that Covid-19 is a disease that mainly affects elderly people and so we shouldn't undermine the capitalist economy or the lives of young people in order to protect them. In Mapuche culture, old people are highly valued as key repositories of knowledge; they are needed to guide younger people and therefore must be accorded the utmost care and protection. For the Mapuche too, there can be no return to normality, as we are the beginning of a very dark cycle in planetary history in which many terrible things will happen.

Source: Quidel Lincoleo, 2020

SUMMARY

- Disasters are not singular, one–off, natural events but are ongoing processes that are embedded in existing relations of power and (colonial) histories. There is no such a thing as a natural disaster.
- We need to understand the unfolding of disasters in the context of diverse forms of coloniality, including debt crisis, austerity, forced migration, and financial incentives for wealthy investors.

- People die and suffer not because of the magnitude of earthquake or hurricane, but because of structural exclusion and negligence and because of the ways in which capitalists and politicians capitalize on human misery (disaster capitalism).
- Disasters frequently unleash important struggles for decolonization by artists, activists, intellectuals, and ordinary citizens.
- There are Indigenous ways of making sense of disasters such as Covid-19 which disrupt the dominant ways in which the pandemic is discussed and represented.

DISCUSSION QUESTIONS

1   Why it is problematic to refer to Hurricane Mitch, Hurricane María, or the 1985 Mexican earthquake as natural disasters?
2   What is meant by the coloniality of disaster? Why is it useful to understand disasters such as Hurricane María or Covid-19 through a decolonial lens?
3   What is disaster capitalism and what form does it take in contemporary Latin America?
4   How might Indigenous knowledges help us to build a better world in the wake of the global pandemic?

NOTE

1   This history of Puerto Rico draws in large part on Banuchi and Morales (2018)

FURTHER READING

Bonilla, Y. and LeBrón, M. (eds) (2019) *Aftershocks of Disaster: Puerto Rico Before and After the Storm.* Chicago: Haymarket Books.
An anthology that brings together many Puerto Rican voices on the disaster – how it was experienced, represented, and used to make profits, as well the many modes of cultural and political resistance it engendered.
Poniatowska, E. (1995) *Nothing, Nobody: The Voices of the Mexico City Earthquake.* Philadelphia: Temple University Press.
A text that captures through the voices of survivors the grassroots organizing in the face of government failure in the 1985 earthquake.
Quijano Valencia, O. and Corredor Jimenez, C. (eds) (2020) *Pandemia al sur.* Buenos Aires: Prometeo Libros.
This book is only available in Spanish, but it contains a set of decolonial reflections on Covid-19 and its consequences for the Global South
Sou, G. (2019) *After María: Everyday Recovery from Disaster*

A graphic novel in English and Spanish on the experiences and consequences of Hurricane María in Puerto Rico Available at www.gemmasou.com

Recommended films

Cuba: Living between Hurricanes (2019), directed by Michael Chanan
A film about the impacts of climate change, commodity trade, and tourism on Cuba.
The Battle for Paradise: Naomi Klein Reports from Puerto Rico (2018), directed by Lauren Feeney
A short film that documents the disaster capitalism that has followed Hurricane María and the community activists that are fighting for a different kind of future on the island and for more sustainable forms of development.

Useful websites

NOAA National Centers for Environmental Information
The website of the National Centers for Environmental Information (NCEI) of the National Oceanic and Atmospheric Administration (NOAA) contains searchable and customizable datasets on hazards and disasters associated with earthquakes, volcanic eruptions, and tsunami.
www.ngdc.noaa.gov/hazel/view/about

Puerto Rico syllabus
A list of resources for teaching and learning about the current crisis and social activism in Puerto Rico.
https://puertoricosyllabus.com/

ReliefWeb
A humanitarian information service run by the United Nations Office for the Coordination of Humanitarian Affairs (OCHA) that contains information on humanitarian emergencies around the world, including maps, reports and infographics.
https://reliefweb.int/

CHAPTER 10

The Latin American city

INTRODUCTION

This chapter focuses on the Latin American city as a key site of decolonial struggle. The intention is not to set up a divide between rural and urban political struggles, as the city and the countryside are closely connected. Rural-urban migration has been a constant feature of Latin America for several decades and many of the people who migrate to cities often leave family members behind to work the land. The city is also not disconnected from the rural social movements against extractivist practices that we discussed in Chapter 4. As Anthony Bebbington (2018: 322) emphasizes, the killings of environmental defenders are "related to global and national flows of commodities and finance, so they are flows that articulate the rural and the urban always." In many urban protests, we witness the articulation of extractivism with gender-based violence as outcomes of the same political system (Cupples and Glynn, 2018; Quiroga, 2019).

Urbanization rates are very high in Latin America and 80% of all Latin Americans live in cities (Duarte, 2019; Franko, 2019). There are now more than 50 cities in Latin America with more than a million inhabitants and five with more than ten million; Mexico City (21 million); São Paulo (21 million); Buenos Aires (13 million), Rio de Janeiro (12 million), and Lima (11 million).

Latin American cities have grown very rapidly since the Second World War. While large Global North cities like New York or London took a long time to reach their current size, Latin American cities did so in the space of a few decades. Rural misery generated successive waves of rural–urban migration, especially in the second half of the 20th century, as large numbers of Latin Americans moved to cities in search of employment, education, health care, and a better standard of living. The rate of urbanization was so rapid that urban authorities could not cope with the influx so informal squatter settlements or shantytowns sprang up on the edges of all major cities. People

DOI: 10.4324/9781003110453-10

built their own homes out of whatever materials they could access and gradually improved them over time. These informal settlements usually lacked access to water, electricity, sanitation, garbage collection, public transport, education, and healthcare, which all became issues around which urban populations mobilized and struggled. One in four Latin Americans still live in irregular settlements (Franko, 2019) and more than 115 million have migrated to the city since the year 2000 (Duarte, 2019). In irregular settlements, the *favelas*, *villas miserias*, or *asentamientos informales*, residents are frequently exposed to substantial health risks associated with inadequate or contaminated drinking water, poor sanitation, indoor woodfuel burning, dumping of waste, and substandard housing, as well as stigmatization, criminalization, and police brutality. Respiratory illnesses are common and residents are often forced to buy drinking water at exorbitant prices from vendors who sell water from trucks. Some urban residents literally live and work on garbage dumps. In the State of Mexico, whole families live on the Neza II landfill site in Bordo Xochiaca, where they work to separate materials that can be recycled from those that can't. It is dirty and dangerous work that involves adults and their children breathing in toxins.

The sheer size, density, and rate of growth of many Latin American cities make transport and mobility particularly challenging. Traffic congestion is serious in most Latin American cities as frequently overcrowded buses used by the poor and the young compete for space with the private air-conditioned vehicles of the wealthy. Public transport is intensely politicized too, and hikes in fares often result in protests as we saw in Venezuela in 1989, Brazil in 2013, and Chile in 2019. In recent years, however, there have been some exciting forms of transport innovation that have substantially enhanced the mobility of the poor and reduced polluting vehicle emissions. These include the implementation of bus rapid transit (BRT) systems such as the Transmilenio in Bogotá or the Metrobús in Mexico City. The creation of cable car systems (*metrocable* or *teleférico*) in La Paz, Medellín, and Rio de Janeiro that connect central business districts to popular *barrios* or *favelas* can make decolonizing contributions to spatial justice (Leibler and Musset, 2010; Brand and Dávila, 2011; Freire-Medeiros and Name, 2017; see Figure 10.1). Bogotá now also has a Ciclovía, a network of bike lanes, and cycling activism has been one of the ways in which the bogotanos assert their right to the city as well as their right to mobility (Castañeda, 2019).

Most Latin American cities are sites of poverty, inequality, and violence, juxtaposed with urban affluence and sophistication. While the city is subject to many top-down pressures, it is also constantly being brought into being by bottom-up modes of resistance; these include high-profile and highly visible forms of political protest as well as the micropractices of everyday life that involve tactical subversions of the unacceptable status quo. These also include gang violence and organized crime that require quite nuanced analysis (see Box 10.1). These practices have socio-economic, political, and cultural

Figure 10.1 The metrocable of Medellín

Source: Photo by author

dimensions as they involve the right to a dignified life and livelihood, the right to a life free from violence, and the right to be in the city. Argentina has a long history of marginalized or oppressed people temporarily taking control of parts of the city to make a political intervention and change the meanings attached to that space. For many years, the weekly protest of the Madres de la Plaza de Mayo put the rights of the disappeared on the political agenda where it still remains (see Chapter 6). In recent years, as Natalie Quiroga (2019) writes, Argentinean women have organized a *tetazo* in Buenos Aires in which they bared their breasts to draw attention to the culture of violence that surrounds women's bodies and LGBTQ+ people in Buenos Aires, Asunción, and Bogotá have participated in a *besatón* (a kissing marathon) to assert their rights to express sexual orientation freely and without discrimination.

BOX 10.1 GANG VIOLENCE IN EL SALVADOR

Since the 1990s, El Salvador has been one of the most violent countries in the Americas. For the past two decades, the homicide rate has consistently greatly exceeded the rate that the UN considers to be an epidemic (more than eight killings per 100,000 people) and in 2015 it reached a peak of 103, higher than most countries at war (Martínez and Martínez, 2019). These horrific statistics are frequently attributed by policymakers and media commentators to gang violence and there is no doubt that gang members have committed many brutal killings that often involve ritualistic forms of group communication (Martínez and Martínez, 2019). Central American gangs or *maras* have transnational origins and their development results in large part from US foreign policy. Many Central American refugees that had fled the US-financed and supported civil wars in El Salvador and Guatemala in the 1970s and 80s were drawn to gangs in cities such as Los Angeles in pursuit of a sense of identity, respect, and belonging. It was in Los Angeles that two main gangs emerged; the 18th Street gang (also known as Barrio 18 or B-18 in Spanish) and a rival group, Mara Salvatrucha, also known as MS-13. In 1996, as a result of the Illegal Immigration Reform and Immigrant Responsibility Act, 50,000 gang members were deported to Central America, where their criminal activities continued and expanded (Rodgers, 2019). In desperate economic circumstances, with hardly any employment opportunities, some of them turned to "contract killings, extortions, drug sales and kidnappings" as a means to survive (Méndez, 2019: 378). In the 2000s, the Salvadoran government began to carry out punitive *mano dura* (hard hand) policies in order to stamp out the gangs, measures that were largely unsuccessful and probably made the problem even worse. Indeed, the measures have involved extrajudicial killings of gang members by state forces and have been accompanied by the automatic attribution of violent crimes to the gangs (Méndez, 2019; Wolf, 2012). Furthermore, Sonja Wolf (2012) states that *mano dura* led to gangs strengthening their recruitment criteria and professionalizing their operations. She notes that thousands of gang members ended up in crowded jails where they continued to plan, strategize, and receive support from members of the outside.

In 2012, the leadership of MS-13 and B-18 issued a joint statement to affirm their commitment to finding a solution. While they recognized the role that had played in the violence that had been afflicted, they also emphasized that it is crucial to understand the role played by US foreign policy, the civil war, forced migration, and neoliberal structural adjustment. The statement says:

> In spite of the mistakes we have made and for which more than 10,000 of our members are serving sentences in the penitentiary centres, it cannot be denied that we are also Salvadorans and that we are a social by-product of the disastrous socioeconomic policies derived from the development models that have been implemented in El Salvador for many years. These models led us into war in the 1980s and we are children of this war, because most of our members lost our parents in that conflict. Others of us come from homes that were broken up because we and our parents were forced to migrate and we experienced the uprootedness that came from being displaced from our places of origin.
> (Los voceros nacionales de la Mara Salvatrucha MS13 y Pandilla 18, 2012, my translation)

And it goes on to call into question much of the colonizing and stigmatizing crisis research (see Tuhiwai Smith, 2012) that social scientists have done on the *maras*:

> To those make a living from doing analysis we invite you to rethink the approach you use to analyze our phenomenon. As long as they continue to analyze us only as a criminal phenomenon, your analyses will be wrong as will your recommendations to resolve it. It is necessary that they understand once and *for all that we are a social phenomenon and that the war that we have been forced to wage has socio-economic causes* and therefore its solution is not only legal and repressive but also involves social and economic measures.
> (Los voceros nacionales de la Mara Salvatrucha MS13 y Pandilla 18, 2012, my translation, emphasis added)

While much research on gang-related violence has indeed been very limited in the ways that the gang members themselves recognize, some research has attempted to paint a more nuanced picture of the *maras* that accounts for the structural conditions of urban marginality and exclusion in which they exist and persist. For example, Wolf (2012) believes that their threat to society as well their transnational reach and cross-border organization have been overstated, while Méndez (2019) characterizes gang violence as work that is situated within a broader political economy that includes the colonial legacies of unequal land tenure, environmental destruction, and conflict-driven migration that exposed young Salvadorans to discrimination in the United States. The violence work of the gang members must therefore be understood alongside the violence work of state forces that have massacred and killed with impunity. As brothers Oscar and Juan José Martínez (2019) state in their account of the life and death of Miguel Ángel Tobar, a MS-13 assassin known as the Hollywood kid, "he was the product of a long series of violent acts" carried out by US and Salvadoran politicians and by land-owning and coffee growing élites.

Sources: Wolf, 2012; Rodgers, 2019; Méndez, 2019; Martínez and Martínez, 2019

URBAN INEQUALITIES AND GENTRIFICATION

Latin America's urban inequalities are clearly visible and the poor and the rich often interact on a daily basis. In Mexico City, popular neighbourhoods with high levels of poverty and precarity such as Colonia Granada and Cerrada Andrómaco are surrounded by the flashy air-conditioned high rises of Polanco, often considered one of the city's swankiest neighbourhoods (Aguayo Ayala, 2016). After Mexico's richest man Carlos Slim created Plaza Carso, a luxury commercial, entertainment, and residential space, the real estate boom took off. Public spaces became privatized and fragmented, the pressures on water and sanitation increased, and it is virtually impossible to get around if you don't have a car. As a result of all the tall buildings, the original inhabitants that remain in their flimsy homes are even deprived of the sunshine that they used to enjoy (Brooks, 2019). But in many ways, they are the fortunate ones who have managed to stay even as the city around them is transformed. In other cities, state-led or state-endorsed modes of gentrification and practices of capital accumulation often displace and marginalize lower-income inhabitants and their economic activities.

Gentrification is the process whereby the original working class and sometimes middle-class inhabitants of an urban neighbourhood are displaced by wealthier outsiders and as a consequence the character of a neighbourhood changes. In the global North, this phenomenon has been partly explained by the existence of a rent gap, the difference between current rents and potential rents (see Smith, 1987; Slater, 2015) and usually involves "active entrepreneurial governance from the state as well as a rise in invested capital" (López-Morales, 2019: 504); in other words, important neoliberalizing alliances between state and capital. Gentrification is often violent, involving the privatization of public space, an increase in securitization, verticalization, and forced evictions (López-Morales, 2019) and it generates sharp divisions between real estate developers and land speculators on the one hand and low-income urban residents on the other. In many parts of Latin America, it is driven by tourism that sometimes results in international capital displacing local businesses or the conversion of affordable homes for locals into expensive short-term rentals for tourists.

As Kate Swanson (2019) writes, street vendors who run food or market stalls are often particularly affected by the processes of gentrification, as city councils, excited by the potential growth in revenue, seek to remove them from city spaces, through criminalization and urban cleansing policies. In the process, they get constructed as a kind of visual pollution. Swanson asserts that such actions are especially harmful, as often those who work on the streets are usually more vulnerable populations, including migrants from rural areas or from other Latin American countries or single mothers. But street vendors organize, resist, and fight back to protect their livelihoods and their right to be in the city. For example, as Verónica Crossa (2009) documents, vendors in Mexico City's historic centre were confronted with an entrepreneurial urban governance (EUG) programme, known simply as *El Programa*, led by local politicians and wealthy businessmen such as Carlos Slim, that involved their removal. But the street vendors mounted an effective resistance to the pro-gramme that enabled them to stay. This resistance has included taking advantage of clientelistic relations of power, forming support networks with formal shop owners, spraying the police with tear gas, or becoming "nomadic agents of resistance" or *toreros* (bullfighters), which means they carry their goods on their body and walk around to sell rather than setting up a stall. This collection of tactics means that "despite powerful forces aligned to remove them (…) street vendors (…) challenged, undermined and even subverted the *Programa*" (p.44). Verónica Gago (2017) also draws on attention to the resistant subjectivities of market traders in Buenos Aires who enact a "neoliberalism from below." While it is mostly harmful, the outcomes of gentrification can at times be somewhat contradictory as it sometimes makes the streets a bit safer or provides sources of employment for local people (if they can afford to stay in the area).

It is clear, however, that gentrification can be understood as a form of urban colonialism (see Atkinson and Bridge, 2005) as original inhabitants are evicted, displaced, or eliminated by state-capital allied forces. The dialogues between decolonial theorizing and urban studies are, however, still quite underdeveloped. Much of the scholarship on the early experiences of gentri-fication has been conducted in Global North contexts and led by scholars of political economy, so it is important to consider the extent to which they are applicable in the Latin American context and whether they adequately con-sider their colonial/decolonial dimensions. As we've emphasized, a decolonial approach means recognizing that theories created to analyze oppressions in the

Global North might not work so well in the Global South. It is also means recognizing that the economic, the political, and the cultural are all entangled, so we should not give primacy to economic drivers without understanding the cultural politics of race and gender that *always* underpin them.

Some (mostly Global North) authors describe gentrification as a global process that they refer to as planetary urbanization (PU). For example, Neil Brenner and Christian Schmid (2012) claimed that the urban condition is now a planetary one, shared by all the spaces and places in the world, including jungles, oceans, and forests. This political economy approach has been very influential and clearly has explanatory power for contemporary urban processes and the modes of injustice that they engender. It has, however, been subject to critique from decolonial, feminist, queer, and postcolonial scholars who believe that the PU thesis overemphasizes capitalist accumulation and exploitation and neglects the role played by other forces such as colonialism and patriarchy (Oswin, 2018; see also Ruddick et al., 2018; Cupples, 2019). Whether gentrification remains a useful concept or not in Latin America (and many think it does), it is important to recognize that Latin American gentrification processes have many differences as well as similarities (López-Morales, 2018) and that not all forms of urban injustice result from profit-seeking state-capital alliances (see Calderón et al., 2019). If we are going to decolonize urban studies more broadly or see gentrification through a decolonial lens, it is imperative that we explore how it is driven by hierarchies of race and gender as well as those of class and also how coloniality means the racial dimensions of gentrification are frequently central but also obscured. Furthermore, PU theorists have not yet engaged with the relational ontologies that we discussed in Chapter 4, as "sentient nonhumans, indigenous cosmovisions, popular religiosities and supernatural beings are all seen as external to the contestation of capitalism, at best as beliefs to be respected rather than as resources for thinking otherwise" (Cupples, 2019: 211, see Box 10.2). Arturo Escobar (2018) proposes one approach to doing so and suggests that we should seek to accelerate a process that he calls *rurbanization* that involves bringing Indigenous and campesino knowledges and cosmovisions back into the city. This approach enables us to address what for Escobar is not so much as a crisis of capitalism (although he dislikes neoliberal discourses of smart cities and resilience just as much as the PU theorists), but rather an epistemological and civilizational one in which certain ways of knowing are deemed to be non-admissable. In many Latin American cities there are territorial movements that draw strength from ancestral knowledges and collective labour that has delinked from the state. As Zibechi (2018) writes, people are forced by marginality and dispossession "to create their own spaces and territories in order to survive" and to "seek the health and education that the system denies them." He refers to the MST-led Pueblo sin Miedo (People without Fear) settlement in São Bernardo do Campo in São Paulo where 30,000 people reside and are organizing their own government and internal security without and beyond the colonial state.

It seems that the PU theorists are still thinking in Eurocentric terms about the one–world world when the planetary actually contains many worlds (Law, 2015; see also Dussel, 2006).

BOX 10.2 SANTA MUERTE: SUPERNATURAL URBAN ACTIVISM

In the late 20th century, the Santa Muerte cult, a form of popular religiosity that involves the worship of a skeleton saint, spread dramatically all over Mexico (see Figure 10.2). It attracts thousands of worshippers who come to ask for protection from violence, incarceration, ill health, adultery, and unemployment. Her devotees are often marginalized urban inhabitants, including prisoners, the LGBTQ+ population, unemployed teenagers, drug addicts, street children, garbage pickers, and migrant and informal sector workers. Like many popular cultural forms, Santa Muerte is both vilified and adored. Adored by her millions of followers, she's ferociously condemned by the Catholic Church, the state, the military, and political and intellectual elites. She is a syncretic figure that draws on Catholic traditions as well as on Mictlantecuhtli, the Aztec God of Death. Altars exist all over Mexico,

Figure 10.2 A Santa Muerte altar in Puebla

Source: Photo by author

but some of the largest ones have emerged in *barrios* that have been hit hard by neoliberal economic change and are often stigmatized by local governments. One of the most famous altars in Mexico City is that of Doña Queta in Tepito, a neighbourhood that is often described in the tabloid and mainstream press as extremely violent. But Tepito is also a locus of working-class resistance and a site of highly creative entrepreneurial activity where low-income inhabitants negotiate the exclusions of globalization by openly selling all kinds of contraband; fake perfume, pirated DVDs, and fake brand label jeans. It is also a neighbourhood that had its social fabric destroyed by the reconstruction carried by local government planners and World Bank funds after the 1985 earthquake when they demolished the Casa Blanca housing project and banned the traditional co-location of work and family life, forcing people out into the streets. While hegemonic forces work to stigmatize and scapegoat devotees and articulate the practice of Santa Muerte worship with organized crime and *narcocultura*, the guardians and devotees work to disarticulate these elements. In the monthly rosary in Tepito, there is sometimes a focus is on the failures of the public health system and the criminal justice system in which the urban poor are neglected or unjustly incarcerated (Roush, 2014). Indeed, the cult has become an important arena of sensemaking around questions of neoliberal urban abandonment. Furthermore, at a time when public space is increasingly privatized and fragmented, during Santa Muerte masses and rosaries, thousands of people turn out on the streets in large numbers for many hours, share food and gifts, disrupt the everyday social order, and produce a different way of being together. They reclaim the city for themselves, albeit temporarily, and destabilize the neoliberal common sense that sees gentrification and urban inequalities as inevitable. Because of its religious associations, Santa Muerte is not generally considered to be an anti-capitalist and decolonial social movement. But de Sousa Santos (2015: 64) believes that pluralist and progressive religions and spiritualities, unlike fundamentalist or reactionary ones, possess an "insurgent humanism" (p.64) that is aligned with anti-capitalist and decolonial struggles. Furthermore, he sees a radical potential in religious practices that's lacking in more secular forms of political mobilization. Santa Muerte's defiance of the status quo is multifaceted and intersectional, adopting not only an anti-market and anti-state stance but also including and accepting those denied access to the social contract, including those inhabiting alternative sexualities and gender identities, the incarcerated, and the addicted.

Sources: Cupples and Glynn, 2019; Roush, 2014; de Sousa Santos, 2015

So perhaps gentrification is just too northern, too colonial to be of much use in the Global South, while others believe it has purchase as long as we remain attentive to geographic specificity. While the Anglocentric term gentrification (*gentrificación*) is not used much in Latin America, urban eviction and displacement have been a key feature of Latin America cities for many decades, as have informal squatter settlements. Indeed, there is an important body of scholarship from the 1960s and 1970s and to which Aníbal Quijano (1972) made a significant contribution, noting how in the 1970s marginality had become a generalized feature of Latin American cities (for more detail, see Vegliò, 2018; Slater, 2019).

The academic literature on Latin American gentrification is growing, but scholarship has been concentrated mostly in Argentina, Chile, Colombia, and Mexico, neglecting other countries in the region where the contexts are quite different (see López-Morales, 2019; Díaz Parra, 2021). As we noted in Chapter 9, inner-city neighbourhoods of Mexico City, especially Roma and Condesa, began to gentrify rapidly after the 1985 earthquake. Despite the extraordinary levels of precarity, informality, and vulnerability that characterize contemporary Mexico City, these parts of the city are filled with luxury condominiums, flashy restaurants, coffee roasters, artisanal bakeries, craft breweries, pet grooming businesses, and beauty salons, including for children. As Ibán Díaz Parra (2021: 473) writes, there is a risk of subsuming many different kinds of urban transformations under the term gentrification understood in terms of the "capitalist production of space." Furthermore, as in the scholarship on the Global North, there has been a tendency to focus on a few specific neighbourhoods where gentrification can be identified, such as Getsemaní in Cartagena (see below) and La Boca in Buenos Aires. These experiences are valuable in their own right but they are not necessarily amenable to generalization and can only give us partial insight into what is happening elsewhere. As we discuss below, in Bolivian cities some quite different trends can be identified. When analyzing processes of gentrification, we should also pay close attention to the geographies at play in any given location and ensure that we are exploring the ways which colonial attitudes or decolonial modes of resistance shape urban transformation.

The neighbourhood of Getsemaní in Cartagena de Indias in Colombia used to be a fairly run-down working-class area with quite high levels of both poverty and crime. Today it is a cool and trendy neighbourhood with vibrant restaurants, bars, brightly coloured houses, as well as highly photogenic graffiti and street art. As you can see from Figure 10.3, the images of popular literary figures such as Gabriel García Márquez on the city's walls show the interaction between the cultural economy and the financial economy and how cultural capital is converted into financial capital (see Bourdieu, 1984). These become part of the conditions of possibility for gentrification which go on to have mostly negative impacts on the locals as outsiders move in and locals can no longer afford to stay. There is significant local opposition to the kinds of

Figure 10.3 Getsemaní's street art: Gabriel García Márquez

Source: Photo by author

changes underway (see Figure 10.4). Melissa Valle's recent study on processes of gentrification in Getsemaní reveals how race and racism are central to urban transformation but are elided from discussions about gentrification that are seen to be largely about class. Valle does not deny that class is central to gentrification, but she asserts that race too "is a critical mediating factor" (p.1236). As we discussed in Chapter 8, discourses of *mestizaje* produce insidious forms of Afro-erasure in Colombia and the stigmatization by the interior of coastal communities that have large Afro populations. Furthermore, Afro-Colombians are much more likely to live in poverty than white-mestizos. Afro-Colombian poverty is particularly acute in Cartagena, which was one of the ports to which enslaved Africans arrived and which today is an important tourist attraction. As a result of tourism, property prices in Getsemaní have increased tenfold in the last decade and original owners are displaced by local tax increases. Valle's fieldwork reveals the ubiquity of racist attitudes and discourses in the city that involve the pigmentocratic valorization of lighter skin, while Blackness is displaced both spatially to the maroon community of Palenque and temporally to the colonial era when slavery still existed. She notes how racist ideologies in Colombia provide a discursive means for gentrifiers to understand the local Afrodescendant population as "undesirable and incapable of contributing to

Figure 10.4 Gentrification in Getsemaní

Note: On a newspaper article that celebrates the changes in Getsemaní, somebody has scrawled "don't sell your houses, this is a paradise"

Source: Photo by author

the enhancement of the place" (p.1247), but do so without mentioning race as the prime force driving their displacement. Race and racism are then central to the cultural fabric of the city, but are detached from the sensemaking practices that surround gentrification.

It is important to recognize that the urban invisibilization of Blackness in Cartagena does not, however, go uncontested. Local Afro-Colombian activist and screen actor, Jhon Narváez, has participated in a series of mediated decolonial interventions to draw attention to anti–Black discrimination in the city. In some of these he appears as a superhero, Capitán Cartagena (see Figure 10.5)[1]. Narváez is also part of a social movement called Pedro Romero vive aquí (Pedro Romero lives here) that seeks to visibilize the Black population in Cartagena and contest the privatization of public space. Pedro Romero was an Afrodescendant from Getsemaní who made a decisive contribution to the independence of Cartagena but who has not been commemorated or recognized in the ways the *criollos* have been. Although since 2007 there has been a statue of him in the Plaza de la Trinidad, he is often referred to

Figure 10.5 Jhon Narváez is Capitán Cartagena

Source: Photo by Fabian Alvarez, reproduced with permission

"the hero without a face" (Tatis Guerra, 2017). This collective assembled an action called *Frágil o embalados* (Fragile or wrapped) in which several colonial statues, including those of Christopher Columbus and Pedro de Heredia, were covered in bubble wrap and packing material to erase them from the urban landscape.[2] Such activism around colonial monuments, that is taking place in much of Latin America as well in many other parts of the world, can, according to Nicholas Mirzoeff (2018) be understood as part of the struggle against invisibilization (see Chapter 11).

A different but also racially underpinned process of gentrification is that taking place in Quibdó on the Pacific Coast of Colombia in Chocó. Here gentrification is fuelled in part by the armed conflict and its legacies. In Quibdó, as Calderon et al. (2019: 26), argue "armed groups have become 'rentiers' since real estate investments have been an option for money laundering through capital fixing." Here we can witness a "mutated version of gentrification" that occurs "when groups outside the law look for a 'spatial fix' that transforms the built environment through extra-legal forms of capital accumulation" (p.26). The state is therefore not present in the way it is in other experiences of gentrification, and it is indeed difficult to identify the class positioning of the gentrifiers.

So gentrification is a useful concept but requires caution with the application of northern theories. In particular, we need to pay close attention to the racial hierarchies constructed in the colonial and postcolonial eras. This requires working with more epistemological pluralism than many scholars of gentrification do and not separating the cultural dimensions from the economic ones. After all, any form of socio-economic urban exclusion is accompanied by a powerful imaginary of what the city should mean and who belongs there and who doesn't. And these imaginaries are always contested by counter-hegemonic ones that celebrate and endorse the city's Black, Indigenous and working class inhabitants who resist neoliberalizing, colonizing, and stigma-tizing forces and remake the city in their own image.

THE REBEL CITIES OF BOLIVIA

While we should extend and disrupt the Eurocentric political economy approach to gentrification, we should do the same with our analyses of urban activism. Contemporary urban social movements increasingly involve the cre-ation of new territories and new communal worlds. As Sian Lazar (2008) and Raúl Zibechi (2005, 2012) write, such movements are collective, leaderless, autonomous, and spontaneous, they enact participatory and horizontal forms of decision-making, and they function beyond and inbetween the state in a way that calls the welfarist premises of the Eurocentric left into question. Furthermore, these movements frequently draw on ancestral knowledges that Eurocentric white-mestizo Latin America condemns to non-existence. They are based on various forms of decolonial refusal, including the mechanisms that seek to separate the economic, the political and the cultural. In their spontaneity and creativity, they often manage to achieve the impossible.

The ancestral and territorial forms of world-making that we see in urban Bolivia are revealing that geographical processes such as gentrification along with neoliberalism more broadly are not inevitable. Just as street vendors in Mexico are able to successfully fight against their elimination, in Bolivia there is a resilient pluri-economy in place that protects non-neoliberal economic forms. As Kate Maclean (2014) notes, the pluri-economy is built on a dis-ruption of neoliberal competitive individualism and a respect for collective, redistributive, and reciprocal Andean labour and livelihood practices. The gentrifying model is not as global as it might seem.

One of the most inspirational urban social movements is that which takes place in the Aymara city of El Alto in Bolivia. Indeed, Zibechi (2005) describes El Alto as the "most significant rebel city in all of Latin America" (see also Lazar, 2008). El Alto is very close to the capital La Paz (but even higher up in the Andes) and it has experienced rapid and dramatic population growth in the past decade. The city had been growing for decades but grew especially rapidly in the wake of the neoliberal structural adjustment package that closed the tin mines and sent newly unemployed Indigenous miners to the cities in search of employment. Rapid population growth put a lot of pressure on the city in terms of infrastructure and services, meaning that new migrants did not have and had to fight for clean water, sanitation, electricity, healthcare, and paved roads. Today, more than 80% of the population defines itself as Indigenous and the vast majority of these as Aymara.

Zibechi (2005) believes that if you look at El Alto with a colonial, Eurocentric or western perspective, you see chaos, but this is because El Alto is organized differently. As he writes:

> This society has its own political and social institutions, its own economy, and a culture that is clearly distinct from the mestizo-white "official" society built on state institutions and the market economy.

El Alto is a self-help city with a strong sense of territorial identity. The city's inhabitants have responded to state neglect by lobbying authorities and by building things themselves. This work is organized through neighbourhood councils, unions and labour organizations, and cultural associations. They have also created their own family-run and collective economy largely built on autonomy, horizontal solidarity, and communal culture rather than on top-down capitalist exploitation. The more than 500 neighbourhood councils in El Alto, that Pedro Mamani Ramírez (2005, 2006) refers to as "*microgobiernos barriales*" (popular neighbourhood microgovernments), function through a bottom–up participatory democracy and decisions are taken in regular assemblies in which all families and households must participate. These councils also often administer justice in the case of conflicts or infractions in ways that do not involve the Bolivian state.

It is also a city with a long rebellious history. In 1781, El Alto was the site in which the rebels led by Tupac Katari established their headquarters (see Box 2.4), in 1899 it was the site of a human blockade during the Federal War, and it played a key role in the 1952 revolution. In more recent years, residents have drawn effectively on ancestral and communal forms of organization. In 2003, they mobilized successfully against the proposed sale of natural gas and a proposed increase in property taxes, setting up effective blockades through a system of rotating shifts while also ensuring that protestors were supported and fed. They were met with brutal repression and protestors were shot and killed, but they also forced the resignation of the president, put an end to the sale of natural gas, and the cancellation of the new tax codes. Two years later, in 2005, as we mentioned in Chapter 4, the people of El Alto mobilized again, calling for the nationalization of hydrocarbons and a new constituent assembly, demands that were both achieved after the 2006 election of Evo Morales, Bolivia's first Indigenous president. The neighbourhood microgovernments were able to deprive the colonizing white-mestizo state of its legitimacy, drawing on collective ancestral knowledges, not only to effectively organize a political protest, but to create safe houses, use radio to communicate, care for the sick, and help with trauma (Mamani Ramírez,

2005). In 2003, events unfolded in an unprecedented but in a deeply terri-torializing manner that exposes the historical conditions of exploitation and discrimination in Bolivia.

> During the uprising, the government and the state disappeared from the city of El Alto. Instead, subaltern symbols were raised, such as the *wiphala*, the pride in knowing how to speak Aymara, Andean religious practices, in addition to a dense distribution of kinships defined according to the logics of action of the *ayllus* and rural communities.
>
> (Mamani-Ramírez, 2005: 12, my translation)

These urban social movements pose important theoretical and political questions. As Zibechi (2012: 50) writes, not only do they "challenge the state and the dominant classes, but also call into question the political and theor-etical framework of the old left." Moreover, they have enabled the survival of the popular sectors, as without such forms of organizing, the brutality of neoliberalism would have condemned them to premature death (see Zibechi, 2012: 119). It was these mobilizations that brought Evo Morales to power. As Hylton and Thompson (2007: 17) put it, "the election of Evo Morales did not bring about a revolution. It was a revolution that brought about the government of Evo Morales."

Under Evo Morales, as Kate Maclean (2018) writes, La Paz and El Alto went on to experience a decade long economic boom that led to substantial infrastructure development, investment, and a rise in real estate prices. As we've noted above, whenever these things coincide, we see changing patterns of mobility. In London, Barcelona, and San Francisco as well as in Mexico City, their combination has led to gentrification in the shape of the disposses-sion and displacement of traditional working-class inhabitants. But in Bolivia, the process has unfolded quite differently. Inequality declined under Morales and local governments created new public transport networks, mostly not-ably cable cars, that substantially improved the mobility of the urban poor. According to Maclean, the main beneficiaries of these recent processes of urban transformation are people of Indigenous or mestizo descent who work in the vast informal markets in the North of La Paz and the city of El Alto, in areas that used to be marginalized and stigmatized. Maclean says that one remarkable aspect is that it is the traditional Aymara women dressed in a layered *pollera* skirt, bowler hat, and an *aguayo* (brightly coloured carrying cloth) who has accumulated sufficient capital by trading in informal markets to displace others. Some of these women have been able to purchase luxurious properties in the affluent, and until now mostly white, Zona Sur area of the city. New Aymara transport and financial mobility also meant, as Maclean writes, that Aymara people developed new modes of consumption and started to frequently visit the Mega-Center shopping mall in the Zona Sur, especially on two-for-one movie nights at the multiplex cinema. Their Indigenizing presence produced a culture shock among non-Indigenous people in the area who responded with racist comments on social media.

This trope of the wealthy Aymara woman with expensive real estate ambitions is represented in a 2009 Bolivian film *Zona Sur* (*Southern District*). The film produced a high level of civic engagement and debate in Bolivia and concern among those who have enjoyed white privilege until now. Maclean (2018) discusses a key scene in which a white family's Aymara godmother, Comadre Remedios, turns up at their home (at a time when they are struggling financially) and offers to buy their home in cash. The challenge to gender, class, and racial privilege that the rich Aymara woman represents means that she is both vilified as corrupt and undeserving or celebrated as the proof of Indigenous empowerment in a decolonizing Bolivia. Either way, "this image of the rich Aymaran woman, and the reversal of expected patterns of urban development that she represents, places the colonial, cultural and gendered dynamics that structure how capital shapes urban space into sharp relief" (p.712).

The urban transformations in Bolivia are aesthetic too (see Box 10.3). As pride in being Indigenous has grown, traditional clothing such as *polleras* and bowler hats, so-called *chola* style, have become high fashion items and are sold in expensive boutiques and displayed in fashion shows (Maclean, 2018, 2019; Salinas Maceda, 2016; Tapia Arce, 2017). Not so long ago, young Aymara people might feel embarrassed if their mother wore a *pollera*, but not anymore (see Figure 10.6). As feminist and Aymara blogger, YouTuber and

Figure 10.6 "Cholas paceñas" wearing traditional dress in La Paz, Bolivia

Source: LBM1948, CC BY-SA 4.0, via Wikimedia Commons

activist Yolanda Mamani calls her blog, it is now fashionable to be a chola – *ser chola está de moda*. As chola fashion has become widespread, Mamani (2015) does, however, express concern at white appropriation of her culture and the absorption of the craft into capitalist networks of production.

BOX 10.3 CHOLET: DECOLONIAL ARCHITECTURE IN EL ALTO

Freddy Mamani is an Aymara civil engineer and architect of humble origins who has built dozens of colourful, psychedelic, and fantastical buildings in the city of El Alto in Bolivia and dozens more outside of El Alto (see Figure 10.8). These buildings function as event halls where the growing wealthy Aymara middle class discussed above can reside and hold lavish parties and banquets. Each building has a small shopping mall at street level and is topped by a rooftop building which is usually the owner's residence. This residence is referred to as a cholet, a word that combines cholo, the once pejorative term for a rural Indigenous or mestizo migrant now reappropriated, with chalet. The floors below the rooftop residence are often composed of apartments that can be occupied by family members or rented out. The buildings pay tribute to a range of traditional Andean colours, styles and symbols, including the geometric architecture at Tiwanaco and Andean textiles and ceramics. They contain images of llamas and condors. Interestingly, these ancestral claims, along with Mamani's qualifications, are often called into question by non-Indigenous architects. At times, the aesthetic value of the buildings is dismissed by making comparisons with Las Vegas . But the buildings are perhaps intended to be over the top as they are primarily spaces in which to have fun and as a result, they are a challenge to white bourgeois society that tends to value moderation. There is no doubt that Mamani's buildings have transformed the urban landscape of El Alto and now there are many copies. A 2018 documentary *Cholet: The Work of Freddy Mamani* directed by Isaac Niemand captures the challenge posed by Mamani to Eurocentric thinking and the practice of architecture, as well as to growing Aymara empowerment, pride and social mobility witnessed in Bolivia in the past two decades and especially since the election of Evo Morales. Daniel Runnels (2019) sees Mamani's work as "Indianizing the urban landscape" and a form of resistance to colonial oppression, while Elisabetta Andreoli (2020) describes the architecture as creating an aesthetic of excess rooted in the strident colours that are central to Bolivian culture. Furthermore, she understands the phenomenon through the lens of the Aymara concept of *ch'ixi* which helps us to identify its political potential. Drawing on the work of Silvia Rivera Cusicanqui Andreoli, she states that *ch'ixi* "represents

Figure 10.7 The architectural work of Freddy Mamani

Source: Grullab, CC BY-SA 3.0, via Wikimedia Commons

how different cultures with different temporalities happen to coexist both in complementary and antagonizing ways. Beyond and below layers of coexistence and complementarity, this residual difference harbours a conflict. It is in this residual difference where the political – a crucial dimension for Rivera – is located" (p.203). Mamani's buildings can be said to embody a form of border thinking or transmodernity, given how they combine modern and futuristic architecture with traditional aesthetic forms, western and Aymara architectural elements and influences, spiritual values with mercantile ones. He incorporates western influences in a way which is also challenging to Eurocentric architects. As Martina Thorne (2019: 83) writes:

> Mamani's daring lies in the appropriation, reinterpretation, and decontextualization of European architectural grammar. Some of his appropriations include columns. Nothing is more foreign to Andean architecture than this element that is preferred by Mamani for the interior decoration of the cholets. The columns do not fulfill a structural support function, but serve a purely decorative role. Placing them one on top of

the other, adorning them with motifs and colours characteristic of Andean textiles, Mamani subverts Western aesthetic canons and translates them into an Aymaran aesthetic language. It is not a simple imitation but rather the translation and reconceptualization of western architectural vocabulary to Aymaran architectural language (my translation).

So Mamani is openly challenging western norms. As he himself said "I've broken the old architectural canons. I'm a transgressor" and he emphasizes how his work differs substantially from what he was taught at university (cited in Thorne 2019). Mamani has received international acclaim for his work and his work has been featured in many exhibitions and trade publications

Sources: Runnels, 2019; Thorne 2019; Andreoli, 2020, *Cholet*

It appears that in more recent years El Alto has experienced some degree of political demobilization. As Angus McNelly (2019) writes, El Alto still has a rebellious character that it can draw on at any time, but it has been subdued in part by co-optation and domestication by state forces. And in spite of the positive changes that the city has witnessed, "infrastructure remains sparse and underfunded, and many residents lack adequate access to education, health care, and basic services" (McNelly, 2019: 339).

But there is no doubt that the cities of La Paz and El Alto are indigenizing and decolonizing in a range of ways – businesses, architecture, fashion, and governance are organized according to Indigenous ways of knowing and being. The wealthy *chola* and Mamani's architecture are evidence of the kinds of Indigenous economic empowerment and cultural renaissance that have taken place in Bolivia under Evo Morales. It shows the ways in which racist terms like *cholo* can be rearticulated and how people who have been kept on the margins of society can move to the centre. Indigenous Bolivians are proud of their Indigenous identity and heritage, and they are claiming their right to be in the city and to make it their own.

SUMMARY

- Latin American cities are large and have grown very rapidly. Many Latin Americans live in informal settlements.
- Latin American cities face many challenges including inadequate access to water or sanitation, respiratory illnesses, substandard housing, police brutality, and gang violence.

- Urban gangs such as MS-13 and B-18 should be understood as social phenomena with origins in colonial violence, including US-supported military intervention and forced deportation.
- Gentrification is a useful concept for analysing the kinds of urban transformation taking place in Latin America, but it is important that we take the decolonial challenge seriously and understand how such processes are driven by race and racism as well as by class.
- Latin America is a site of public transport innovation that has enhanced the mobility of the poor in important ways.
- The forms of urban mobilization and transformation we see in Bolivia and elsewhere in Latin America are sometimes based on anti-capitalist and anti-colonial forms of organization and on ancestral traditions.
- The popular economies and resistant tactics of market traders and street vendors in Argentina, Bolivia and Mexico are remaking cities in ways that confound the politics of neoliberalism.

DISCUSSION QUESTIONS

1   Are urban political struggles different from rural ones? If so, how? How are cities and the countryside connected?
2   What role have rural–urban migration and squatter movements played in making contemporary Latin American cities?
3   What is the role of colonialism and imperialism in bringing about gang membership and violence in Central America?
4   Are the theoretical frameworks used to analyse cities in the global North suitable for analysing cities in the global South? How can we decolonize urban studies?
5   How do you feel about the enrolment of Santa Muerte in struggles to counter urban marginality?
6   How have urban social movements changed over the past decades? To what extent are these struggles decolonial ones?
7   How do the inhabitants in El Alto mobilize against oppression? What makes this mobilization so different from conventional left-wing politics?
8   Does the work of Freddy Mamani encompass what might be called border thinking or transmodernity? How does it challenge conventional architecture rooted in Eurocentrism? How does it also challenge essentialist understandings of Indigenous identity?
9   How might we analyse the rise of a new Aymara bourgeoisie in a city such as El Alto? To what extent can Mamani's *salones* and the lavish activities that take place in them be understood as a form of *ch'ixi*?

NOTES

1 See https://cumbiafilms.com/project/capitan-cartagena/
2 A video of the action can be seen here https://vimeo.com/channels/pedroromeroviveaqui/43307873

FURTHER READING

HBLAD Chapters 30 (Swanson), 42 (López-Morales), 43 (Rodgers), 44 (Lombard), 45 (Duarte), 46 Lopes de Sousa, 47 (Quiroga)

Cupples, J. and Slater, T. (eds) (2019) *Producing and Contesting Urban Marginality: Interdisciplinary and Comparative Dialogues*. London: Rowman and Littlefield.

An edited collection that explores urban phenomena in Latin America and beyond.

Gago, V. (2017) *Neoliberalism from Below: Popular Pragmatics and Baroque Economies*. Durham, NC: Duke University Press.

An ethnographic study conducted in an urban market in Buenos Aires reveals the agency of market traders to remake neoliberalism from the ground up.

Lazar, S. (2008) *El Alto, Rebel City: Self and Citizenship in Andean Bolivia*. Durham, NC: Duke University Press.

A book that explores citizenship practices, collective organization and popular protest in El Alto, Bolivia and that also deploys ethnographic methods.

Recommended films

Cholet: The Work of Freddy Mamani (2018), directed by Isaac Niemand

A film about the lavish buildings in El Alto Bolivia that have been created by Aymara architect Freddy Mamani that draws on ancestral traditions and are appealing for the growing number of wealthy Aymara businesspeople.

Roma (2018), directed by Alfonso Cuarón

A film set in Mexico City in the 1970s focused on the life of an Indigenous maid for a middle-class Mexican family. It deals with the complexities of gender, race, and class.

Zona Sur (*Southern District*) (2009), directed by Juan Carlos Valdivia

A film about urban transformation in contemporary Bolivia and the ways in which the cultural politics of gender, race and class intervene to produce new, surprising and decolonizing outcomes.

Useful website

The blog of feminist Aymara activist Yolanda Mamani (in Spanish)
https://sercholaestademoda.blogspot.com/

CHAPTER 11

Making the decolonial turn

LOS 43

In September 2014, in Iguala in the Mexican state of Guerrero, 43 students were kidnapped and have never been seen again. The students all attended the Raúl Isidro Burgos Normal School in the town of Ayotzinapa, part of a network of schools known as *Normales* that were created after the Mexican revolution to train teachers to work in the most deprived parts of Mexico. The school provides residential tuition for male students who come from poor, Indigenous, and campesino backgrounds and is well known for its political activism. Students and teachers at the school have been important voices in contesting the inadequacy of state funding to public education as well as neoliberal education reforms. All but one of those taken were in their first year of study at the school.

What exactly happened to the 43 students or *normalistas* is still not completely clear, but it appears that they were kidnapped by the police in collaboration with drug traffickers. On the night of the attack, the students were on their way to temporarily seize buses so that they could participate in the commemoration of the 1968 Tlatelolco student massacre in Mexico City, an event that takes place every year. Eyewitness accounts vary but it appears that a police ambush on the night of 26 September resulted in the deaths of six people, including people in other passing vehicles killed by unidentified shooters, and the kidnapping of the 43 students who were seen being rounded up in police vehicles. A soccer team in a similar looking bus to that of the *normalistas* also got attacked. It appears too that over the course of the night of 26 September, there was not just one attack on the students but several and no police force arrived to respond to the attacks (Hernández, 2018). According to the Attorney General, the students were handed over to an organized criminal organization known as the Guerreros Unidos and were murdered and burned in a garbage dump, an account that has been contested

DOI: 10.4324/9781003110453-11

by international investigators appointed by the Inter-American Commission on Human Rights (IACHR) (Meyer and Hinojosa, 2019). This was not the first time that students from the school were exposed to state-led and police brutality. Indeed, just prior to the kidnappings, then President Peña Nieto had deemed the students a security threat. It is still not known if the students are dead or alive, although bodies believed to belong to three of them have been located and identified.

Investigations have led to the arrest and imprisonment of a small number of police officers and gang members. Federal and state authorities accused the mayor of Iguala, José Luis Abarca Velázquez, and his wife, María de los Angeles Pineda, of ordering the attacks. They went into hiding for more than a month but were later arrested in Mexico City in November 2014. Both were imprisoned for money laundering and links to organized crime, but by 2021 they had still not yet been convicted of masterminding the kidnapping and murdering of the Ayotzinapa 43. Investigative journalist Anabel Hernández (2018), who has published a book on the Ayotzinapa 43 and had to leave Mexico because of threats to her life, believes that the official story is a total fabrication. Her book reveals a world of official complicity, cover-up, lies, and tampering with evidence "where the government tortured people to manufacture suspects and extract confessions" (p.40). Hernández believes that two of the buses that the students had taken contained a hidden $2 million dollars' worth of heroin unbeknownst to them.

The disappearances outraged many people and produced hundreds of angry protests and marches, many peaceful and some violent all over Mexico, as well as in many other countries (see Figure 11.1). Forced disappearances are common in Mexico but have not attracted the level of mobilization that the Ayotzinapa case did. According to Tomasso Gravante (2020), the Ayotzinapa disappearances have produced a collective cultural trauma, producing unprecedented levels of identification with the students and their parents. Ayotzinapa became articulated to other forms of colonial violence, including economic marginality. All over Mexico you see banners and posters in solidarity with the disappeared and their families (see Figure 11.2). As Melissa Wright (2019: 236) shows, the case has resulted in the mass production of political poster art, "shouts from the walls" that "advocate militant politics around the intersected themes of state terror, neoliberal exploitation, and social justice."

Many of these posters and banners contain a slogan that originated in the 1970s in the context of forced disappearances: "vivos se los llevaron, vivos los queremos" ("they took them alive, we want them back alive"). As Luis Fernando Méndez Franco (2015) writes, this slogan is particularly significant given the existing conjuncture in Mexico, that he describes as one of a weakened nation-state that expresses its sovereign power necropolitically and routinizes premature death. He notes that this slogan "puts life at the centre of social protest" (p.67) and it is expressed by a spontaneous, indignant, and

Figure 11.1 Protest to demand the return of the 43 missing *normalistas*

Source: PetrohsW, CC BY-SA 4.0, via Wikimedia Commons

heterogeneous collective. The slogan "Fue el Estado/#FueElEstado" ("It was the State") is also frequently used online and offline. Nobody imagines that the students are actually still alive but the demand for life, the demand for their safe return, refuses to let the state off the hook. Furthermore, it is an idea of life that "is beyond and even above death" (p.72). Both slogans have been used in association with human rights atrocities, politically motivated assassinations, and extrajudicial killings in Honduras, Nicaragua, and Colombia, as well as in Mexico, and are used right across the continent to express solidarity with the Ayotzinapa 43.

Activism led by the families of the missing students and other victims of Mexico's drugs war have led to the discovery of many mass graves in the state of Guerrero, underscoring the endemic nature of organized crime. The case reveals the extent of impunity and corruption in Mexico as well as the close relationship between the state and the cartels. It also reveals how those that threaten the colonial-capitalist status quo through their activism and critical education are deemed expendable. For Catherine Walsh (2019: 223), the struggles for liberation taking place in Ayotzinapa and the indignation created by the crimes committed against the 43 students are part of an ongoing struggle to "assemble a collective pedagogy" in order to "reclaim (…) that which has been negated, eliminated, violated, and despoiled, including subjectivities, knowledges, bodies, territories, existence, and life."

Figure 11.2 Expression of solidarity with the 43 missing *normalistas*, Coyoacán, México

Note: The banner reads: "We are missing 43!!! They took them alive, we want them back alive! We are all Ayotzinapa!!!"

Source: Photo by author

When President Andrés Manuel López Obrador came to power in 2018, he passed a presidential decree that promised to bring justice to the Ayotzinapa families and announced it was reopening the investigation. In June 2020, Guerreros Unidos leader José Ángel Casarrubias Salgado, otherwise known as "El Mochomo," was arrested in connection with the missing students. In June 2021, Humberto Velázquez Delgado, one of the police officers believed to be a key player in the case was assassinated at his workplace (Flores Contreras, 2021). Little progress has been made, and the *normalistas* are still demanding the safe return of their classmates.

THE POLITICS OF LIFE AND DEATH

This book has emphasized that studying contemporary Latin America in relation to questions of development and decolonization involves understanding struggles for hegemony between Eurocentric and ancestral cosmovisions, between capitalism and socialism, between extractivism and sustainability, between imperialism and solidarity, between the left in power and the organized left in resistance that is out on the streets, and between different modes of doing gender. Latin America is a continent built on trauma and injustice where the brutality of colonialism in the 15th, 16th, and 17th centuries gave way to coloniality after independence. In the 20th and 21st centuries, coloniality produced horrendous forms of economic, political, and epistemic violence – military dictatorships that tortured and disappeared their political opponents; structural adjustment policies that destroyed campesino livelihoods, accelerated environmental destruction, and threw millions of people into poverty; a violent drugs war; and state-multinational partnerships that have pursued neocolonial forms of extractivism and been prepared to eliminate those that stand in its path. These forms of violence have led to myriad collective movements and struggles in defence of land, territory, human rights, and dignity that operate beyond the state (see Zibechi, 2018). All of these struggles are as much cultural and epistemic as they are political, as they involve the contestation of dominant modern, colonial, patriarchal, and capitalist imaginaries and epistemes.

Particularly at stake here is a struggle that has been a constant since 1492 between *the politics of life and the politics of death*. In the words of Gabriel García Márquez, uttered when he accepted the Nobel Peace Prize in 1982, "to oppression, plundering, and abandonment, we respond with life" (cited in Camacho de Schmidt and Schmidt, 1995: ix). Latin America is the site where human rights atrocities and injustice co-exist daily with courageous forms of political struggle in defence of life and dignity. Coloniality brings racism, suffering, and premature death, but also the will to survive against the odds. The Mexican state reproduces its politics of death, and the people respond overwhelmingly with a politics of life. "Vivos los queremos" "We want them (back) alive" is Latin America's popular and affective modus operandi. As Emilia Reyes (2020: 263) puts it, "life has been seriously threatened by this capitalistic and neoliberal system" and it is these lives in the Global South, the lives of people "whose identities were anthropogenically conceived in a notion of alterity by this system for the very purpose of their exclusion" that are those most at risk. But when we refer to life, we need to extend beyond human life, as the system, as Reyes underscores, has also threatened "the healthy life of the Planet, the survival of biodiversity and the integrity of ecosystems" (p.23).

Indeed, we see this assertion of human and non-human life over and over again in the deaths of Latin Americans who are killed with impunity while fighting for social justice. The Ayotzinapa case has galvanized people in Mexico and beyond because it reveals both the persistence and intensification of the colonial matrix of power, or what becomes of those who reside in what Frantz Fanon (1967) called the Zone of Non-Being (ZNB), the hell to which racialized and colonized subjects are confined and denied their humanity. Ayotzinapa is its own hell, but it stands in for the experience of everyone in the ZNB that lives with constant insecurity, precarity, and the threat, not just of premature death, but of things that are even worse than death, such as torture and rape (see Maldonado-Torres, 2016). It speaks to experiences in Guerrero, but also in Wallmapu, Ciudad Juárez, Cali, the favelas of Rio de Janeiro, and in the desert between the United States and Mexico. But just like the Ayotzinapa 43, those eliminated – Berta Cáceres, Oscar Cazorla, Vicky Hernández, Marielle Franco, the children of the Madres de la Plaza de Mayo, the maquiladora workers in Ciudad Juárez, the Ixil people massacred by their own US-funded government, the students who fought in the Nicaraguan uprising of 2018, the undocumented migrants in the desert – do not die but give rise to life. They live on in collective memory, they are rehumanized through collective and artistic acts of commemoration, and they shape the contours of decolonial struggle.

While rebellion, resistance, and life politics have been constant features of the past five centuries, these have taken on an urgent quality more recently. Today, in their public defence of life, the people come out onto the streets also with the slogan *Nos quitaron tanto que nos quitaron hasta el miedo* – They took so much from us, they even took away our fear, and they promise too to neither forgive nor forget – *Ni perdón, ni olvido*. Collectively they constitute a demand for dignity, and as Chileans Soledad Falabella Luco and Carolina Solar Andrade (2020) put it, the struggle will continue until *la vida vale la pena* – life is worth it. As a video that reflects back on the April 2018 uprisings in Nicaragua reveals, amidst the fear, terror, confusion, the noise of gunfire, spilled blood, homes under siege, the forced exile, and political incarceration, people collectively generate hope, convictions, spontaneous mutual aid, friendship, solidarity, and new ways of understanding the world and being in the world (La Digna Rabia, 2021). Through this rebirth, meanings of flags, revolution, and dictatorship get rearticulated, old histories are rewritten through feminist and decolonial lenses, silenced histories are heard, and new histories are made. Activists recognize that liberation is not linear but is subject to constant setbacks, and they are taking a long-term historical perspective to negotiate the multiple contradictions that racial heteropatriarchal capitalism produces.

A number of decolonial scholars, including de Sousa Santos, Grosfoguel, and Maldonado-Torres, have extended Fanon's (1967) analysis of the ZNB to explore the ways in which epistemic and political violence co-exist and mutually constitute one another with a view to producing analyses that might disrupt the forms of colonial violence at work. De Sousa Santos (2007: 45) writes:

> Modern Western thinking is an abyssal thinking. It consists of a
> system of visible and invisible distinctions, the invisible ones being the
> foundation of the visible ones. The invisible distinctions are established

> through radical lines that divide social reality into two realms, the realm of "this side of the line" and the realm of "the other side of the line."

So an abyssal line divides people in the world, placing some of us in the Zone of Being (ZB), where there might be oppressions of class, gender or sexuality, but because of racial privilege, humanity is not in question, and some of us in the Zone of Non-Being (ZNB), where humanity is in question and where classism, sexism, or homophobia are exacerbated by racial oppression (de Sousa Santos, 2007; Grosfoguel, 2018). Abyssal thinking assumes that below the line, there is no intellectuality or knowledge production; the knowledges that exist in this realm are "produced as non-existent" which means "not existing in any relevant or comprehensible way of being" (de Sousa Santos, 2007: 46). The zone in which we are located influences our chances of making it in life, our ability to find political redress in moments of injustice, whether we get heard when we say no to something that brings us (or the planet) harm, and whether our saying no results in us being tortured or killed. It is race, rather than gender or class, that determines in which zone one resides. Grosfoguel (2018: 266) puts it as follows:

> In the zone of being, subjects, for reasons of being racialized as superior, do not live racial oppression but racial privilege. (…) In the zone of non-being, due to the subjects being racialized as inferior, they live racial oppression in place of racial privilege. Therefore, the oppression of class, sexuality, and gender that exist in the zone of non-being is qualitatively different and more devastating from those oppressions that exist in the zone of being.

What this means, as Grosfoguel clarifies, is that the ZB is characterized by regulation and emancipation, the ZNB by appropriation and violence. In the former, there are legal and usually non-violent means through which to address oppressions and fight for rights. In the latter, while there are moments of regulation in which rights are secured, as we've seen throughout this book, the norm is that violent dispossession proceeds with impunity. There is a ZB and ZNB in both the Global North and the Global South and both zones are heterogeneous. Men have advantages denied to women, cis women have advantages denied to trans women, professional Black men have advantages not enjoyed by poor Black men, and white women in Europe are generally better off than Black men in Latin America. As Grosfoguel (2018) puts it, in the ZNB there is police brutality, in the ZB there is police protection. There is a ZNB in Minneapolis as well as in Rio de Janeiro. For Nelson Maldonado-Torres (2016), the ZNB is where dehumanization and endless war are the order of the day.

Ayotzinapa is a Zone of Non-Being. Those disappeared were young, mostly Indigenous, and from poor rural backgrounds, training to be teachers to educate people like them. They were politically aware and active students, who drew on a long-term collective memory of historical colonial violence in Mexico. They were trying to make a better world, both through their studies and through their activism. They were standing up to forms of power that routinely condemn them and those like them to premature death. Not only were they kidnapped without trace, but the state cannot but act with impunity.

The ZNB needs more conventional development – it needs resources and investment in education, healthcare, housing, transport, and water systems, and it might continue to benefit from the application of some conventional social or environmental science. But this will not be enough, because the ZNB is not just a site of inadequate infrastructure, weak governance and so on but a site of dehumanization and epistemic violence. It also needs respect for its territorial integrity, for Indigenous, Afrodescendant and campesino ways of knowing, as well as respect for the rights of nature, but such respect is *not* compatible with colonial-capitalist business as usual. Decolonization is not about giving some people in the ZNB access to the forms of citizenship available in the ZB, but rather it involves bringing about new understandings of the human, in which any doubts about "the full humanity of the colonized" are eliminated (Maldonado-Torres, 2017: 129; see also Wynter, 2003).

CIVILIZATIONAL, CONJUNCTURAL, AND EPISTEMOLOGICAL CRISIS

This analysis enables us to understand that Ayotzinapa, horrific and horrendous though it is, is not an aberration. It is an outcome not just of the colonial matrix of power, but also of the apocalyptic phase of capitalism into which we are moving. Capitalist accumulation is reaching its limits, it is collapsing under the weight of its own contradictions, so those in pursuit of profit and the policy elites that support them have accelerated their activities and the elimination of those that stand in their way is no longer a last resort, if indeed it ever was. Indeed, the presence of militarized police in riot gear has become ubiquitous in Latin America. They are there to challenge not only those who would bring down the government through extra-legal means such as armed conflict, but those who want free and fair elections, those who want to protect forests, birds, and butterflies, those who want an end to gender-based violence, and those who want state institutions that are not corrupt. The thing they have in common is that they challenge the modern-colonial status quo. Riot police and racialized forms of police brutality are one of the many violent responses that are arrayed against those that have tried to mobilize Eurocentric forms of international law and put them to decolonial purposes to protect life. The authoritarian elites send out the riot police because they are frightened, as the hegemony of modern–colonial thought is being tangibly undermined. While impunity leaves most of the perpetrators unpunished, the violence they wield must be understood as the loss of legitimation. Furthermore, while this violence instils terror, it fails to eliminate critical thinking. As Maldonado-Torres (2016: 25) writes: "If anything, suffering gives further rise to thought." As we witnessed during the dictatorships and military governments of the 1980s, repression does not work "as a form of long-term control" but instead creates "a presence of protest for ongoing life" (Schirmer, 1989: 24, 25).

The civilizational and epistemological crisis at work here is part of both a global as well as a Latin American conjuncture. Eurocentric governments and institutions, the colonial-capitalist power bloc, are struggling – they must state their support for the SDGs and ILO C169, they must express their commitment to tackling climate change, and they must get behind initiatives for gender and racial equality. But the same entities also work to keep colonial and Eurocentric discourses in circulation. Consequently, these contradictions can be observed in the way they assert their commitment to the next round of climate change talks, while simultaneously engaging in culture wars over the question of free speech, gender studies, critical race theory, colonial monuments, and what should and shouldn't be on school and university curricula. Discourses of cancel culture and free speech are being mobilized by governments and Eurocentric academics dishonestly to protect colonial structures of power and white privilege. As Maldonado-Torres (2016: 8) puts it, when we raise the question of decolonization, we are up against "a decadent and genocidal modern/colonial attitude of indifference" but one that is couched in liberal values cloaked in a veneer of neutrality. Colonial discourses get reworked and recirculated to prevent decolonial movements gaining traction and to silence and dismiss those who put them on the agenda. As Maldonado-Torres (2016) emphasizes, it is achieved by drawing on the "objective" knowledge enshrined in established Eurocentric disciplines.

Seen from this perspective, it is clear that the attempt in many parts of the world to eliminate critical thought is part of this same colonial matrix of power. The growing forms of oppression and repression that we witness across Latin America are because this critical questioning is undermining the hegemony of modern–colonial thought. It is also why critical race theory and gender studies, as well as critical analyses of colonialism, are under attack from governments and elite academics the world over, why so many white feminists are putting their energies into attacking trans rights, and why so many people are defending statues of white supremacists who committed or endorsed genocide (see Vergès, 2021; Phipps, 2020). Statues have been removed or brought down repeatedly throughout history the world over without complaints from the power bloc (see Figure 11.3), but at the current time, the statues of racists and colonizers are subject to disproportionate defence and protection. The statue of Cecil Rhodes or Winston Churchill comes to symbolize the racial privilege perceived to be under threat. In other words, many of those who

Figure 11.3 What remains of the statue of former Nicaraguan dictator Somoza, brought down on 19 July 1979

Note: It is now at the Loma de Tiscapa in Managua

Source: Photo by author

have enjoyed racial privilege until now are not prepared to relinquish it. It certainly applies to those that have become billionaires, as well as to some middle-class gentrifiers, scientists, historians, and media professionals, but it also applies to some of the white working poor who are marginalized by capitalism but cling onto their white privilege and turn against migrants and refugees, instead of forming alliances with racialized populations. It is interesting to note that while Oxford University keeps trying to find ways not to remove the statue of Cecil Rhodes, in 2020 and 2021 the Misak people of Colombia removed several statues of colonizers, including the statues of Sebastián de Belalcázar in Popayán and Cali and the statue of Gonzalo Jiménez de Quesada in Bogotá. Their actions inspired others who went on to remove the statue of Misael Pastrana Borrero and Diego de Ospina y Medinilla in Neiva and Gilberto Alzate Avendaño in Manizales (see Oquenda, 2021). The authorities in Bogotá went on to pre-emptively remove the statues of Christopher Columbus and Isabel de Castilla in Bogotá, as they were the Misak's next target. A few days later people brought down the statue of Christopher Columbus in Barranquilla. The removal of Gonzalo Jiménez de Quesada in Bogotá was followed by a symbolic funeral ritual in June 2021 led by the Muisca people, whose ancestors were tortured by Jimenez de Quesada, to allow him to depart and promote healing and forgiveness in the face of colonial pain and anger (Martínez Ante, 2021, see Figure 11.4) In the midst of pandemic, a national strike and widespread protests, the Misak and Muisca people have opened a debate about colonial heritage and the colonial wound, articulated with the generalized rejection of the death politics of the government of Iván Duque.

Figure 11.4 The Muisca funeral ritual for the statue of Gonzalo Jiménez de Quesada

Note: The tweet reads: The Muisca community performed a funeral ritual for the statue of Gonzalo Jiménez de Quesada. They wanted to "let him go and heal his scars." Empty pedestals and an urban landscape that is trying to be resignified, another effect of the protests

Source: https://twitter.com/cataoquendo/status/1407733275125235719?s=20

DECOLONIZING TEACHING AND RESEARCH

This situation has profound implications for education as our campuses have become ideological and epistemological battlegrounds. Decolonial struggle and scholarship have revealed that things that were thought of as social or environmental problems that could be resolved through the application of scientific knowledge turn out to be civilizational crises that require much more radical solutions. To quote Escobar (2016: 15, my translation), it has forced a recognition that "we are facing modern problems for which there are no modern solutions." In other words, we need to make the decolonial turn, as Eurocentric physical and social science and mainstream development have clearly not been able to solve the global challenges we face and have in many cases made things worse.

While not everything that Europeans do or say is colonial, many Europeans draw on coloniality in order to construct their (superior) identities and advance a position. As a result, as noted above, the humanity of colonized others remains in question. We must therefore take on board that "the project of Western modernity as a whole is inherently a colonial one" (Maldonado-Torres, 2016: 11). We see this position in US and European universities, often built on the wealth of slavery and in some cases actually built by slave labour, who seek to attract international students from the Global South where they will purportedly receive a superior (Eurocentric) education. When they arrive, these students are subject to differential and discriminatory treatment and don't find their worldviews, or authors or professors who look like them, present in their courses. We dehumanize and reproduce coloniality when we pretend we are helping. Furthermore, students and faculty of colour are expected to embrace the liberal values of the institution. Ideals such as "objectivity" and "excellence" are mobilized to defend "the boundaries between those who claim to be in the zone of being human and those condemned to the zone of dehumanization" (Maldonado-Torres, 2016: 14). These ideals are not just wrong, they are dangerous.

We are then living through a moment of profound conjunctural crisis, described once by Antonio Gramsci (1971: 276) as a time when "the old is dying, but the new cannot be born." This conjunctural analysis explains why the UK government has embarked on economically and logistically disastrous initiatives such as Brexit, because they facilitate the revitalization of the affects of empire and the renewal of racial capitalism. Moving to Latin America, these conjunctural politics can help to explain why an anti-imperialist revolution aimed at the liberation of the subordinated, such as the Nicaraguan one, can be appropriated by a corrupt and heteropatriarchal leadership willing to turn its weapons against its own people in order to reproduce coloniality and enable capitalist accumulation through extractivist projects. It helps us to explain in spite of political progress made in interrogating racial democracy, reducing poverty, and revealing the racism that drives police brutality and other forms of violence in Brazil, the election of Jair Bolsonaro as president. Bolsonaro is a fascist racist misogynist who doesn't care about the environment, a man "who openly declared his misogyny, homophobia, Negrophobia, and contempt for Indigenous people" just a few months after the murder of Marielle Franco (Vergès, 2021:12). De Santana Pinho (2021) believes that this reactionary wave results from the progress that Brazil has made to lift Black populations out of poverty that has enabled them to access spaces of consumption, such as luxury shopping malls. It produces what she calls "injured whiteness," the reaction of those not used to sharing such spaces with the poor Black populations. So even though capitalism is making life harder for everyone except for the wealthiest 10% of the world's population, and is literally destroying the planet on which all of our lives depend, some of those who still enjoy relative racial privilege are doing everything to hang onto it and are interpellated by politicians such as Bolsonaro. Those who don't enjoy racial privilege are putting their bodies in front of the riot police.

EMBRACING LIFE

So let us end by exploring some of the ways in which we can make the decolonial turn by embracing a politics of life, rather than a politics of death. There are a number of things we can do. One is to turn our attention to the intellectual production in the ZNB, which Fanon (1967) saw as a site of truth, in order to rehumanize what has been dehumanized. This doesn't mean we can't read or cite Foucault or Gramsci anymore; we can, but it does mean we need to be vigilant about our reading and citation practices. It might mean avoiding the application of theoretical insights developed in the ZB to the ZNB without modification. We must pay close attention to the specific historical experiences in the ZNB, to the body-politics and geopolitics of knowledge, to the forms of intellectual production produced by people who live in and come from that location (Grosfoguel, 2018). If we continue to silence and invisibilize these experiences and this intellectuality, we run the risk of recolonization. We need to learn with and speak from a decolonial locus of enunciation, regardless of our own social location. So as well as reading and citing Silvia Rivera Cusicanqui, Beatriz Nascimento, Sylvia Wynter, Fausto Reinaga, and Linda Tuhiwai Smith, we must also act in solidarity with the Zapatistas, COPINH, the MST, and other territorial political and artistic movements that are remaking the world. To do so ethically means developing horizontal forms of knowledge exchange that do not replicate the paternalism or sense of superiority often displayed by aid workers, development experts, or western scientists. We must also learn about and confront our own complicity, as well as the complicity of the university in which we work or study, with the colonial matrix of power. Finally, we have to work with

Indigenous knowledges and decolonial struggles without romanticization or essentialism and in a way that recognizes how Indigenous lives are entangled with capitalism and extractivism and these entanglements can produce quite contradictory outcomes (Copeland, 2018). We can as Maldonado-Torres (2016: 21) suggests, intervene philosophically with "intersubjective love and understanding", to radically reconfigure modes of human recognition. Decolonial artists and activists including environmental defenders are crucially important in creating the right political conditions for such a thing to occur.

De Sousa Santos (1995, 2002) provides us with three useful metaphors that we can use to adopt a decolonial locus of enunciation: the frontier, the baroque and the South. The frontier is focused on the fluid, plural, and non-hierarchical social and political relations to be found in the margins. The baroque is a mode of utopian subjectivity and sociability that emerges from hybridity and *mestizaje* and is characterized by subversion and laughter. The South refers to suffering that is caused by capitalism and colonialism. He proposes an epistemology of the South that involves three elements "to learn that the South exists; to learn how to go to the South; to learn from and with the South" (de Sousa Santos, 2002:16; 2014). He notes how José Martí's *Nuestra América* which provides a subaltern and situated view of the American continent possesses a baroque ethos in which *Nuestra América* is Caliban in constant struggle against Prospero, characters from Shakespeare's *The Tempest*. A number of Latin American and Caribbean intellectuals, including José Enrique Rodó, Aimé Césaire, and Darcy Ribeiro, have extended and reworked the Caliban/Prospero metaphor and have turned also to the spirit Ariel as a site of decolonial potentiality. De Sousa Santos (2014) draws on this work to present Ariel as a baroque angel subject to three transfigurations. The first is one that posits a decolonial and liberatory *mestizaje* that, unlike dominant *mestizaje*, reveals rather than conceals racism. The second transfiguration is Ariel's transformation into a reflexive organic intellectual that is clear about whose side he is on. The third and final one is epistemological and involves a fundamental shift in the way that knowledge is produced, challenging the epistemological dominance of northern theories and visibilizing instead knowledges that emerge from suffering and decolonial struggles – *the epistemologies of the South*. As de Sousa Santos goes on to note, embracing the epistemologies of the South means being vigilant about one's own practices, to be sure that our work – our practices in the field and our modes of interaction, dissemination, and representation – are not doing more harm than good.

While decolonial thinking cannot be fixed or codified, and like other knowledges, it is culturally and geographically specific, dynamic, and always in the process of construction and emergence, Table 11.1 is an attempt to articulate *some* of the ways that decoloniality differs from Eurocentrism. It is intended as a flexible starting point that you can add to through your own reading and political activism. The terms are not equivalent. As the table

Table 11.1 Eurocentric thought vs. decolonial thought

	Colonial/Eurocentric	Decolonial
1492	Discovery of America	Conquest of America
Land	Resource, private property	Communal, collective entity, laden with meaning
Territory	Piece of land	A collective space in which to reproduce life, and where life becomes "world" (Escobar)
Humanity of the colonized	Humanity in question	Humanity affirmed
Nature	Separate from culture	Cosmovision that include nature and culture, humans and nonhumans as entangled and inseparable
Volcanoes, mountains, lakes, rivers	As resources or hazards to be managed	As alive, sentient, agentic beings that require human care and respect
Earth-beings and water spirits	Non-admissable and non-existent	Fully fledged and alive political actors
Corn	A staple food stuff	A sacred staple food stuff and ancestor
Gender and sexuality	Binary and hierarchically organized. Patriarchy and heterosexuality normalized	Fluid, non-binary, depatriarchalization, multiple genders and sexualities
Civilization	Europe as site of civilization and as civilizing agent	Pre- and post-conquest Latin America as site of multiple civilizations
Colonialism	Quite bad, but had some positives and is now in the past	Brutal, violent, and genocidal, and continues today in the form of coloniality
Capitalism	Not perfect, but the only political system that is thinkable. Nothing to do with colonialism	A political system that is entangled with and inseparable from colonialism, racism, and heteropatriarchy. A system that has no future (de Sousa Santos). Needs to be dismantled, there are many alternatives
Free market	A system that gives individuals the freedom to buy and sell and create wealth and that by itself eliminates any distortions created by economic processes	An economic model that benefits a small minority, naturalizes dispossession, and is subject to constant crises

Table 11.1 Cont.

	Colonial/Eurocentric	Decolonial
Feminism	A liberal variety that incorporates women into the neoliberal social order. Provides white cis women with opportunities and forms of mobility denied to women of colour and to people with non-dominant gender identities. Complicit with racial capitalism and transphobia	A decolonial and intersectional feminism that decentres white feminism and its colonizing underpinnings. Accords epistemological privilege to Black and Indigenous women and to LGBTQ+ people. Refuses complicity with racial capitalism and transphobia
Modernity	A time of scientific and industrial progress and an end to religious tyranny. Led by European Enlightenment thinkers who were able to think rationally. There is one modernity	A colonial project. Not something that exists in addition to coloniality, but something created by it and from which it cannot be separated. Something that can be forged by different people in different places, in opposition to or instead of Euro-centered modernity. There are multiple modernities or transmodernity
Slavery	A colonial system of coerced labour that was brought to an end thanks to the benevolence and statesmanship of parliamentarians such as William Wilberforce	A violent colonial system of coerced labour that lasted for 200 years and made many Europeans extraordinarily wealthy, including as a result of the compensation some slaveowners received when slavery was abolished. It was ended as a result of the rebellions of the enslaved and their descendants are still waiting for reparations
The human	Natural organism	Hybrid organism made of organic matter and symbolic content (Wynter)

(continued)

Table 11.1 Cont.

	Colonial/Eurocentric	Decolonial
Methodology	Research on or about, extractivist, aimed at understanding	Research with, in solidarity, aimed at understanding and transforming
Positionality	Feigns objectivity and balance, disguises its politics	Takes sides, overtly and explicitly on the side of the oppressed and dehumanized. Anti-colonial, anti-capitalist, anti-patriarchal
Time	Linear and singular, moves forward according to the clock	Circular, spiral, and multiple. Moves at different speeds. The past-future is contained in the present (Rivera Cusicanqui)
Eurocentric knowledge	Universal, objective, disembodied, scientific and superior to other knowledges	Like all knowledges, local, particularistic, embodied, socially located. Not superior, and sometimes used to legitimize oppression or dismiss other ways of knowing. Should be treated horizontally not hierarchically
Indigenous knowledge	Superstition, belief, often wrong – to be dismissed or tolerated	Knowledge that provides useful insight, documents history, reproduces life, and defends territory
Art	Something apolitical (or only mildly political) to be enjoyed as part of leisure or aesthetic pleasure. Elite, hard to access for those without cultural capital	Laden with political meaning and can contribute to a decolonial aesthetics. Political and participatory, for and of the people World-making
Climate change	An environmental problem that can be managed through technocratic means within capitalism and for which all of humanity is responsible	A civilizational crisis caused by rich white people in the ZB that is causing misery for people in the ZNB. It can only be addressed if we dismantle colonial racial heteropatriarchal capitalism and put an end to extractivism

Table 11.1 Cont.

	Colonial/Eurocentric	Decolonial
Covid-19	Unprecedented, can be managed through Eurocentric and biopolitical modes of public health and "following the science" so that we can return to (colonial-capitalist) normality	The expected and easily predicted outcome of capitalist coloniality. There was no normality especially for those in the ZNB before and we should not be seeking to return to the same system that produced it

indicates, there is a Eurocentric feminism and a decolonial feminism, but there is no decolonial capitalism, as capitalism by its very nature is colonial.

If we decide to reside in the frontier, adopt a baroque subjectivity, and take on board the suffering of the South, the suffering that is Ayotzinapa, it becomes hard to be optimistic about mainstream development and initiatives such as the SDGs. It's clear, however, that the epistemological crisis outlined above – and the growing awareness that we are accelerating our own extinction as a species – is beginning to dismantle the epistemic authority of modern-colonial development. In 2020, the leadership of ECLAC announced that all of the development strategies that have been implemented in Latin American have failed. Gudynas (2020) called this event a confession, noting that it was surprising that such a statement did not produce much of a response from governments, the media, or citizens. As he writes, for decades, ECLAC has both produced and engaged with, albeit sometimes critically, with mainstream development thinking. In particular, they failed to abandon a belief in economic growth as the engine of development and rather than denouncing extractivisms, supported them while arguing in favour of forms of technological innovation that could make them cleaner and greener and of the reinvestment of profits in local communities to reduce inequalities and social conflicts. Their recognition that extractivisms bring harm and that there has been hardly any technological innovation that has diminished the harms or expanded the benefits is significant. Gudynas recommends that ECLAC heeds the words of its founder Raúl Prebisch (1963: 20, my translation) who wrote:

> The tendency to import ideologies is still very strong in Latin America, and it is as strong as the tendency of the centres to export them. (…) It is not a question of closing the intellect to what is thought and done in other countries. Fortunately, there is a growing interest in development theory and problems in the large centres, and it would be a serious mistake not to take advantage of the valuable contribution thus made

> to us. But nothing exempts us from the intellectual obligation to analyze our own phenomena and find our own image in the effort to transform the existing order of things. Let us make intelligent use of outside thought and experiences, but only as a formative element of our own thinking.

There is abundant evidence of this endogenous thinking, but it is found mostly within anti-extractivist and territorial social movements and in movements such as Zapatismo and in intercultural universities such as URACCAN or Universidad de la Tierra (see Cupples and Glynn, 2014b; Esteva, 2014, 2017, 2019; Mato, 2019). The challenge is how to inject this thinking into the westernized university (see Cupples, 2018; Grosfoguel, 2012), as well as into the institutional spaces of development in order to bring an end to the politics of death carried out in the name of development.

We must recognize that our contemporary universities are not adequately organized for decolonization, and endeavours that claim to decolonize must be accompanied by the return of stolen land, the end of extractivism, and reparations for slavery. We should, however, focus on the epistemic. Land dispossession is made possible by epistemic violence, in which the people deprived of their lands are also deprived of their epistemic autonomy and political self-determination. All of us, students, researchers, and teachers, confront, whether we wish to or not, the question of how we come to know what we know and take up a position in which there is no neutrality. Either we reproduce myths of universality and objectivity to sustain Europe's problematic and colonizing "claim to sole epistemological authority" (Gopal, 2021: 880) or we begin to transform our disciplines, our methodologies, our intellectual hierarchies, the ways in which we are together with human and nonhuman others, and our practices of citation, and start to ask different questions posed from different epistemic locations. As we do so, we must try to mobilize a politics of hope, love, and life. It is June 2021; the Zapatistas have just arrived in Europe.

SUMMARY

- Studying contemporary Latin America in relation to questions of development and decolonization involves understanding struggles for hegemony between Eurocentric and ancestral cosmovisions, between capitalism and socialism, between extractivism and sustainability, and especially between the politics of life and death.
- Colonial violence has led to myriad collective movements and struggles in defence of land, territory, human rights, and dignity that often operate beyond the state.

- Activists who are murdered with impunity do not die but give rise to life, as they live on in collective memory and in acts of commemoration and they continue to shape the contours of decolonial struggle.
- The case of the Ayotzinapa 43 is not an aberration, it is an outcome of the colonial matrix of power and the apocalyptic phase of capitalism into which we are moving.
- The epistemological and civilizational conjunctural crisis we are living through requires radical solutions and a politics of hope, love, and life.

DISCUSSION QUESTIONS

1 Why did Ayotzinapa have such a big impact on Latin America and what are its implications for understanding the Latin American conjuncture?
2 Explain why the debates about free speech and colonial monuments are so fraught at the current time. Why do so many people want to defend the colonial–capitalist status quo?
3 Study Table 11.1. Based on your reading and studies so far, can you think of other ways that a decolonial approach differs from a Eurocentric one?
4 How can students and academics decolonize the university? What things will you do or approach differently now?
5 What role do love, hope, and life play in these struggles and in your learning?

FURTHER READING

de Sousa Santos, B. (2014) *Epistemologies of the South: Justice Against Epistemicide.* Boulder, CO, and London: Paradigm Publishers.
A book that reveals how struggles for justice are epistemic as well as political. It emphasizes how there are ways of knowing that exceed Eurocentric understanding and it is these knowledges that are central to emancipation.

Recommended films
Ayotzinapa, el paso de la tortuga (2018), directed by Enrique García Meza
A documentary about the struggle for justice by the students of Ayotzinapa
The 43 (2019), directed by Matías Gueilburt
A two-part docuseries on the missing 43 students in Iguala, Mexico

A playlist for rebellion and decolonization

Lila Downs (feat. Juanes) – La Patria Madrina (Mexico)
Victor Jara – Te recuerdo Amanda (Chile)
Gotan Project – El capitalism foráneo (Argentina)
Calle 13 – Latinoamérica (Puerto Rico)
Rebeca Lane – Alma mestiza (Guatemala)
Amparo Ochoa – La maldición del Malinche (Mexico)
Silvio Rodríguez – Te doy una canción (Cuba)
Mercedes Sosa – Sólo le pido a Dios (Argentina)
Juan Luis Guerra – Ojalá que llueva café (Dominican Republic)
Los Olimareños – Milonga del Fusilado (Uruguay)
Luis Enrique Mejía Godoy – Somos hijos del maíz (Nicaragua)
Carlos Mejía Godoy – El zenzontle pregunta por Arlen (Nicaragua)
Guardabarranco – Guerrero del amor (Nicaragua)
Manu Chao – EZLN para tod@s todo (Spain)
Maná – ¿Dónde jugarán los niños? (Mexico)
Going to Bocas – Walter Ferguson (Costa Rica)
Joan Baez – La llorona (US)
Caetano Veloso – Cucurrucucu Paloma (Brazil)
Soul Vibrations – Rock down Central America (Nicaragua)
Los Bukis – Tu cárcel (Mexico)
Joe Arroyo – Rebelión (Colombia)
Héctor Lavoe – El periódico de ayer (US/Puerto Rico)
Grupo Niche – Etnia (Colombia)
Ana Tijoux y MC Millaray – Rebelión de octubre (Chile)
Carlos Vives – La gota fría (Colombia)
Rita Indiana – Pa' Ayotzinapa (Dominican Republic)
Atahualpa Yupanqui – Hermanos (Argentina)
Inti-Illimani – Exiliada del Sur (Chile)
Celia Cruz – Guantanamera (Cuba)

Ali Primera – La patria es el hombre (Venezuela)
Aventura – Obsesión (Dominican Republic)
Daymé Arocena – Negra caridad (Cuba)
Sara Curruchich – Hija de la tierra (Guatemala)
David Rudder – Haiti I am sorry (Trinidad and Tobago)

References

Abad Faciolince, H. (2006) *El olvido que seremos*. Bogotá: Editorial Planeta Colombia.

Adorno, R. (2000) *Guaman Poma: Writing and Resistance in Colonial Peru*, 2nd ed. Austin: University of Texas Press.

Afolabi, N. (2013) Reversing dislocations: African contributions to Brazil in the works of Manuel Querino, 1890–1920. *History Compass* 11(4): 259–267.

Afro-descendant Women Human Rights Defenders Project (2012) *Defeating Invisibility: A Challenge for Afro-Descendant Women in Colombia*. Bogotá: PCN.

Aguayo Ayala, A. (2016) Nuevo Polanco: Renovación urbana, segregación y gentrificación en la Ciudad de México. *Iztapalapa. Revista de Ciencias Sociales y Humanidades*, 37(80): 101–123.

Ahmed, S. (2012). *On Being Included*. Durham, NC: Duke University Press.

Ahmet, A. (2020) Who is worthy of a place on these walls? Postgraduate students, UK universities, and institutional racism. *Area* 52:678–686.

Alarcón-González, D. and McKinley, T. (1999) The adverse effects of structural adjustment on working women in Mexico. *Latin American Perspectives* 26(3): 103–117.

Aman, R. (2018) Other knowledges, Other interculturalities: The colonial difference in intercultural dialogue. In J. Cupples and R. Grosfoguel (eds) *Unsettling Eurocentrism in the Westernized University*. London: Routledge, pp. 171–186.

Anderson, M. (1996) The quilombo of Palmares: A new overview of a maroon state in seventeenth-century Brazil." *Journal of Latin American Studies* 28: 545–66.

——— (2009) *Black and Indigenous: Garífuna Activism and Consumer Culture in Honduras*. Minneapolis: University of Minnesota Press.

——— (2011) *Disaster Writing: The Cultural Politics of Catastrophe in Latin America*. Charlottesville and London: University of Virginia Press.

Anderson, J. and Gomes, P. (2021) Blackness, race, and politics in Argentina. *Oxford Research Encyclopedia of Politics* 21 February https://oxfordre.com/politics/view/10.1093/acrefore/9780190228637.001.0001/acrefore-9780190228637-e-1703 (Accessed 1 June 2021).

Andolina, R., Laurie, N. and Radcliffe, S. (2009) *Indigenous Development in the Andes: Culture, Power, and Transnationalism*. Durham, NC: Duke University Press.

André, R. (2011) The invisible war against Afro-Colombians. *Americas Quarterly Blog* 16 March www.americasquarterly.org/node/2322 (Accessed 4 June 2021).

Andreoli, E. (2020) Architecture in El Alto: The politics of excess. In W Van Acker and T Mical (eds) *Architecture and Ugliness: Anti-Aesthetics and the Ugly in Postmodern Architecture*. London: Bloomsbury Visual Arts, pp. 193–208.

Andrews, G. R. (1980) *The Afro-Argentines of Buenos Aires*. Madison: University of Wisconsin Press.

——— (1996) Brazilian racial democracy, 1900–90: An American counterpoint. *Journal of Contemporary History* 31(3): 483–507.

——— (2018) Afro-Latin America by the numbers: The politics of the census. *ReVista: Harvard Review of Latin America* 4 January https://revista.drclas.harvard.edu/afro-latin-america-by-the-numbers/ (Accessed 29 May 2021).

Andrews, K. (2016) The black studies movement in Britain. *The Black Scholar* 6 October www.theblackscholar.org/black-studies-movement-britain/ (Accessed 3 March 2021).

Ángel, A. (2016) La construcción retórica de la corrupción. *Chasqui* 132: 309–327.

Antón, J., Bello, A., Del Popolo, F., Paixão, M. and Rangel, M. (2009*) Afrodescendientes en América Latina y el Caribe: Del reconocimiento estadístico a la realización de derechos*. Santiago: Naciones Unidas.

Anzaldúa, G. (1999) *Borderlands/La Frontera: The New Mestiza*. San Francisco: Aunt Lute Books.

Aparicio, F. R. (1998) *Listening to Salsa: Gender, Latin Popular Music, and Puerto Rican Cultures*. Hanover: Wesleyan University Press.

Appelbaum, N. P., Macpherson, A. S. and Rosemblatt, K. A. (2003) Introduction: Racial nations. In N. P Appelbaum, A. S. Macpherson and K. A. Rosemblatt (eds) *Race and Nation in Modern Latin America*. Chapel Hill: University of North Carolina Press, pp. 1–31.

Araujo, A. (2014) Coca-Codo Sinclair confirmó la muerte de 13 personas en el proyecto hidroeléctrico. *El Comercio* 14 December www.elcomercio.com/actualidad/muerto-heridos-accidente-cocacodosinclair-hidroelectrica.html (Accessed 14 April 2021).

Araujo, A. L. (2015) The mythology of racial democracy in Brazil. *Open Democracy* 22 June www.opendemocracy.net/en/beyond-trafficking-and-slavery/mythology-of-racial-democracy-in-brazil/ (Accessed 2 June 2021).

Arias, A. (2016) Violence, Indigeneities and human rights. In S. A. McClennen and A. Schulteis Moore (eds) *The Routledge Companion to Literature and Human Rights*. New York: Routledge, pp. 326–332.

Arteta, I. (2017) "¿Es usted negro?", una pregunta mal planteada o ausente en los censos. *El País* 25 January https://elpais.com/internacional/2017/01/24/actualidad/148526 5647_183464.html (Accessed 2 June 2021).

Asher, K. (2017) Spivak and Rivera Cusicanqui on the dilemmas of representation in post-colonial and decolonial feminisms. *Feminist Studies* 43(3): 512–524.

——— (2020) Fragmented forests, fractured lives: Ethno-territorial struggles and development in the Pacific Lowlands of Colombia. *Antipode* 52(4): 949–970.

Assies, W. (2003) David versus Goliath in Cochabamba: Water rights, neoliberalism, and the revival of social protest in Bolivia. *Latin American Perspectives* 30(3): 14–36.

Atkinson, R. and Bridge, G. (eds) (2005) *Gentrification in a Global Context: The New Urban Colonialism*. Abingdon: Routledge.

Babb, F. E. (2010) *After Revolution: Mapping Gender and Cultural Politics in Neoliberal Nicaragua*. Austin: University of Texas Press.

———— (2019) LGBTQ sexualities and social movements. In J. Cupples, M. Palomino-Schalsha and M. Prieto (eds) *Routledge Handbook of Latin American Development*. London: Routledge, pp. 308–318.

Baca, D. (2010) Rhetoric, interrupted: La Malinche and Nepantlisma. *Rhetorics of the Americas: 3114 BCE to 2012 CE*. New York: Palgrave Macmillan, pp. 143–151.

Bacigalupo, A. M. (2003) Rethinking identity and feminism: Contributions of Mapuche Women and Machi from Southern Chile. *Hypatia* 18(2):32–57.

———— (2007) *Shamans of the Foye Tree: Gender, Power, and Healing among Chilean Mapuche*. Austin: University of Texas Press.

Bairros, L. (2008) A community of destiny: New configurations of racial politics in Brazil. *Souls: A Critical Journal of Black Politics, Culture, and Society* 10(1): 50–53.

Ballvé, T. and McSweeney, K. (2020) The 'Colombianisation' of Central America: Misconceptions, mischaracterisations and the military-agroindustrial complex. *Journal of Latin American Studies, 52*(4), 805–829.

Baltodano, M. (2008) Mujeres sandinistas para la historia. *Mujeres en Red* Agosto www.mujeresenred.net/spip.php?article1585 (Accessed 10 February 2021).

Banuchi, O. and Morales, E. (2018) A cartoon history of colonialism in Puerto Rico: A primer on how the island became the last colony. *The Village Voice* 19 March www.villagevoice.com/2018/03/19/a-cartoon-history-of-colonialism-in-puerto-rico/ (Accessed 30 November 2020).

Bartlett, J. (2021) 'A new Chile': political elite rejected in vote for constitutional assembly. *The Guardian* 18 May www.theguardian.com/world/2021/may/18/a-new-chile-political-elite-rejected-in-vote-for-constitutional-assembly (Accessed 30 June 2021).

Barton, J. R. (1997) *A Political Geography of Latin America*. London: Routledge.

BBC (2011) Bolivia's Evo Morales says no to DEA agents' return. 4 March www.bbc.co.uk/news/world-latin-america-12643404 (Accessed 8 May 2021).

Bebbington, A. (2019) Rural social movements: Conflicts over the countryside. In J. Cupples, M. Palomino-Schalsha and M. Prieto (eds) *Routledge Handbook of Latin American Development*. London: Routledge, pp. 321–331.

Belli, G. (2001) *El país bajo mi piel*. Barcelona and Managua: Plaza y Janés and Anamá.

Benería, L., Floro, M., Grown, C. and MacDonald, M. (2000) Introduction: Globalization and gender. *Feminist Economics*. 6(3): vii–xviii.

Benjamin, A. (2000) A time of reconquest: History, the Maya revival, and the Zapatista rebellion in Chiapas. *The American Historical Review* 105(2): 417–450.

Beristain, M., Páez Rovira, D. and Fernández, I. (2009) *Las palabras de la selva: Estudio psicosocial del impacto de las explotaciones petroleras de Texaco en las comunidades amazónicas de Ecuador*. Bilbao: Hegoa.

Bhambra, G. (2014) Postcolonial and decolonial dialogues. *Postcolonial Studies* 17(2): 115–212.

Bhambra, G., Gebrial, D. and Nişancıoğlu, K. (2018). *Decolonising the University*. London: Pluto Press.

Bidaseca, K. (2015) Feminicide and the pedagogy of violence: An essay on exile, coloniality and nature in third feminism. In G. Melville and C. Ruta (eds) *Thinking the Body as a Basis, Provocation, and Burden of Life*. Berlin: Walter de Gruyter, pp. 224–234.

Bird, A. (2011) Biofuels, mass evictions and violence build on the legacy of the 1978 Panzos massacre in Guatemala. *Upside Down World* 23 March https://upsidedownworld.org/archives/guatemala/biofuels-mass-evictions-and-violence-build-on-the-legacy-of-the-1978-panzos-massacre-in-guatemala/ (Accessed 5 June 2021).

Blackwell, M., Boj Lopez, F. and Urrieta, L. (2017) Special issue: Critical Latinx indigeneities. *Latino Studies* 15: 126–137.

Blanco, R. and Teixeira Delgado, A. C. (2019) Problematising the ultimate Other of Modernity: The crystallisation of coloniality in international politics. *Contexto Internacional* 41(3): 599–619.

Blasberg, M., Glüsing, J. and Kollenbroich, B. (2021) "It's the biggest genocide in our history." *Spiegel International* 30 March www.spiegel.de/international/world/ex-president-lula-on-brazil-s-corona-disaster-it-s-the-biggest-genocide-in-our-history-a-ada6e391-14a5-4231-9d9f-212194ecf914 (Accessed 3 April 2021).

Blaser, M. (2014) Ontology and indigeneity: On the political ontology of heterogeneous assemblages. *cultural geographies* 21(1): 49–58.

Blaser, M. and de la Cadena, M. (2018) Introduction: Pluriverse: Proposals for a world of many worlds. In M. Blaser and M. de la Cadena (eds) *A World of Many Worlds*. Durham, NC: Duke University Press, pp1–22.

Boggs, V. (1992) *Salsiology: Afro-Cuban Music and the Evolution of Salsa in New York*. New York: Excelsior Music Publishing.

Bonilla, Y. (2020) The coloniality of disaster: Race, empire, and the temporal logics of emergency in Puerto Rico, USA. *Political Geography* 78 (online early).

Bonilla, Y. and LeBrón, M. (2019a) Introduction: Aftershocks of disasters. In Y. Bonilla and M. LeBrón (eds) *Aftershocks of Disaster*. Chicago: Haymarket Books, pp. 1–20.

——— (eds) (2019b) *Aftershocks of Disaster*. Chicago: Haymarket Books.

Bonilla, A. and Mancero, M. (2020) "Venimos a luchar por el pueblo, no por el poder": el levantamiento indígena y popular en Ecuador 2019. In C. Parodi and N. Sticotti (eds) *Ecuador: La insurrección de octubre*. Buenos Aires: CLACSO, pp. 271–280.

Booth, R. (2020) UK more nostalgic for empire than other ex-colonial powers. *The Guardian* 11 March www.theguardian.com/world/2020/mar/11/uk-more-nostalgic-for-empire-than-other-ex-colonial-powers (Accessed 28 July 2020).

Borja, D. and Buitrón Cañadas, V. (2020) Sí, la normalidad es el problema: Inequidad, exclusión y fuerza estatal en la crisis del Covid-19 en Guayaquil. *Journal of Latin American Geography* 19(3): 224–233.

Bourdieu, P. (1984) *Distinction: A Social Critique of the Judgement of Taste*, translated by R. Nice. London: Routledge.

Boyle, C. M. (1993) Touching the air: The cultural force of women in Chile. In S.A. Radcliffe and S. Westwood (eds) *Viva: Women and Popular Protest in Latin America*. London and New York: Routledge, pp. 156–172.

Bradshaw, S. (2001) Reconstructing roles and relations: Women's participation in reconstruction in post- Mitch Nicaragua. *Gender and Development* 9(3): 79–87.

Brand, P. and Dávila, J. D. (2011) Mobility innovation at the urban margins. *City* 15(6): 647–661.

Brenner, N. and Schmid, C. (2012) Planetary urbanization. In M. Gandy (ed) *Urban Constellations*. Berlin: Jovis, pp. 10–13.

Brooks, D. (2019) México: "El contraste es brutal", cómo es vivir en Cerrada Andrómaco, el callejón que quedó atrapado entre los edificios de "Ciudad Slim." *BBC Mundo* 5 April www.bbc.com/mundo/noticias-america-latina-47620849 (Accessed 13 June 2021).

Bryan, J. (2009) Where would we be without them? Knowledge, space, and power in indigenous politics. *Futures* 41(1): 24–32.

———— (2019) Counter-mapping development. In J. Cupples, M. Palomino-Schalsha and M. Prieto (eds) *Routledge Handbook of Latin American Development*. London: Routledge, pp. 263–272.

Buchenau, J. and Johnson, L.L. (eds) (2009) *Aftershocks: Earthquakes and Popular Politics in Latin America*. Albuquerque: University of New Mexico Press, pp. 184–221.

Burki, T. (2021) Behind Cuba's successful pandemic response. *The Lancet* 21(4): 465–466.

Bush, D., Bush, S., Cayasso-Dixon, K., Cupples, J., Gleghorn, C., Glynn, K., Henríquez Cayasso, G., Lee, D., Moreno Rojas, C., Perea Lemos, R., Ribeiro, R. and Valencia, Z. (2019) Afro-Latino-América: Black and Afro-descendant rights and struggles. In J Cupples, M Prieto and M Palomino-Schalscha (eds) *The Routledge Handbook of Latin American Development*. London: Routledge, pp. 236–251.

Butler, J. (1990) *Gender Trouble: Feminism and the Subversion of Identity*. London: Routledge.

Cadena-Roa, J. (2002) Strategic framing, emotions, and Superbarrio: Mexico City's masked crusader. *Mobilization: An International Journal* 7(2): 201–216.

Calderón F. and Castells, M (2020) *The New Latin America*, translated by R. McGlazer. Cambridge: Polity Press.

Calderón, E., Gray, N., Kallin, H. and Soytemel, E. (2019) An explanatory or mystifying concept? The use value of gentrification theory. In J Cupples and T Slater (eds) *Producing and Contesting Urban Marginality: Interdisciplinary and Comparative Dialogues*. London: Rowman and Littlefield, pp. 19–42.

Calvo, L. and Rueda Esquibel, C. (2015) *Decolonize Your Diet*. Vancouver: Arsenal Pulp Press.

Camacho de Schmidt, A. and Schmidt, A. (1995) Foreword: The shaking of a nation. In E. Poniatowska (1995) *Nothing, Nobody: The Voices of the Mexico City Earthquake*. Philadelphia: Temple University Press, pp. ix–xxix.

Cannon, T. (1994) Vulnerability analysis and the explanation of 'natural' disasters. In A. Varley (ed) *Disasters, Development and Environment*. Chichester: Wiley, pp. 13–30.

Carey, M. (2008) The politics of place: Inhabiting and defending glacier hazard zones in Peru's Cordillera Blanca. In B. Orlove, E. Wiegandt and B. H. Luckman (eds) *Darkening Peaks: Glacier Retreat, Science and Society*. Berkeley: University of California Press, pp. 229–240.

Carmagnani, M. (2011) *The Other West: Latin America from Invasion to Globalization*. Berkeley: University of California Press.

Carneiro, S. (2011) *Racismo, Sexismo e Desigualdade no Brasil*. São Paulo: Selo Negro Edições.

Casey, N. and Krauss, C. (2018) It doesn't matter if Ecuador can afford this dam. China still gets paid. *New York Times* 24 December www.nytimes.com/2018/12/24/world/americas/ecuador-china-dam.html? (Accessed 17 April 2021).

Castañeda, P. (2019) From the right to mobility to the right to the mobile city: Playfulness and mobilities in Bogotá's cycling activism. *Antipode* 52(1): 58–77.

Castillo, I. and Largaespada, M. (2010) La insurrección de los niños en Matagalpa. In M. Baltodano (ed) *Memorias de la lucha sandinista*. Managua: IHNCA-UCA, pp. 153–168.

Castro Buzon, N. (2020) Gladys Tzul Tzul "Las mujeres indígenas reivindicamos una larga memoria de lucha por la tierra." *Revistas Amazonas* 3 April www.revistaamazonas.com/2020/04/03/gladys-tzul-tzul-las-mujeres-indigenas-reivindicamos-una-larga-memoria-de-lucha-por-la-tierra/ (Accessed 22 May 2021).

Cepek, M. (2012) The loss of oil: Constituting disaster in Amazonian Ecuador. *The Journal of Latin American and Caribbean Anthropology* 17(3): 393–412.

Chaguaceda, A. (2019) Russia and Nicaragua: Progress in bilateral cooperation. *Global Americans* 28 March https://theglobalamericans.org/2019/03/russia-and-nicaragua-progress-in-bilateral-cooperation/ (Accessed 16 April 2021).

Chalhoub, S. (2018) Slavery and precarious freedom: A strange co-existence in 19th-century Brazil. *ReVista: Harvard Review of Latin America* 17(2) https://revista.drclas.harvard.edu/slavery-and-precarious-freedom/ (Accessed 25 June 2021).

Chambers, S. C. (2001) Republican friendship: Manuela Sáenz writes women into the nation, 1835–1856. *The Hispanic American Historical Review* 81(2): 225–257.

Chamosa, O. (2010) Criollo and Peronist: The Argentine folklore movement during the first Peronism, 1943–1955. In M. B. Karush and O. Chamosa (eds) *The New Cultural History of Peronism: Power and Identity in Mid-Twentieth-Century Argentina*. Durham, NC: Duke University Press, pp. 113–142.

Cheng, Y. (2007). Sino-Cuban relations during the early years of the Castro Regime, 1959–1966. *Journal of Cold War Studies* 9(3): 78–114.

Chuchryk, P. (1991) Women in the revolution. In T. W. Walker (ed) *Revolution and Counter-Revolution in Nicaragua*. Boulder, CO: Westview Press, pp. 143–166.

COHA (2011) A victory for gay rights in Latin America. *Council of Hemispheric Affairs* 25 May www.coha.org/latin-america-progresses-forward-a-victory-for-gay-rights/ (Accessed 28 June 2021).

Coffey C., Espinoza Revollo, P., Harvey, R. Lawson, M., Parvez Butt, A., Piaget, K., Sarosi, D. and Thekkudan, J. (2020) *Time to Care: Unpaid and Underpaid Care Work and the Global Inequality Crisis*. Oxford: Oxfam.

Collier, G. A. (1994) The new politics of exclusion: Antecedents to the rebellion in Mexico. *Dialectical Anthropology* 19: 1–44.

Composto, C. and Navarro, M.L. (2014) Presentación. In C. Composto and M.L. Navarro (eds) *Territorios en disputa: Despojo capitalista, luchas en defensa de los bienes comunes naturales y alternativas emancipatorias para América Latina*. México: Bajo Tierra Ediciones, pp. 13–15.

Conferencia Nacional de Organizaciones Afrocolombianas (2015) Cuéntame – Yo cuento en este cuento – CNOA [YouTube] www.youtube.com/watch?v=vVEiG9qrrzU (Accessed 4 June, 2021).

Copeland, N. (2019) Linking the defence of territory to food sovereignty: Peasant environmentalisms and extractive neoliberalism in Guatemala. *Journal of Agrarian Change* 19: 21–40.

COPINH (2015) Discurso de Berta Cáceres en el Opera House, San Francisco California al recibir el Premio Ambiental Goldman, el 20 de abril, 2015. *COPINH* 22 April https://copinh.org/2015/04/discurso-de-berta-caceres-en-el-opera-house-san-francisco-california-al-recibir-el-premio-ambiental-goldman-el-20-de-abril-2015/ (Accessed 2 August 2020).

Corrales, J. (2010) Latin American gays: The post-left leftists. *Americas Quarterly* 19 March www.americasquarterly.org/gay-rights-Latin-America (Accessed 30 June 2021).

Corrales, J. and Pecheny, M. (2010) Introduction: The comparative politics of sexuality in Latin America. In J. Corrales and M. Pecheny (eds) *The Politics of Sexuality in Latin America: A Reader on Lesbian, Gay, Bisexual, and Transgender Rights*. Pittsburgh: University of Pittsburgh Press, pp. 1–30.

Correa-Cabrera, G. (2019) The War on Drugs in Latin America from a development perspective. In J. Cupples, M. Palomino-Schalsha and M. Prieto (eds) *Routledge Handbook of Latin American Development*. London: Routledge, pp. 132–144.

Correia, J. E. (2019) Soy states: Resource politics, violent environments and soybean terri-torialization in Paraguay. *The Journal of Peasant Studies* 46: 316–336.

Cortés, H. (2018[1521]) *Five Letters of Cortes to the Emperor: 1519–1526.* Mount Pleasant, SC: Arcadia Press.

Cotto Morales, L. (2020) Social movements, crises, and mobilizations: A look at summer 2019. *Latin American Perspectives* 47(3): 129–137.

Crossa, V. (2009) Resisting the entrepreneurial city: Street vendors' struggles in Mexico City's historic center. *International Journal of Urban and Regional Research* 33: 43–63.

Cuadra, P. A. (1997) *El Nicaragüense.* Hispamer: Managua.

Cultural Survival (2020) In memoriam: 28 Indigenous rights defenders murdered in Latin America in 2019. 28 January www.culturalsurvival.org/news/memoriam-28-indigenous-rights-defenders-murdered-latin-america-2019 (Accessed 15 August 2020).

Cumbre Internacional Mujeres Afro (2017) Campaña Visibilidad Afrodescendiente en Censo 2017 Nicaragua [YouTube] www.youtube.com/watch?v=bzkldrvehvY (Accessed 4 June 2021).

Cupples J (2002) The field as a landscape of desire: Sex and sexuality in geographical field-work. *Area* 34(4): 382–390.

——— (2004) Rural development in El Hatillo, Nicaragua: Gender, neoliberalism and environmental risk. *Singapore Journal of Tropical Geography* 25(3): 343–357.

——— (2006) Between maternalism and feminism: Women in Nicaragua's counter-revolu-tionary forces. *Bulletin of Latin American Research* 26(1): 83–103.

——— (2007) Gender and Hurricane Mitch: Reconstructing subjectivities after disaster. *Disasters: The Journal of Disaster Studies, Policy and Management* 31(2): 155–175.

——— (2012) Wild globalization: The biopolitics of climate change and global capitalism on Nicaragua's Mosquito Coast. *Antipode* 44(1): 10–30.

——— (2018) Introduction: Coloniality resurgent, coloniality interrupted. In J Cupples and R Grosfoguel (eds) *Unsettling Eurocentrism in the Westernized University.* London: Routledge, pp. 1–22.

——— (2019) Conclusion: Urban research and the pluriverse: Analytical and pol-itical lessons from scholarship in varied margins. In J. Cupples and T. Slater (eds) *Producing and Contesting Urban Marginality: Interdisciplinary and Comparative Dialogues.* London: Rowman and Littlefield, pp. 205–226.

——— (2020) Love in the time of Covid-19: Or, Nicaragua, the strange country where children still go to school. *Journal of Latin American Geography* 19(3): 323–333.

Cupples, J. and Glynn, K. (2013) Postdevelopment television? Cultural citizenship and the mediation of Africa in contemporary TV drama. *Annals of the Association of American Geographers* 103(4): 1003–1021.

——— (2014a) The mediation and remediation of disaster: Hurricanes Katrina and Felix in/and the new media environment. *Antipode* 46(2): 359–381.

——— (2014b) Indigenizing and decolonizing higher education on Nicaragua's Atlantic Coast. *Singapore Journal of Tropical Geography* 35(1): 56–71.

——— (2018) *Shifting Nicaraguan Mediascapes: Authoritarianism and the Struggle for Social Justice.* Cham: Springer.

——— (2019) Popular religiosity and struggles for urban justice in Mexico: A decolonial analysis of Santa Muerte. In J. Cupples and T. Slater (eds) *Producing and Contesting Urban Marginality: Interdisciplinary and Comparative Dialogues.* London: Rowman and Littlefield, pp. 117–139.

Cupples, J., Glynn, K. and Larios, I. (2007) Hybrid cultures of postdevelopment: The struggle for popular hegemony in rural Nicaragua. *Annals of the Association of American Geographers* 97(4): 786–801.

Cupples, J and Grosfoguel, R. (eds) (2018) *Unsettling Eurocentrism in the Westernized University*. London: Routledge.

Cupples, J., Palomino-Schalsha, M. and Prieto, M. (2019a) Latin American development: Editors' introduction. In J. Cupples, M. Palomino-Schalsha and M. Prieto (eds) *Routledge Handbook of Latin American Development*. London: Routledge, pp. 1–11.

Cupples, J., Palomino-Schalscha, M. and Prieto, M. (2019b) (eds) *Routledge Handbook of Latin American Development*. London: Routledge.

Davies, J. (2010) Zapatista supporters attacked in retaliation for building an autonomous school. *Upside Down World* 14 September https://upsidedownworld.org/archives/mexico/zapatista-supporters-attacked-in-retaliation-for-building-an-autonomous-school/ (Accessed 30 June 2021)

de Certeau, M. (1984) *The Practice of Everyday Life*, translated by S. Rendall. Berkeley: University of California Press.

———— (1988) *The Writing of History*, translated by T. Conley. New York: Columbia University Press.

Declercq, N. F., Degrieck, J., Briers, R. and Leroy, O. (2004) A theoretical study of special acoustic effects caused by the staircase of the El Castillo pyramid at the Maya ruins of Chichen-Itza in Mexico. *The Journal of the Acoustical Society of America* 116: 3328–3335.

Declet-Barreto, J. (2020) #coloniavirus, cambio climático y colonialismo: La construcción colonial de la precariedad en Puerto Rico. *Journal of Latin American Geography* 19(3): 289–295.

DeGuzmán, M. (2017) Latinx: ¡Estamos aquí!, or being "Latinx" at UNC-Chapel Hill. *Cultural Dynamics* 29(3) 214–230.

de la Cadena, M. (2010) Indigenous cosmopolitics in the Andes: Conceptual reflections beyond "politics." *Cultural Anthropology* 25(2): 334–370.

———— (2014) The politics of modern politics meets ethnographies of excess through ontological openings. *Society for Cultural Anthropology* 13 January https://culanth.org/fieldsights/the-politics-of-modern-politics-meets-ethnographies-of-excess-through-ontological-openings (Accessed 3 July 2021).

———— (2015) *Earth Beings. Ecologies of Practice Across Andean Worlds*. Durham, NC: Duke University Press.

de las Casas, B. (1992[1552]) *A Short Account of the Destruction of the Indies*. London: Penguin.

de Landa, D. (1978 [1937] [1552]) *Yucatan Before and After Conquest*, translated by W. Gates. New York: Dover Publications.

Democracia Abierta (2020) As the pandemic continues to accelerate, so does the deforestation of the Amazon. *Open Democracy*, 20 June www.opendemocracy.net/en/democraciaabierta/se-acelera-la-pandemia-y-se-acelera-la-deforestacion-del-amazonas-en/ (Accessed 12 April 2021).

Democracy Now (2020) We are in danger daily: Honduran Afro-Indigenous Garífuna demand return of kidnapped land defenders. 17 August www.democracynow.org/2020/8/17/garifuna_land_defenders_honduras (Accessed 14 April 2021).

Dennis, P. A. (2004) *The Miskitu People of Awastara*. Austin: University of Texas Press.

de Onís, C. (2017) What's in an "x"? An exchange about the politics of "Latinx. *Chiricú Journal: Latina/o Literatures, Arts, and Cultures* 1(2): pp. 78–79.

de Santana Pinho, P. (2010) *Mama Africa: Reinventing Blackness in Bahia.* Durham, NC: Duke University Press

———— (2021) Whiteness has come out of the closet and intensified Brazil's reactionary wave. In B. Junge, S. T. Mitchell, A. Jarrin and L. Cantero (eds) *Precarious Democracy: Ethnographies of Hope, Despair, and Resistance in Brazil.* Brunswick, NJ: Rutgers University Press, pp. 62–76.

desinformémonos (2014) Subcomandante Marcos is no more. 26 May https://upside downworld.org/news-briefs/news-briefs-news-briefs/subcomandante-marcos-is-no-more/ (Accessed 19 May 2021).

de Sousa Santos, B. (1995) Three metaphors for a new conception of law: The frontier, the baroque and the South. *Law and Society Review* 29: 569–584.

———— (2002) Between Prospero and Caliban: Colonialism, postcolonialism, and inter-identity. *Luso-Brazilian Review* 39(2): 9–43.

———— (2004) The World Social Forum: Towards a counter-hegemonic globalization (Part I). In J. Sen, A. Anand, A. Escobar and P. Waterman (eds) *World Social Forum: Challenging Empires.* New Delhi: Viveka Foundation, pp. 235–245.

———— (2007) Beyond abyssal thinking: From global lines to ecologies of knowledges. *Review (Fernand Braudel Center)* 30(1): 45–89.

———— (2014) *Epistemologies of the South: Justice Against Epistemicide.* Boulder, CO, and London: Paradigm Publishers.

———— (2015) *If God Were a Human Rights Activist.* Stanford: Stanford University Press.

———— (2018a) *The End of the Cognitive Empire: The Coming of Age of Epistemologies of the South.* Durham, NC: Duke University Press.

———— (2018b) *Las bifurcaciones del orden: Revolución, ciudad, campo e indignación.* Madrid: Trotta.

———— (2020a) Ecuador: Del centro al fin del mundo. In C. Parodi and N. Sticotti (eds) *Ecuador: La insurrección de octubre.* Buenos Aires: CLACSO, pp. 17–20.

———— (2020b) *La cruel pedagogía del virus.* Madrid: Ediciones Akal.

Díaz del Castillo, B. (1973[1800/1576]) *The Conquest of New Spain,* translated by J.M. Cohen. London: Penguin.

Díaz-Parra I. (2021) Generating a critical dialogue on gentrification in Latin America. *Progress in Human Geography.*45(3): 472–488.

Dixon, K. (2016) *Afro-Politics and Civil Society in Salvador da Bahia, Brazil.* Gainesville: University Press of Florida.

do Nascimento, A. (1980) Quilombismo: An Afro-Brazilian political alternative. *Journal of Black Studies* 11(2): 141–178.

———— (1989) *Brazil, Mixture or Massacre? Essays in the Genocide of a Black People,* translated by E. Larkin Nascimento. Dover, MA: The Majority Press.

Dorfman, A. and Mattelart, A. (1971) *How to Read Donald Duck: Imperialist Ideology in the Disney Comic.* New York: International General.

Dos Santos, T. (1978) *Imperialismo y dependencia.* Mexico City: Ediciones Era.

Draper, N. (2010) *The Price of Emancipation: Slave-Ownership, Compensation and British Society at the End of Slavery.* Cambridge: Cambridge University Press.

Duarte, F. (2019) Urban mobility in Latin America. In J. Cupples, M. Palomino-Schalsha and M. Prieto (eds) *Routledge Handbook of Latin American Development.* London: Routledge, pp. 539–548.

Dubois, L. (2012) *Haiti: The Aftershocks of History*. New York: Picador.

Dudley, S. S. (2010) *Drug Trafficking Organizations in Central America: Transportistas, Mexican Cartels and Maras*. Working Paper Series on U.S.–Mexico Security Collaboration Washington DC and San Diego: Woodrow Wilson International Center for Scholars and University of San Diego.

Durban-Albrecht, E. (2017) Postcolonial disablement and/as transition: Trans★ Haitian narratives of breaking open and stitching together. *Transgender Studies Quarterly* 4(2): 195–207.

Dussel, E. (1976) *Filosofía de la liberación*. Mexico: Editorial Edicol.

――― (1993) Eurocentrism and modernity (Introduction to the Frankfurt Lectures). *Boundary* 20(3): 65–76.

――― (2006) Globalization, organization and the ethics of liberation. *Organization* 13(4): 489–508.

――― (2014) Anti-Cartesian meditations: On the origin of the philosophical anti-discourse of modernity. *Journal for Cultural and Religious Theory* 13(1): 11–53.

Eckstein, S. (2001) Poor people versus the state and capital: Anatomy of a successful community mobilization for housing in Mexico City. In M. A. Garretón (ed) *Power and Popular Protest: Latin American Social Movements*. Berkeley: University of California Press, pp. 329–349.

ECLAC (2014) *Guaranteeing Indigenous People's Rights in Latin America: Progress in the Past Decade and Remaining Challenges*. Santiago: United Nations.

――― (2016) *Horizon 2030: Equality at the Centre of Sustainable Development*. Santiago: United Nations.

Ellner, S. (ed) (2019) *Latin America's Pink Tide: Breakthroughs and Shortcomings*. Lanham, MD: Rowman and Littlefield.

Elson, D. and Pearson, R. (1981) Nimble fingers make cheap workers: An analysis of women's employment in third world export manufacturing. *Feminist Review*. 7: 87–107.

Encarnación, O. (2011) Latin America's gay rights revolution. *Journal of Democracy* 22(2): 104–119.

Engler, Y. and Fenton, A. (2006) *Haiti in Canada: Waging War on the Poor Majority*. Vancouver: Red and Fernwood Publishing.

Enríquez, L. J. and Page, T. L. (2019) The rise and fall of the Pink Tide. In J. Cupples, M. Palomino-Schalsha and M. Prieto (eds) *Routledge Handbook of Latin American Development*. London: Routledge, pp. 87–97.

Escobar, A. (1995) *Encountering Development: The Making and Unmaking of the Third World*. Princeton: Princeton University Press.

――― (2015) Territorios de diferencia: la ontología política de los "derechos al territorio." *Cuadernos de Antropología Social* 41: 25–38.

――― (2016) Sentipensar con la tierra: Las luchas territoriales y la dimensión ontológica de las epistemologías del Sur. *AIBR: Revista de Antropología Iberoamericana* 11(1): 11–32.

――― (2018) *Designs for the Pluriverse: Radical Interdependence, Autonomy, and the Making of Worlds*. Durham, NC: Duke University Press.

Escribano, G. (2013) Ecuador's energy policy mix: Development versus conservation and nationalism with Chinese loans. *Energy Policy* 57: 152–159.

Esguerra Muelle, C. (2019) Coloniality, colonialism, and decoloniality: Gender, sexuality, and migration. In J. Cupples, M. Palomino-Schalsha and M. Prieto (eds) *Routledge Handbook of Latin American Development*. London: Routledge, pp. 54–63.

Esquivel, L. (2005) *Malinche*. México: Suma de Letras.

Esteva, G. (1985) Beware of participation, and Development: metaphor, myth, threat, *Development: Seeds of Change* 3: 77–79.

——— (2014) La libertad de aprender. *Revista Interuniversitaria de Formación del Profesorado* 28(2): 39–50.

——— (2017) La construcción del saber histórico de lucha. In C. Walsh (ed) *Pedagogías decoloniales. Prácticas insurgentes de resistir, (re)existir y (re)vivir, Tomo II*. Quito: Ediciones Abya-Yala, pp. 77–103.

——— (2019) Constructing knowledge in the Latin American University. *Compare: A Journal of Comparative and International Education* 49(3): 501–506.

Estupiñan Vitero, T. (1998) *Volcán Pinchincha: Erupciones, destrucciones e invenciones*. Quito: Abya Yala and Ediciones Banco Central de Ecuador.

EZLN (1996) Cuarta declaración de la selva lacandona. *Enlace Zapatista* 1 January https://enlacezapatista.ezln.org.mx/1996/01/01/cuarta-declaracion-de-la-selva-lacandona/ Available in English at https://radiozapatista.org/?p=20287&lang=en (Accessed 19 March 2021).

——— (2020) Una montaña en alta mar. *Enlace Zapatista* 5 October https://enlacezapatista.ezln.org.mx/2020/10/05/sexta-parte-una-montana-en-alta-mar/ (Accessed 8 May 2021).

Fabricant, N. and Postero, N. (2015). Sacrificing indigenous bodies and lands: The political-economic history of Lowland Bolivia in light of the recent TIPNIS debate. *The Journal of Latin American and Caribbean Anthropology* 20: 452–474.

Falabella Luco, S. and Solar Andrade, C. (2020) No más abusos: Hasta que vivir valga la pena. *Nomadías* 29: 301–327.

Falconí Trávez, D. (2012) Entrevista a Julieta Paredes. *Lectora* 18: 179–195.

Fals Borda, O. (2009) *Una sociología sentipensante para América Latina*. Buenos Aires: CLACSO.

Fanon, F. (2004 [1963]) *The Wretched of the Earth*, translated by R. Philcox. New York: Grove Press.

——— (1967) *Black Skin, White Masks*, translated by C. Lam Markmann. New York: Grove Press.

Farthing, L. C. and Kohl, B. H. (2014) *Evo's Bolivia: Continuity and Change*. Austin: University of Texas Press.

Fernández, R. A. (2006) *From Afro-Cuban Rhythms to Latin Jazz*. Berkeley: University of California Press.

Ffrench-Davies, R. (2010) *Economic Reforms in Chile from Dictatorship to Democracy*, 2nd ed. New York: Palgrave Macmillan.

Ficek, R. E. (2018). Infrastructure and colonial difference in Puerto Rico after Hurricane María. *Transforming Anthropology* 26(2): 102–117.

Fisher, J. (1989) *Mothers of the Disappeared*. London: Zed Books.

——— (1993) *Out of the Shadows: Women, Resistance and Politics in South America*. London: Latin America Bureau.

Fleuriet, K. J. and Castellano, M. (2020) Media, place-making, and concept-metaphors: the US-Mexico border during the rise of Donald Trump. *Media, Culture and Society* 42(6): 880–897.

Flores Contreras, E. (2021) Ejecutan a comandante considerado pieza clave en caso Ayotzinapa. *Proceso* 16 June www.proceso.com.mx/nacional/estados/2021/6/16/ejecutan-comandante-considerado-pieza-clave-en-caso-ayotzinapa-266060.html (Accessed 21 June 2021).

Franko, P. (2019) *The Puzzle of Latin American Economic Development*. 4th ed. Lanham, MD: Rowman and Littlefield.

Fregoso, R. L and Bejarano, C. (2010) Introduction: A cartography of feminicide in the Américas. In R. L. Fregoso and C. Bejarano (eds) *Terrorizing Women: Feminicide in the Americas*. Durham, NC: Duke University Press, pp. 1–42.

Freire-Medeiros, B. and Name, L. (2017) Does the future of the favela fit in an aerial cable car? Examining tourism mobilities and urban inequalities through a decolonial lens. *Canadian Journal of Latin American and Caribbean Studies Revue canadienne des études latino-américaines et caraïbes* 42(1): 1–16.

Froehling, O. (1999) Internauts and guerrilleros: The Zapatista rebellion in Chiapas and its extension into cyberspace. In M. Crang, P. Crang and J. May (eds) *Virtual Geographies: Bodies, Space and Relations*. London: Routledge, pp. 164–177.

Fuerzas Armadas Revolucionarias de Colombia and Ejército de Liberación Nacional (2017) Starting points for the FARC and the ELN. In A. Farnsworth-Alvear, M. Palacios and A. M. Gómez López (eds). *The Colombia Reader: History, Culture, Politics*. Durham, NC: Duke University Press, pp. 367–376.

Gaffield, J. (2015) *Haitian Connections in the Atlantic World: Recognition after Revolution*. Chapel Hill: University of North Carolina Press.

——— (2020) Haiti was the first nation to permanently ban slavery. *The Washington Post*, 12 July www.washingtonpost.com/outlook/2020/07/12/haiti-was-first-nation-permanently-ban-slavery/ (Accessed 13 July 2020).

Gago, V. (2017) *Neoliberalism from Below: Popular Pragmatics and Baroque Economies*. Durham, NC: Duke University Press.

Galeano, E. (1971) *Las Venas Abiertas de América Latina*. México: Siglo XXI.

Galindo, M. (2006) Indias, putas y lesbianas, juntas revueltas y hermanadas. ¡Un libro sobre Mujeres Creando! In E. Monasterios (ed) *No pudieron con nosotras: El desafío del feminismo autónomo de Mujeres Creando*. La Paz: Plural, 27–59.

Ganchev, I. (2020) China pushed the Pink Tide and the Pink Tide pulled China: Intertwining economic interests and ideology in Ecuador and Bolivia (2005–2014). *World Affairs* 183(4): 359–388.

García, C. 1996. *The Making of the Miskitu people of Nicaragua: The Social Construction of Ethnic Identity*. Uppsala: Uppsala University.

García Canclini, N. (2001) *Consumers and Citizens: Globalization and Multicultural Conflicts*. Minneapolis: University of Minnesota Press.

——— (2002) *Latinoamericanos buscando lugar en este siglo*. Buenos Aires: Paidós.

García de León, A. (2001) *Fronteras interiores: Chiapas, una modernidad particular*. México: Océano.

García Pérez, A. (2007) El síndrome del perro del hortelano. *El Comercio* 28 October. Available at www.chs-peru.com (Accessed 29 June 2021).

García-Salgado, T. (2010) The sunlight effect of the Kukulcán Pyramid or the history of a line. *Architecture, Mathematics and Perspective* 12(1): 113–130.

Gardini, G. L. (2021) Analysis and 'normalization' of the surge of external powers in Latin America. In G. L. Gardini (ed) *External Powers in Latin America: Geopolitics between Neo-extractivism and South-South Cooperation*. Abingdon: Routledge, pp. 1–14.

Garvey, M. (1925) African fundamentalism [front page editorial]. *The Negro World* 6 June.

Garvey, M. and Jacques Garvey, A. (2013[1923]) *Philosophy and Opinions of Marcus Garvey: Or Africa for the Africans*. London: Routledge.

Gates Jr., H. L. (2011) *Black in Latin America*. New York: New York University Press.

Gill, L. (2004) *The School of the Americas: Military Training and Political Violence*. Durham, NC: Duke University Press.

Ginsburg, F. 2002. Screen memories: Resignifying the traditional in indigenous media. In F. D. Ginsburg, L. Abu-Lughod, and B. Larkin (eds) *Media Worlds: Anthropology on New Terrain*. Berkeley: University of California Press, pp. 39–57.

Global Witness (2021) The Last Line of Defence: The Industries Causing the Climate Crisis and Attacks against Land and Environmental Defenders www.globalwitness. org/documents/20191/Last_line_of_defence_-_high_res_-_September_2021.pdf (Accessed 21 October 2021).

Glynn, K. and Cupples, J. (2011) Indigenous mediaspace and the production of (trans) locality on Nicaragua's Mosquito Coast. *Television and New Media* 12(2): 101–135.

Gomis, C. (2018) Dismantling Eurocentrism in the French history of chattel slavery and racism. In J. Cupples and R. Grosfoguel (eds) *Unsettling Eurocentrism in the Westernized University*. London: Routledge, pp. 235–247.

Goñi, U. (2021a) The hidden history of Black Argentina. *New York Review* 8 February www-nybooks-com.ezproxy.is.ed.ac.uk/daily/2021/02/08/the-hidden-history-of-black-argentina/ (Accessed 1 June 2021).

———— (2021b) Time to challenge Argentina's white European self-image, black history experts say. *The Guardian* 31 May www.theguardian.com/world/2021/may/31/argentina-white-european-racism-history (Accessed 1 June 2021).

González, A. (2011) Landscape and settlement patterns. In J. Knippers Black (ed) *Latin America: Its Problems and Its Promise: A Multidisciplinary Introduction*. Boulder: Westview Press, pp. 23–38.

González, A. L. (2020) How Afro-Colombians are standing up against racial violence. *Remezcla* 19 June https://remezcla.com/features/culture/colombia-protests-black-lives-matter-racial-violence-in-country/ (Accessed 19 March 2021).

Gonzalez, L. (1982) A mulher negra na sociedade brasileira. In L. Mandel (ed) O Lugar da Mulher: Estudos sobre a Condição Feminina. Edições Graal.

———— (1988) A categoria político-cultural de amefricanidade. *Tempo Brasileiro*. 92/ 93: 69–82.

Gonzalez-Rivera, V. (2020) Why my Nicaraguan father did not "see" his Blackness and how Latinx anti-Black racism feeds on racial silence. *Medium* 10 June https://medium.com/@victoriagonzalezrivera/why-my-nicaraguan-father-did-not-see-his-blackness-and-how-latinx-anti-black-racism-feeds-on-738249ddd100 (Accessed 3 June 2021).

Goodman, J. (2007) Greenwashing energy crops: Biofuels, the biggest scam going. *Counterpunch* 28 December www.counterpunch.org/2007/12/28/biofuels-the-biggest-scam-going/ (Accessed 5 June 2021).

Gopal, P. (2021) On decolonisation and the university. *Textual Practice* 35(6): 873–899.

Gordon, E. T. (1998) *Disparate Diasporas: Identity and Politics in an African Nicaraguan Community*. Austin: Texas University Press.

Gosine, A. (2018) Rescue and real love: same-sex desire in international development. In C. Mason (ed) *Routledge Handbook of Queer Development Studies*. London: Routledge, pp. 193–208.

Graham-Harrison, E. (2020). Poverty, not just populists, to blame for Covid-19's impact on Latin America. *The Guardian* 5 July www.theguardian.com/world/2020/jul/05/poverty-not-just-populists-to-blame-for-covid-19s-impact-on-latin-america (Accessed 5 July 2020)

Gramsci, A. (1971) *Selections from the Prison Notebooks*. London: Lawrence and Wishart.

Grant, W. (2021) *¡Populista! The Rise of the Latin America's 21st Century Strongman*. London: Head of Zeus.

Gravante T. (2020) Forced disappearance as a collective cultural trauma in the Ayotzinapa Movement. *Latin American Perspectives* 47(6):87–102.

Green, D. (2003) *Silent Revolution: The Rise and Crisis of Market Economics in Latin America*. London: Latin America Bureau.

———— (2006) *Faces of Latin America*, 3rd ed. London: Latin America Bureau.

Greenfield, P. (2021) Indigenous peoples face rise in rights abuses during pandemic, report finds. *The Guardian* 18 February www.theguardian.com/environment/2021/feb/18/ indigenous-peoples-face-rise-in-rights-abuses-during-covid-pandemic-report-aoe (Accessed 12 April 2021).

Grosfoguel, R. (2011) Decolonizing post-colonial studies and paradigms of political-economy: Transmodernity, decolonial thinking, and global coloniality. *Transmodernity: Journal of Peripheral Cultural Production of the Luso-Hispanic World* 1(1): np http:// escholarship.org/uc/item/21k6t3fq

———— (2012) The dilemmas of Ethnic Studies in the United States: Between liberal multiculturalism, identity politics, disciplinary colonization, and decolonial epistemologies. *Human Architecture: Journal of The Sociology of Self-Knowledge* 10(1): 81–90.

———— (2013) Epistemic racism/sexism, westernized universities and the four genocides/ epistemicides of the long sixteenth century. *Human Architecture: Journal of the Sociology of Self-Knowledge* 11(1): 73–90.

———— (2018) What is racism? Zone of Being and Zone of Non-Being in the work of Frantz Fanon and Boaventura de Sousa Santos. In J. Cupples and R. Grosfoguel (eds) *Unsettling Eurocentrism in the Westernized University*. London: Routledge, pp. 264–273.

Guamán Poma de Ayala, F. (2005[1978]) Officials and messengers. In O. Starn, R. Kirk and C. I. Degregori (eds) *The Peru Reader: History, Culture, Politics*. Durham, NC: Duke University Press, pp. 76–81.

Gudynas, E. (2010) The new extractivism of the 21st century: Ten urgent theses about extractivism in relation to current South American progressivism. *Americas Program Report* 21 January http://postdevelopment.net/2010/02/19/new-extractivism-of-the-21st-century-10-urgent-theses/ (Accessed 14 April, 2021).

———— (2011) Buen vivir: Today's tomorrow. *Development* 54(4): 441–447.

———— (2015a) Friendly colonialism and the "Harvey" fashion, translated by D. Andreucci. *Entitle* 15 October https://entitleblog.org/2015/10/15/friendly-colonialism-and-the-harvey-fashion (Accessed 12 April 2021).

———— (2015b) *Derechos de la naturaleza: Ética biocéntrica y políticas ambientales*. Buenos Aires: Tinta Limón.

———— (2020) El agotamiento del desarrollo: La confesión de la CEPAL. *Economía Sur* 14 Febuary http://economiasur.com/2020/02/el-agotamiento-del-desarrollo-la-confesion-de-la-cepal/ (Accessed 27 June 2021).

Guerrero, J. (2015) U.S. deportations strain Tijuana's mental health infrastructure. *KPBS* 28 April www.kpbs.org/news/2015/apr/28/us-deportations-strain-tijuanas-mental-health-i/ (Accessed 1 May 2021).

———— (2021) 3 million people were deported under Obama. What will Biden do about it? *The New York Times* 23 January www.nytimes.com/2021/01/23/opinion/sunday/ immigration-reform-biden.html (Accessed 1 May 2021).

Guerrón Montero, C. (2014) Multicultural tourism, demilitarization, and the process of peace building in Panama. *The Journal of Latin American and Caribbean Anthropology* 19(3): 418–440.

Gutierrez, G. (1974) *A Theology of Liberation: History, Politics, and Salvation*. London: SCM Press.

Gutiérrez, E. R. and Fuentes, L. (2010) Population control by sterilization: The cases of Puerto Rican and Mexican-Origin women in the United States. *Latino/a Research Review* 7(3): 85–100.

Guzmán-Concha, C. (2020) When the pandemic meets the insurrection. Santiago, Chile. *International Journal of Urban and Regional Research* [Spotlight On series] www.ijurr.org/spotlight-on/urban-revolts/when-the-pandemic-meets-the-insurrection/ (Accessed 8 June 2021).

Hale, C. R. (1994) *Resistance and Contradiction: Miskitu Indians and the Nicaraguan State 1894–1987*. Stanford: Stanford University Press.

——— (2002) Does multiculturalism menace? Governance, cultural rights and the politics of identity in Guatemala. *Journal of Latin American Studies* 34: 485–524.

——— (2004) Rethinking indigenous politics in the era of the "indio permitido". *NACLA Report on the Americas* 38(2): 16–21.

——— (2005) Neoliberal multiculturalism: The remaking of cultural rights and racial dominance in Central America. *Polar* 28(1): 10–28.

——— (2019) Neoliberal multiculturalism. In J. Cupples, M. Palomino-Schalsha and M. Prieto (eds) *Routledge Handbook of Latin American Development*. London: Routledge, pp. 75–86.

Hale, C. R. and Mullings, L. (2020) A time to recalibrate: Analyzing and resisting the Americas-wide project of racial retrenchment. In J. Hooker (ed) *Black and Indigenous Resistance in the Americas: From Multiculturalism to Racist Backlash*. Lanham: Lexington Books, pp. 21–65.

Hall, S. and Schwarz, B. (2017) *Familiar Stranger: A Life Between Two Islands*. London: Penguin.

Halleck, D. (1994) Zapatistas on-line. *NACLA Report on the Americas* 38(2): 30–32.

Harpelle, R. N. (2001) *The West Indians of Costa Rica: Race, Class, and the Integration of an Ethnic Minority*. Montreal: McGill-Queens University Press.

——— (2003) Cross currents in the Western Caribbean: Marcus Garvey and the UNIA in Central America. *Caribbean Studies* 31(1): 35–73.

Harten, S. (2011) *The Rise of Evo Morales and the MAS*. London: Zed Books.

Hawley, S. (1997) Protestantism and indigenous mobilisation: The Moravian Church among the Miskitu Indians of Nicaragua. *Journal of Latin American Studies* 29(1): 111–129.

Henne, N. C. (2020) *Reading Popol Wuj: A Decolonial Guide*. Tucson: University of Arizona Press.

Hernández, A. (2018) *A Massacre in Mexico: The True Story Behind the Missing 43 Students*. London and Brooklyn: Verso.

Hetherington, K. (2020) *The Government of Beans: Regulating Life in the Age of Monocrops*. Durham, NC: Duke University Press.

Hewitt, K. (1995) Excluded perspectives in the social construction of disaster. *International Journal of Mass Emergencies and Disasters* 13(3): 317–339.

Hickel, J. (2019) The contradiction of the sustainable development goals: Growth versus ecology on a finite planet. *Sustainable Development* 27:873–884.

Hogenboom, B. (2019) Latin America and China. In J. Cupples, M. Palomino-Schalsha and M. Prieto (eds) *Routledge Handbook of Latin American Development*. London: Routledge, pp. 179–191.

Hooker J (2005) Indigenous inclusion/black exclusion: Race, ethnicity, and multicultural citizenship in contemporary Latin America. *Journal of Latin American Studies* 37(2): 285–310.

hooks, b. (1984) *Feminist Theory from Margin to Centre*. Boston: South End Press.

Hope, J. (2020) Globalising sustainable development: Decolonial disruptions and environmental justice in Bolivia. *Area* (online early).

Howard, A. and Dangl, B. (2007) The multinational beanfield war: Soy cultivation spells doom for Paraguayan campesinos. *In These Times* 12 April https://upsidedown world.org/archives/paraguay/the-multinational-beanfield-war-soy-cultivation-spells-doom-for-paraguayan-campesinos/ (Accessed 5 June 2021).

Humane Borders (2021) 2020 was deadliest year for migrants crossing unlawfully into US via Arizona. *Humane Borders/Fronteras Compasivas* 12 February https://humanebord ers.org/2020-was-deadliest-year-for-migrants-crossing-unlawfully-into-us-via-ariz ona/ (Accessed 2 May 2021).

Hutchison, E., Miller Klubock, T., Milanich, N. B. and Winn, P. (2014) The Pinochet dictatorship: Military rule and neoliberal economics. In E. Hutchison, T. Miller Klubock, N.B. Milanich and P. Winn (eds) *The Chile Reader: History, Culture, Politics*. Durham, NC, Duke University Press, pp. 443–441.

Hylton, F. and Thomson, S. (2007) *Revolutionary Horizons: Past and Present in Bolivian Politics*. London: Verso.

ISHR (2020) ISHR intervenes in regional legal proceedings against Honduras in case of transfemicide of defender Vicky Hernández. 11 November www.ishr.ch/news/ lgbti-rights-ishr-intervenes-regional-legal-proceedings-against-honduras-case-transfemicide (Accessed 15 May 2021).

Jenkins, R. (2009) The Latin American case. In R. Jenkins and E. Dussel (eds) *China and Latin America: Economic Relations in the Twenty-First Century*. Bonn: Deutsches Institut für Entwicklungspolitik, pp. 21–63.

Kampwirth, K. (2002) *Women and Guerrilla Movements: Nicaragua, El Salvador, Chiapas, Cuba*. University Park: Pennsylvania State Press.

Karush, M. B. (2012) Blackness in Argentina: Jazz, tango and race before Perón. *Past and Present* 216(1): 215–245.

Kaufman,T. (1994a) The native languages of Meso-America. In C. Moseley and R.E. Asher (eds) *Atlas of the World's Languages*. London: Routledge, pp. 34–41.

——— (1994b) The native languages of South America. In C. Moseley and R.E. Asher (eds) *Atlas of the World's Languages*. London: Routledge, pp. 46–76.

Kay, C. (1989) *Latin American Theories of Development and Underdevelopment*. London: Routledge.

——— (2019a) Modernization and dependency theory. In J. Cupples, M. Palomino-Schalsha and M. Prieto (eds) *Routledge Handbook of Latin American Development*. London: Routledge, pp. 15–28.

——— (2019b) Theotonio Dos Santos (1936–2018): The revolutionary intellectual who pioneered Dependency Theory. *Development and Change* 51(2): 599–630.

Kennon, I. (2020) Costa Rica legalized same-sex marriage. Where does the rest of Latin America stand on marriage equality? *Atlantic Council* 2 June www.atlanticcouncil.org/ blogs/new-atlanticist/costa-rica-legalized-same-sex-marriage-where-does-the-rest-of-latin-america-stand-on-marriage-equality/ (Accessed 22 June 2021).

Kishore, N., Marqués, D., Mahmud, A., Kiang, M.V., Rodriguez, I., Fuller, A., Ebner, P., Sorensen, C., Racy, F., Lemery, J., Maas, L., Leaning, J., Irizarry, R. A., Balsari, S. and Buckee, C.O. (2018). Mortality in Puerto Rico after Hurricane Maria. *The New England Journal of Medicine* 379: 162–170.

Klein, H. S. (1966) Peasant communities in revolt: The Tzeltal republic of 1712. *Pacific Historical Review* 35(3): 247–263.

Klein, N. (2007) *The Shock Doctrine: The Rise of Disaster Capitalism*. New York: Metropolitan Books.

——— (2018) *The Battle for Paradise: Puerto Rico Takes on the Disaster Capitalists*. Chicago: Haymarket Books.

Koch, A., Brierley, C., Maslin, M.M. and Lewis, S. L. (2019) Earth system impacts of the European arrival and Great Dying in the Americas after 1492. *Quaternary Science Reviews* 207: 13–36.

Koop, F. (2019) Belt and Road: The new face of China in Latin America. *Diálogo Chino* 25 April https://dialogochino.net/en/infrastructure/26121-belt-and-road-the-new-face-of-china-in-latin-america/ (Accessed 30 April 2021).

Koopman, S. (2020) Building an inclusive peace is an uneven socio-spatial process: Colombia's differential approach. *Political Geography* 83 (online early).

Krenak, A. (2020) El mañana no está a la venta. In O. Quijano Valencia and C. Corredor Jimenez (eds) (2020) *Pandemia al sur*. Buenos Aires: Prometeo Libros, pp. 23–29.

La Digna Rabia (2021) Nada está olvidado. https://nadaestaolvidado.com/#voces (Accessed 26 June 2021).

Lagarde, M. (2006) Del femicidio al feminicidio. *Desde el Jardín de Freud* 6: 216–225.

Laing, A. F. (2019) Subaltern geographies in the plurinational state of Bolivia: The TIPNIS affair. In T. Jazeel and S. Legg (eds) *Subaltern Geographies*. Athens: University of Georgia Press, pp. 167–190.

Lakhani, N. (2020) *Who Killed Berta Cáceres: Dams, Death Squads, and an Indigenous Defender's Battle for the Planet*. London: Verso.

Lancaster, Roger N. (1992) *Life is Hard: Machismo, Danger and the Intimacy of Power in Nicaragua*. Berkeley: University of California Press.

Lane, J. (2003) Digital Zapatistas. *The Drama Review* 47(2): 129–144.

Laurie, N., Andolina, R. and Radcliffe, S. (2002) The excluded 'indigenous'? The implications of multi-ethnic politics for water reform in Bolivia. In R. Sieder (ed) *Multiculturalism in Latin America: Indigenous Rights, Diversity and Democracy*. Basingstoke: Palgrave Macmillan, pp. 252–276.

Laurie, N., Dwyer, C., Holloway, S. and Smith, F. (1999) *Geographies of New Femininities*. Essex: Longman.

Law, J. (2015) What's wrong with a one-world-world? *Distinktion: Scandinavian Journal of Social Theory* 16(1): 126–139.

Lawler M (2005) *Marcus Garvey: Black Nationalist Leader*. Philadelphia: Chelsea House.

Lazar, S. (2008) *El Alto, Rebel City: Self and Citizenship in Andean Bolivia*. Durham, NC: Duke University Press.

Leibler, L. and Musset, A. (2010) Un transporte hacia la justicia espacial? El caso del metrocable de Medellín. *Scripta Nova: Revista Electrónica de Geografía y Ciencias Sociales* 14 https://raco.cat/index.php/ScriptaNova/article/view/200029 (Accessed 13 June 2021).

Levine, D. H. (1988) Assessing the impacts of liberation theology in Latin America. *The Review of Politics* 50(2): 241–263.

Lewis, S. E. (2004) The Zapatistas in context. *Latin American Perspectives* 31(6): 107–109.

Li, F. (2015) *Unearthing Conflict: Corporate Mining, Activism, and Expertise in Peru*. Durham, NC: Duke University Press.

Lind, A. (2005) *Gendered Paradoxes: Women's Movements, State Restructuring, and Global Development in Ecuador*. University Park: Pennsylvania State University Press.

Linz, J. (2021) Where crises converge: the affective register of displacement in Mexico City's post-earthquake gentrification. *cultural geographies* 28(2): 285–300.

Llenín-Figueroa, B. (2019) This was meant to be a hurricane diary. In Y. Bonilla and M. LeBrón (eds) *Aftershocks of Disaster*. Chicago: Haymarket Books, pp. 96–100.

Lloréns, H. (2018). Ruin Nation: In Puerto Rico, Hurricane Maria laid bare the results of a long-term crisis created by dispossession, migration, and economic predation. *NACLA Report on the Americas* 50(2): 154–159.

López-Morales, E. (2019) Just another chapter of Latin American gentrification. In J. Cupples, M. Palomino-Schalsha and M. Prieto (eds) *Routledge Handbook of Latin American Development*. London: Routledge, pp. 503–516.

Los voceros nacionales de la Mara Salvatrucha MS13 y Pandilla 18 (2012) Comunicado de la Mara Salvatrucha MS13 y Pandilla 18. *Aresteguí Noticias* 23 March https://aristeguin oticias.com/2303/mundo/comunicado-de-la-mara-salvatrucha-ms13-y-pandilla-18/ (Accessed 15 June 2021).

Lovell, W. G. (1992) "Heavy shadows and black night": Disease and depopulation in colonial Spanish America. *Annals of the Association of American Geographers* 82: 426–443.

——— (2020) From Columbus to Covid-19: Amerindian antecedents to the global pandemic. *Journal of Latin American Geography* 19(3): 177–185.

Lozano Lerma, B. (2021) Colombia: "The government is killing us." *Territorio de Ideas* 27 May https://territoriodeideas.com/2021/05/27/colombia-the-government-is-kill ing-us/ (Accessed 25 June 2021).

Lugones, M (2007) Heterosexualism and the colonial/modern gender system. *Hypatia* 22(1): 186–209.

——— (2010) Toward a decolonial feminism. *Hypatia* 25(4): 742–759.

——— (2016) The coloniality of gender. In W. Harcourt (ed) *The Palgrave Handbook of Gender and Development*. Basingstoke: Palgrave Macmillan, pp. 13–33.

Lyons, K. M. (2016) Decomposition as life politics: Soils, selva, and small farmers under the gun of the U.S.-Colombian War on Drugs. *Cultural Anthropology* 31(1): 56–81.

Maclean, K. (2014) Evo's jumper: identity and the used clothes trade in 'post-neoliberal' and 'pluri-cultural' Bolivia. *Gender, Place and Culture* 21(8): 963–978.

——— (2018) Envisioning gender, Indigeneity and urban change: The case of La Paz, Bolivia. *Gender, Place and Culture* 25(5): 711–726.

——— (2019) Fashion in Bolivia's cultural economy. *International Journal of Cultural Studies* 22(2): 213–228.

Maldonado-Torres, N. (2016) Outline of ten theses on coloniality and decoloniality. *Fondation Frantz Fanon*. Available at https://fondation-frantzfanon.com/outline-of-ten-theses-on-coloniality-and-decoloniality/ (Accessed 19 June 2021).

——— (2017) On the coloniality of human rights. *Revista Crítica de Ciências Sociais* 114: 117–136.

Mamani, Y. (2015) Como cholas debemos repensarnos y revalorarnos y revalorar nuestra cultura. *Ser chola está de moda* [blog] 29 January https://sercholaestademoda.blogspot. com/2015/01/ser-chola-esta-de-moda.html (Accessed 13 June 2021)

Mamani Ramírez, P. (2005) *Microgobiernos barriales: Levantamiento de la Ciudad de El Alto*. El Alto: Centro Andino de Estudios Estratégicos.

———— (2006) Territory and structures of collective action: Neighborhood micro-governments. *Ephemera: Theory and Politics in Organization* 6(3): 276–286.

Manrique, L. (2016) Dreaming of a cosmic race: José Vasconcelos and the politics of race in Mexico, 1920s–1930s. *Cogent Arts and Humanities* 3(1): 1–13.

Marcella, G. (2012) China's military activity in Latin America. *Americas Quarterly* 20 January www.americasquarterly.org/fulltextarticle/chinas-military-activity-in-latin-america/ (Accessed 30 April 2021).

Mariátegui, J. C. (1970[1927]) *Peruanicemos el Perú*. Lima: Biblioteca Amauta.

Marquéz, R. (2018) What's in the 'x' of Latinx? *Medium* 9 July https://medium.com/center-for-comparative-studies-in-race-and/whats-in-the-x-of-latinx-9266ed40766a (Accessed 10 April 2021).

Martí, J. (1977[1891]) Our America. In P. S. Foner (ed) *Our America by José Martí: Writings on Latin America and the Struggle for Cuban Independence*. New York: Monthly Review Press, pp. 84–94.

Martí i Puig, S. (2010) The emergence of indigenous movements in Latin America and their impact on the Latin American political scene: Interpretive tools at the local and global levels. *Latin American Perspectives* 37(6): 74–92.

Martínez Ante, O.L. (2021) El 'funeral' muisca de la estatua de Gonzalo Jiménez de Quesada. *El Tiempo* 25 June www.eltiempo.com/cultura/arte-y-teatro/el-funeral-de-gonzalo-jimenez-de-quesada-598116 (Accessed 28 June 2021).

Martín Barbero, J. (2004) *De los medios a las mediaciones: Comunicación, cultura y hegemonía*. Barcelona: Editorial Gustavo Gili.

Martínez Novo, C. (2018) Ventriloquism, racism and the politics of decoloniality in Ecuador. *Cultural Studies* 32(3): 389–413.

Martínez, O. and Martínez, J.J. (2019) *The Hollywood Kid: The Violent Life and Violent Death of an MS-13 Hitman*. London: Verso.

Mato, D. (2019) Intercultural universities. In J. Cupples, M. Palomino-Schalsha and M. Prieto (eds) *Routledge Handbook of Latin American Development*. London: Routledge, pp. 213–224.

McCaughan, E. J. (2020) "We didn't cross the border, the border crossed us": Artists' images of the US-Mexico border and immigration. *Latin American and Latinx Visual Culture* 2(1): 6–31.

McClenaghan, S. (1997) Women, work and empowerment: Romanticizing the reality. In E. Dore (ed) *Gender Politics in Latin America: Debates in Theory and Practice*. New York: Monthly Review Press, pp. 19–35.

McCoy-Torres, S. (2016) "Cien porciento tico tico": Reggae, belonging, and the Afro-Caribbean ticos of Costa Rica. *Black Music Research Journal* 36(1): 1–29.

McGee, M. (2018) Queer paradise: Development and recognition in the Isthmus of Tehuantepec. In C. Mason (ed) *Routledge Handbook of Queer Development Studies*. London: Routledge, pp. 209–222.

McGuinness, A. (2003) Searching for "Latin America": Race and sovereignty in the Americas in the 1850s. In N. P Appelbaum, A. S. Macpherson and K. A. Rosenblatt (eds) *Race and Nation in Modern Latin America*. Chapel Hill: University of North Carolina Press, pp. 87–107.

McIlwayne, C. and Ryburn, M (2019) Diversities of international and transnational migration in and beyond Latin America. In J. Cupples, M. Palomino-Schalsha and M. Prieto (eds) *Routledge Handbook of Latin American Development*. London: Routledge, pp. 145–155.

McNeish, A. (2013) Extraction, protest and indigeneity in Bolivia: The TIPNIS effect. *Latin American and Caribbean Ethnic Studies* 8(2): 221–242.

McNelly, A. (2019) The highs and lows of Bolivia's rebel city. *NACLA: Report on the Americas* 51(4): 333–340.

McSherry, J.P. (2002) Tracking the origins of a state terror network: Operation Condor. *Latin American Perspectives* 29(1):38–60.

Medina, N. (2009) *Mestizaje: Remapping Race, Culture, and Faith in Latina/o Catholicism.* New York: Orbis.

Melgarejo, V. and Bucholtz, M. (2020) "Oh, I don't even know how to say this in Spanish" The linguistic representation of Latinxs in "Jane the Virgin." *Spanish in Context* 17(3): 488–510.

Menchú, R. and Burgos-Debray, E. (1984) *I, Rigoberta Menchú: An Indian Woman in Guatemala.* London: Verso.

Méndez, M. J. (2018) "The river told me": Rethinking intersectionality from the world of Berta Cáceres. *Capitalism, Nature, Socialism* 29(1): 7–24.

——— (2019) The violence work of transnational gangs in Central America. *Third World Quarterly* 40(2): 373–388.

Méndez Franco, L. F. (2015) La vida en el imaginario de la resistencia popular por Ayotzinapa: La comunidad en contextos de terrorismo de Estado. *El Cotidiano,* 189: 67–72.

Mendoza, B. (2016) Coloniality of gender and power: From postcoloniality to decoloniality. In L. Disch and M. Hawkesworth (eds) *The Oxford Handbook of Feminist Theory.* Oxford: Oxford University Press, pp. 100–121.

Mezzadra, S. and Neilson, B. (2019) *The Politics of Operations: Excavating Contemporary Capitalism.* Durham, NC: Duke University Press.

Meyer, M. and Hinojosa, G. (2019) Five years on, still no justice for Mexico's 43 disappeared Ayotzinapa students. *WOLA* 24 September www.wola.org/analysis/five-year-anniversary-ayotzinapa-mexico/ (Accessed 13 July 2020).

Mignolo, W. D. (2000) *Local Histories/Global Designs: Coloniality, Subaltern Knowledges, and Border Thinking.* Princeton: Princeton University Press.

——— (2005) *The Idea of Latin America.* Malden, MA: Blackwell.

——— (2007) Delinking: The rhetoric of modernity, the logic of coloniality and the grammar of de-coloniality. *Cultural Studies* 21(2–3): 449–514.

——— (2011) *The Darker Side of Western Modernity: Global Futures, Decolonial Options.* Durham, NC: Duke University Press.

Mignolo, W. D. and Hoffman, A. (2017) Interview – Walter Mignolo/Part 2: Key Concepts. *E-International Relations* 21 June www.e-ir.info/2017/01/21/interview-walter-mignolopart-2-key-concepts/ (Accessed 21 March 2021).

Milan, S., Treré, E. and Masiero, S. (2021) Introduction: Covid-19 seen from the land of otherwise. In S. Milan, E. Treré and S. Masiero (eds) *Covid-19 from the Margins: Pandemic Invisibilities, Policies and Resistance in The Datafied Society.* Amsterdam: Institute of Network Cultures, Amsterdam, pp. 14–23.

Miles, R. (2021) Russia in Latin America. In G. L. Gardini (ed) *External Powers in Latin America: Geopolitics between Neo-extractivism and South-South Cooperation.* London: Routledge, pp. 59–74.

Milian, C. (2019) *Latinx.* Minneapolis: University of Minnesota Press.

Mínguez García, H. and Zamarripa Nungaray, J. (2016) Tácticas artivistas frente a la violencia en Ciudad Juárez. *Kult-Ur* 3(5): 211–28.

Miranda. M. and Vallejo, G. (2006) Sociodarwinismo y psicología de la inferioridad de los pueblos hispanoamericanos. Notas sobre el pensamiento de Carlos O. Bunge. *Frenia* 6: 57–77.

Mirzoeff, N. (2018) The politics of seeing within the global city. *Hyperallergic* 29 May https://hyperallergic.com/444073/the-politics-of-seeing-within-the-global-city/ (Accessed 8 June 2021).

Mochkofsky, G. (2020) Who are you calling Latinx? *The New Yorker* 5 September www.newyorker.com/news/daily-comment/who-are-you-calling-latinx (Accessed 22 June 2021).

Mohanty, C. T. (1991) Under Western eyes: Feminist scholarship and colonial discourses. In C. Mohanty, A. Russo and L. Torres (eds) *Third World Women and the Politics of Feminism*. Bloomington: Indiana University Press, pp. 51–80.

Moldano, A. (2017) A portrait of drug "mules" in the 1990s. In A. Farnsworth-Alvear, M. Palacios and A. M. Gómez López (eds) *The Colombia Reader: History, Culture, Politics*. Durham, NC: Duke University Press, pp. 501–504.

Mollet, S. (2017) Celebrating critical geographies of Latin America: Inspired by an NFL quarterback. *Journal of Latin American Geography* 16(1): 165–171.

Molyneux, M. (1986) Mobilization without emancipation? Women's interests, state and revolution in R.F. Fagan, C.D. Deere and J. L. Coraggio (eds) *Transition and Development: Problems of Third World Socialism*. New York: Monthly Review Press. 280–302.

Moser, C. O. N. (1993) Adjustment from below: Low-income women, time and the triple role in Guayaquil, Ecuador. In S. A. Radcliffe and S. Westwood (eds) *Viva: Women and Popular Protest in Latin America*. London and New York: Routledge, pp. 173–196.

Motta, S. (2019) Feminising our revolutions. *Soundings* 71: 15–27.

Muñoz, C. (2020) Brazil suffers its own scourge of police brutality. *Americas Quarterly* 3 June www.americasquarterly.org/article/brazil-suffers-its-own-scourge-of-police-brutality/ (Accessed 19 March 2021).

Muñoz, W. (2015). De la página a la calle: Los "performances" de Mujeres Creando en Bolivia. *Letras Femeninas*, 41(1): 182–194.

Murray, P. S. (2001) 'Loca' or 'libertadora'?: Manuela Sáenz in the eyes of history and historians, 1900-c.1990. *Journal of Latin American Studies* 33(2): 291–310.

Nascimento, B. (2021a[1988]) For a (new) existential and physical territory. *Antipode* 53(1): 279–316.

———— (2021b[1985]) The concept of quilombo and Black cultural resistance. *Antipode* 53(1): 279–316.

Nash, D. (2012) The First New Chronicle and Good Government: On the history of the world and the Incas up to 1615. *Ethnohistory*. 59(1): 202–205.

Nathanson, M. (2017) Damming or damning the Amazon: Assessing Ecuador/China cooperation. *Mongabay* 22 November https://news.mongabay.com/2017/11/damming-or-damning-the-amazon-assessing-ecuador-china-cooperation/ (Accessed 21 May 2021).

Nations, J.D. (1994) The ecology of the Zapatista Revolt. *Cultural Survival* March www.culturalsurvival.org/publications/cultural-survival-quarterly/ecology-zapatista-revolt (Accessed 23 June 2021).

Nederveen Pieterse, J. (2000) After post-development. *Third World Quarterly* 21(2): 175–191.

Nelson, D. M. (2019) Low intensities. *Current Anthropology* 60 (Supplement 19): 122–133.

Noriega, C. (2020) 'Black Lives Matter is seen as a trend – it's time to wake up.' A view from Colombia. *Huck* 27 July www.huckmag.com/perspectives/activism-2/colombia-black-lives-matter-trend-racism/ (Accessed 19 March 2021).

OAS (2019) IACHR brings Honduras case before IA Court. 9 May www.oas.org/en/iachr/media_center/PReleases/2019/112.asp (Accessed 15 May 2021).

O'Boyle, B. (2019) The rise and fall of CICIG: LatAm's biggest stories of the 2010s. *Americas Quarterly* 11 December www.americasquarterly.org/article/the-rise-and-fall-of-cicig-latams-biggest-stories-of-the-2010s/ (Accessed 5 July 2020).

Observatório de Mortes Violentas de LGBTI+ no Brasil (2021) *Mortes Violentas de LGBTI+ no Brasil: Relatório 2020*. Florianópolis: Editora Acontece Arte e Política LGBTI+ e Grupo Gay da Bahia (GGB).

O'Keefe, P., Westgate, K. and Wisner, B. (1976) Taking the naturalness out of natural disasters. *Nature* 260: 566–567.

Oliver-Smith, A. (1986) *The Martyred City: Death and Rebirth in the Andes*. Albuquerque: University of New Mexico Press.

——— (1994) Peru's five hundred year earthquake: Vulnerability in historical context. In A. Varley (ed) *Disasters, Development and Environment*. Chichester: Wiley. 31–48.

Olmo, G. D. (2020) Irán y Venezuela: Megasis, el inusual supermercado que el país del Golfo Pérsico abrió en Caracas (y qué dice de las tensiones con Estados Unidos). *BBC* 31 July www.bbc.com/mundo/noticias-america-latina-53604561 (Accessed 3 May 2021).

Olusoga, D. (2018) The Treasury's tweet shows slavery is still misunderstood. *The Guardian* 12 February www.theguardian.com/commentisfree/2018/feb/12/treasury-tweet-slavery-compensate-slave-owners (Accessed 13 July 2020).

Open Society Justice Initiative (2013) *Judging a Dictator: The Trial of Guatemala's Ríos Montt*. New York: Open Society Justice Initiative.

Oquendo, C. (2021) Colombia carga contra los monumentos a los conquistadores españoles. *El País* 23 June https://elpais.com/internacional/2021-06-23/colombia-carga-contra-los-monumentos-a-los-conquistadores-espanoles.html?ssm=TW_CC (Accessed 28 June 2021).

Orellana, D. (2021) A 11 años del asesinato de Vicky Hernández, su madre reclama justicia. *Washington Blade* 7 May www.washingtonblade.com/2021/05/07/a-11-anos-del-asesinato-de-vicky-hernandez-su-madre-reclama-justicia (Accessed 15 May 2021).

Oswin, N. (2018) Planetary urbanization: A view from outside." *Environment and Planning D: Society and Space* 36(3): 540–546.

Pacheco, M. (2020) Alerta en la Corporación Eléctrica del Ecuador por erosión del río Coca. *El Comercio* 4 May www.elcomercio.com/actualidad/alerta-celec-erosion-rio-coca.html (Accessed 14 April 2021).

Page, J. (2019) China counts the costs of its big bet on Venezuela. *Wall Street Journal* 1 February www.wsj.com/articles/china-counts-the-costs-of-its-big-bet-on-venezuela-11549038825 (Accessed 29 April 2021).

Paley, D. (2014) *Drug War Capitalism*. Oakland, CA: AK Press.

Palmer. P. (2004[1993]) West Indian Limón. In S. Palmer and I. Molina (eds) *The Costa Rica Reader: History, Culture and Politics*. Durham, NC: Duke University Press, pp. 237–242.

Panizza, F. (2009) *Contemporary Latin America: Development and Democracy Beyond the Washington Consensus*. London: Zed Books.

Papyrakis, E. and Pellegrini, L. (2019) The resource curse in Latin America. *Oxford Research Encyclopedia of Politics* https://doi.org/10.1093/acrefore/9780190228637.013.1522 (Accessed 13 April 2021).

Paredes, J. (2008) *Hilando fino: Perspectives from Communitarian Feminism.* La Paz: Comunidad Mujeres Creando.

Patterson, E. M. (2016) Reconciling Indigenous peoples with the judicial process: An examination of the recent genocide and sexual slavery trials in Guatemala and their integration of Mayan culture and customs. *Quebec Journal of International Law* 29(2): 225–252.

Paton, D. (2020) Proper channels. *History Workshop* 15 June www.historyworkshop.org.uk/proper-channels/ (Accessed 16 June 2020).

Paz, O. (1950) *El laberinto de soledad.* México: Cuadernos Americanos.

Pedersen, A. (2014). Landscapes of resistance: Community opposition to Canadian mining operations in Guatemala. *Journal of Latin American Geography* 13(1): 187–214.

Peet, R. (2003) *Unholy Trinity: The IMF, the World Bank and the WTO.* London: Zed Books.

Perreault, T. (2001) Developing identities: Indigenous mobilization, rural livelihoods, and resource access in Ecuadorian Amazonia. *Ecumene* 8(4): 381–413

——— (2005) State restructuring and the scale politics of rural water governance in Bolivia. *Environment and Planning A* 37(2): 263–284.

Petropoulou, C. (2018) Social resistances and the creation of another way of thinking in the peripheral "Self-Constructed Popular Neighborhoods": Examples from Mexico, Argentina, and Bolivia. *Urban Science* 2(1): 1–22.

Phillips, D. (2018) Their forefathers were enslaved. Now, 400 years later, their children will be landowners. *The Guardian* 5 March www.theguardian.com/world/2018/mar/05/descendants-of-slaves-celebrate-brazil-land-rights-victory (Accessed 2 June 2021).

——— (2020) 'We are facing extermination': Brazil losing a generation of indigenous leaders to Covid-19. *The Guardian* 20 June www.theguardian.com/global-development/2020/jun/21/brazil-losing-generation-indigenous-leaders-covid-19 (Accessed 7 June 2021).

Phipps, A. (2020) *Me, Not You: The Trouble with Mainstream Feminism.* Manchester: Manchester University Press.

Picq, M. (2011) Indigenous resistance is the new 'terrorism.' *Al Jazeera* 10 July www.aljazeera.com/opinions/2011/7/10/indigenous-resistance-is-the-new-terrorism (Accessed 14 April 2021).

Pilkington, E. (2020) Parents of 545 children still not found three years after Trump separation policy. *The Guardian* 20 October www.theguardian.com/us-news/2020/oct/21/trump-separation-policy-545-children-parents-still-not-found (Accessed 1 May 2021).

Pineda, B. (2001) The Chinese Creoles of Nicaragua: Identity, economy, and revolution in a Caribbean Port City. *Journal of Asian American Studies* 4(3): 209–233.

——— (2006) *Shipwrecked Identities: Navigating Race on Nicaragua's Mosquito Coast.* New Brunswick, NJ: Rutgers University Press

Pinkoski, K. (2012) Maoism in South America: Comparing Peru's Sendero Luminoso with Mexico's PRP and PPUA. *Constellations* 4(1): 232–247.

Piñon, J. (2017) Jane the Virgin: The pursuit of new Latina/o representations. *ReVista* Fall 23–26.

Poniatowska, E. (1995) *Nothing, Nobody: The Voices of the Mexico City Earthquake.* Philadelphia: Temple University Press.

Postero, N. G. (2004) Articulation and fragmentation: Indigenous politics in Bolivia. In N. G. Postrero and L. Zamosc (eds) *The Struggle for Indigenous Rights in Latin America.* Eastborne: Sussex Academic Press, pp. 189–216.

Prebisch, R. (1950) *The Economic Development of Latin America and its Principal Problems.* New York: United Nations.

——— (1963) *Hacia una dinámica del desarrollo latinoamericano.* México: Fondo de Cultura Económica.

Quidel Lincoleo, J. (2020) Fey ga akuy ti aht'ü. Entonces el día llegó. Una lectura de la pandemia desde un mapuche rakizuam. *Comunidad de Historia Mapuche* 21 April www.comunidadhistoriamapuche.cl/fey-ga-akuy-ti-ahtu-entonces-el-dia-llego-una-lectura-de-la-pandemia-desde-un-mapuche-rakizuam/ (Accessed 27 May 2020).

Quijano, A. (1972) La constitución del "mundo" de la marginalidad urbana. *Revista Eure* 2(5): 89 106.

——— (2000) Coloniality of power, Eurocentrism and Latin America. *Nepantla,* 1(3): 533–580.

——— (2005) The challenge of the "indigenous movement" in Latin America. *Socialism and Democracy* 19(3): 55–78.

——— (2007) Coloniality and modernity/rationality. *Cultural Studies* 21(2–3): 168–178.

Quijano Valencia, O. and Corredor Jimenez, C. (eds) (2020) *Pandemia al sur.* Buenos Aires: Prometeo Libros.

Quiroga Díaz, N. (2019) Rethinking the urban economy: Women, protest, and the new commons. In J. Cupples, M. Palomino-Schalsha and M. Prieto (eds) *Routledge Handbook of Latin American Development.* London: Routledge, pp. 560–569.

Radcliffe, S. A. (1999) *Gender, Migration and Domestic Service.* London: Routledge.

Radcliffe, S. A and Radhuber, I. M. (2020) The political geographies of D/decolonization: Variegation and decolonial challenges of /in geography. *Political Geography* 78: 1–12.

Rahnema, M. (1997) Towards post-development: searching for signposts, a new language and new paradigms. In M. Rahnema and V. Bawntree (eds) *The Post-Development Reader.* Cape Town: David Philip, pp. 377–403.

Ramírez, S. (2007) *Tambor Olvidado.* San José: Aguilar.

Ramos, A. R., Guerreiro Osório, R. and Pimenta, J. (2009) Indigenising development. *Poverty in Focus* 17: 3–5.

Randall, M. (1981) *Sandino's Daughters: Testimonies of Nicaraguan Women in Struggle.* Vancouver: New Star.

Ranis, P. (2006) Factories without bosses: Argentina's experience with worker-run enterprises. *Labor: Studies in Working-Class History of the Americas* 3(1): 11–23.

Ranta, E. M (2016) Toward a decolonial alternative to development? The emergence and shortcomings of *Vivir Bien* as state policy in Bolivia in the era of globalization, *Globalizations* 13(4): 425–439.

Rautner, M. and Cuffe, S. (2020) Nicaraguan beef, grazed on deforested and stolen land, feeds global demand. *Mongabay* 10 June https://news.mongabay.com/2020/06/nicaraguan-beef-grazed-on-deforested-and-stolen-land-feeds-global-demand/ (Accessed 14 April 2021).

Rebollo Gil, G. (2018) *Writing Puerto Rico: Our Decolonial Moment.* Cham: Palgrave.

Recinos, A. (1950) *Popol Vuh: The Sacred Book of the Ancient Quiché Maya*, translated by D. Goetz and S. G. Morley. Norman: University of Oklahoma Press.

Reiter, B. (2015) Palenque de San Basílio: Citizenship and republican traditions of a maroon village in Colombia. *Journal of Civil Society* 11(4): 333–347

Reid Smith, T. (2020) Mexican public backs same-sex marriage and would support LGBT+ family members. *Gay Star News* 30 June www.gaystarnews.com/article/mexican-public-backs-same-sex-marriage-and-would-support-lgbt-family-members/ (Accessed 4 July 2021).

Restall, M. (2004) *Seven Myths of the Spanish Conquest*. Oxford: Oxford University Press.

Reuters (2021) Uproar after Argentina president says 'Brazilians came from the jungle.' *The Guardian* 9 June www.theguardian.com/world/2021/jun/09/argentina-president-comments-uproar (Accessed 11 June 2021).

Reyes, E. (2020) Body politics in the Covid-19 era from a feminist lens. *Development* 63: 262–269.

Reyes Cruz, M. (2018). Por quiénes esperamos. *80 grados*. 2 Febuary www.80grados.net/por-quienes-esperamos/ (Accessed 7 June 2021).

Riofrancos, T. (2019) What comes after extractivism? *Dissent* www.dissentmagazine.org/article/what-comes-after-extractivism (Accessed 14 April 2021).

——————— (2020) *Resource Radicals: From Petro-Nationalism to Post-Extractivism in Ecuador*. Durham, NC: Duke University Press.

Rivera Andía, J. J. and Vindal Ødegaard, C. (2019) Introduction: Indigenous peoples, extractivism, and turbulences in South America. In C. Vindal Ødegaard and J. J. Rivera Andía (eds) (2019) *Indigenous Life Projects and Extractivism: Ethnographies from South America*. Cham: Palgrave Macmillan, pp. 1–50.

Rivera Cusicanqui, S. (1991) Aymara past, Aymara future. *Report on the Americas* 25(3):18–45.

——— (2004) La noción de "derecho" o las paradojas de la modernidad post-colonial: indígenas y mujeres en Bolivia. *Aportes andinos* 11: 1–15.

——— (2012) *Ch'ixinakax utwixa*: A reflection on the practices and discourses of decolonization. *South Atlantic Quarterly* 111(1): 95–109.

Rivero, Y. M. (2003) The performance and reception of televisual "ugliness" in *Yo soy Betty la fea*. *Feminist Media Studies* 3(1): 65–81.

Robinson, S. (2010) The Chinese of Central America: Diverse beginnings, common achievements. In W. Look Laid and T. Chee-Beng (eds) *The Chinese in Latin America and the Caribbean*. Boston: Brill, pp. 103–128.

Rocha, L. O. (2020) Estamos em marcha! Anti-racism, political struggle, and the leadership of Black Brazilian women. In J. Hooker (ed) *Black and Indigenous Resistance in the Americas: From Multiculturalism to Racist Backlash*. Lanham: Lexington Books, pp. 159–187.

Rodgers, D. (2019) Gang violence in Latin America. In J. Cupples, M. Palomino-Schalsha and M. Prieto (eds) *Routledge Handbook of Latin American Development*. London: Routledge, pp. 517–527.

Rodríguez Madera, S. L. (2020) From necropraxis to necroresistance: Transgender experiences in Latin America. *Journal of Interpersonal Violence* (online early)

Roldán, M. (2002) *Blood and Fire: La Violencia in Antioquia, Colombia, 1946–1953*. Durham, NC: Duke University Press.

Rosales, O. and Kuwayama, M. (2012) *China and Latin America and the Caribbean: Building a Strategic Economic and Trade Relationship*. Santiago: ECLAC.

Rostow, W. (1960) *The Stages of Economic Growth: A Non-Communist Manifesto.* Cambridge: Cambridge University Press.

Roush, L. (2014) Santa Muerte, protection, and *desamparo:* A view from a Mexico City altar. *Latin American Research Review* 49: 129–48.

Routledge, P. (1998) Going globile: Spatiality, embodiment and mediation in the Zapatista insurgency. In G. Ó'Tuathail and S. Dalby (eds) *Rethinking Geopolitics.* London: Routledge, pp. 240–260.

Rowe, W. and Shelling, V. (1991) *Memory and Modernity: Popular Culture in Latin America.* London: Verso.

Ruddick, S., Peake, L., Tanyildiz, G S. and Patrick, D. (2018) Planetary urbanization: An urban theory for our time?" *Environment and Planning D: Society and Space* 36(3): 387–404.

Runnels, D. (2019) Cholo aesthetics and mestizaje: Architecture in El Alto, Bolivia. *Latin American and Caribbean Ethnic Studies* 14(2): 138–150.

Saini, A. (2019) *Superior: The Return of Race Science.* London: Fourth Estate.

Salgado, S. (2003) *The Teatro Solís: 150 Years of Opera, Concert and Ballet in Montevideo.* Middletown: Wesleyan University Press.

Salinas Maceda, V. (2016) Cholas fashionistas. Cuando la identidad se porta en la pollera, la manta y el sombrero. Paper presented to Festival Internacional de la Imagen, Manizales, Colombia, 3–13 May. Available at http://festivaldelaimagen.com/wp-cont ent/uploads/2017/07/Valeria_Salinas.pdf (Accessed 13 June, 2021).

Salt, K. (2019) *The Unfinished Revolution: Haiti, Black Sovereignty and Power in the Nineteenth-Century Atlantic World.* Liverpool: Liverpool University Press.

Salzinger, L. (2000) Manufacturing sexual subjects: 'Harassment', desire and discipline on a maquiladora shopfloor. *Ethnography* 1(1): 67–92.

Santos de Araújo, F. (2018) Marielle, presente! *Meridians: Feminism, Race, Transnationalism* 17(1): 207–211.

Sarmiento, D. F. (2004[1845]) *Facundo: Civilization and Barbarism,* translated by K. Ross. Berkeley: University of California Press.

Sawyer, M. Q. (2005) Du Bois's Double Consciousness versus Latin American exceptionalism: Joe Arroyo, salsa, and negritude. *Souls: A Critical Journal of Black Politics, Culture, and Society* 7(3–4): 85–95.

Sawyer, S. (1997). The 1992 Indian Mobilization in Lowland Ecuador. *Latin American Perspectives* 24(3): 65–82.

——— (2004) *Crude Chronicles: Indigenous Politics, Multinational Oil and Neoliberalism in Ecuador.* Durham, NC: Duke University Press.

Scauso, M.S., FitzGerald, G., Tickner, A.B., Chadha Behera, N. Pan, C., Shih, C. and Shimuzu, K. (2020) Covid-19, democracies, and (de)colonialities. *Democratic Theory* 7(2): 82–93.

Schirmer, J. G. (1989) "Those who die for life cannot be called dead": Women and human rights protest in Latin America. *Feminist Review* 32: 3–29.

Schiwy, F. (2003) Decolonizing the frame: Indigenous video in the Andes. *Framework* 44(1): 116–132.

Sealey-Huggins, L. (2018). The climate crisis is a racist crisis: Structural racism, inequality and climate change. In A. Johnson, R. Joseph-Salisbury and B. Kamunge (eds) *The Fire Now: Anti-Racist Scholarship in Times of Explicit Racial Violence.* London: Zed Books, pp. 1–12.

Segato, R.L. (2010) Territory, sovereignty, and crimes of the second state: The writing on the body of murdered women. In R. L. Fregoso and C. Bejarano (eds) *Terrorizing Women: Feminicide in the Americas*. Durham, NC: Duke University Press, pp. 70–92.

——— (2016) *La guerra contra las mujeres*. Madrid: Traficantes de Sueños.

——— (2020) Coronavirus: Todos somos mortales. In O.Quijano Valencia and C. Corredor Jimenez (eds) (2020) *Pandemia al sur*. Buenos Aires: Prometeo Libros, pp. 11–21.

Segovia, M. A. (2021) A dream deal with China that ended in nightmarish debt for Venezuela. *Diálogo Chino* 14 February https://dialogochino.net/en/trade-investment/40016-a-dream-deal-with-china-that-ended-in-nightmarish-debt-for-venezuela/ (Accessed 29 April 2021).

Sepúlveda, J. G. de (1941[1547]) *Tratado sobre las justas causas de la guerra contra los indios*. Mexico City: Fondo de Cultura Económica.

Serafini, P. (2020) 'A rapist in your path': Transnational feminist protest and why (and how) performance matters. *European Journal of Cultural Studies*, 23(2): 290–295.

Silva, D. N. and Won Lee, J. (2020) "Marielle, presente": Metaleptic temporality and the enregisterment of hope in Rio de Janeiro. *Journal of Sociolinguistics* 25(2): 179–197.

Silva Loureiro, G. (2020) To be Black, Queer and Radical: Centring the epistemology of Marielle Franco. *Open Cultural Studies* 4: 50–58.

Silverblatt, I. (1987) *Moon, Sun, Witches: Gender Ideologies and Class in Inca and Colonial Peru*. Princeton: Princeton University Press.

Skidmore, T. E. (1993) Bi-racial U.S.A. vs. multi-racial Brazil: is the contrast still valid?'. *Journal of Latin American Studies*, 25(2): 373–86.

Skidmore, T. E. and Smith, P. H. (2005) *Modern Latin America*. 6th edition. Oxford: Oxford University Press.

Slater, T. (2015) Planetary rent gaps. *Antipode* 49(S1): 114–137.

——— (2019) Introduction: Producing and contesting urban marginality. In J. Cupples and T. Slater (eds) *Producing and Contesting Urban Marginality: Interdisciplinary and Comparative Dialogues*. London: Rowman and Littlefield, pp. 1–16.

Smith, C. A. (2016) Towards a Black feminist model of Black Atlantic liberation: Remembering Beatriz Nascimento. *Meridians* 14(2): 71–87.

Smith, C., Davies, A. and Gomes, B. (2021) "In Front of the World": Translating Beatriz Nascimento. *Antipode* 53(1): 279–316.

Smith, N. (1987) Gentrification and the rent gap. *Annals of the Association of American Geographers* 77(3): 462–465.

——— (2006) There's no such thing as a natural disaster. *Items: Insights from the Social Sciences* 11 June https://items.ssrc.org/understanding-katrina/theres-no-such-thing-as-a-natural-disaster/ (Accessed 7 June 2021).

Solnit, R. (2010) *A Paradise Built in Hell: The Extraordinary Communities That Arise in Disaster*. New York: Penguin.

Stensrud, A. B. (2019) Water as resource and being: Water extractivism and life. In C. Vindal Ødegaard and J. J. Rivera Andía (eds) (2019) *Indigenous Life Projects and Extractivism: Ethnographies from South America*. Cham: Palgrave Macmillan, pp. 143–164.

Stephen, L. (2002) Sexualities and genders in Zapotec Oaxaca. *Latin American Perspectives* 29(2): 41–59.

Subcomandante Marcos (2005) *Conversations with Durito: Stories of the Zapatistas and Neoliberalism*. Brooklyn: Autonomedia.

Subcomandante Marcos and Ponce de León, J. (2001) *Our Word Is Our Weapon: Selected Writings*. New York: Seven Stories.

Sudré, L. (2021) Brazil: transgender murders increased 41% in 2020. *Brasil de Fato* 29 January www.brasildefato.com.br/2021/01/29/brazil-transgender-murders-increased-41-in-2020 (Accessed 15 May 2021).

Sue, C. A. (2013) *Land of the Cosmic Race: Race Mixture, Racism and Blackness in Mexico*. Oxford: Oxford University Press.

Sullivan, M. and Lum, T. (2020) China's engagement with Latin America and the Caribbean. *Congressional Research Service In Focus* 12 November https://crsreports.congress.gov (Accessed 29 June).

Sunkel, O. (1973) Transnational capitalism and national disintegration in Latin America. *Social and Economic Studies* 22(1): 132–176.

Susman, P., O'Keefe, P. and Wisner, B. (1983) Global disasters: A radical interpretation. In K. Hewitt (ed) *Interpretations of Calamity from the Viewpoint of Human Ecology*. London: Allen and Unwin, pp. 263–283.

Svampa, M. (2019) *Neo-Extractivism in Latin America: Socio-environmental Conflicts, the Territorial Turn, and New Political Narratives*. Cambridge: Cambridge University Press.

Swanson, K. (2019) Street vendors. In J. Cupples, M. Palomino-Schalsha and M. Prieto (eds) *Routledge Handbook of Latin American Development*. London: Routledge, pp. 355–363.

Sweig, J. (2009) *Cuba: What Everyone Needs to Know*. Oxford: Oxford University Press.

Tapia Arce, A. (2017) Making beauty: The wearing of polleras in the Andean altiplano. *Portal* 28 August https://llilasbensonmagazine.org/2017/08/28/making-beauty-the-wearing-of-polleras-in-the-andean-altiplano/ (Accessed 13 June 2021).

Tatis Guerra, G. (2017) Pedro Romero, el héroe sin rostro. *El Universal* 12 November www.eluniversal.com.co/suplementos/facetas/pedro-romero-el-heroe-sin-rostro-265980-KUEU379232 (Accessed 12 June 2021).

Taylor, C. (2005) Salsa. In L. Shaw and S. Dennison (eds) *Pop Culture Latin America! Media, Arts, and Lifestyle*. Santa Barbara: ABC Clio, pp. 10–12.

Teheran Times (2021) Cuban coronavirus vaccine to start third clinical trial phase in Iran. 18 April www.tehrantimes.com/news/459979/Cuban-coronavirus-vaccine-to-start-third-clinical-trial-phase (Accessed 3 May 2021).

teleSUR (2020) 10 datos importantes sobre Manuela Sáenz. 27 December www.telesurtv.net/news/manuela-saenz-heroina-libertadora-america-latina-20191123-0002.html (Accessed 5 March 2021).

TGEU (2020) TMM Update Trans Day of Remembrance 2020, 20 November https://transrespect.org/en/tmm-update-tdor-2020/ (Accessed 15 May 2021).

Thorne, M. (2019) Cuando el subalterno construye: Freddy Mamani y la emergencia del cholo power boliviano. *Lógoi Revista de Filosofía* 35: 75–86.

Tiano, S. (1990) Maquiladora women: A new category of workers. In K. Ward (ed) *Women Workers and Global Restructuring*. New York: ILR Press, pp. 193–223.

Tiano, S. and Ladino, C. (1999) Dating, mating, and motherhood: Identity construction among Mexican *Maquila* Workers. *Environment and Planning A* 31(2): 305–325.

Tischler, S. (2019) Zapatismo: Reinventing Zapatismo. In J. Cupples, M. Palomino-Schalsha and M. Prieto (eds) *Routledge Handbook of Latin American Development*. London: Routledge, pp. 252–262.

Torres, L. (2018) Latinx? *Latino Studies* 16: 283–285.

Torres López, J. E. (2020) *Reporting on Colombia: Essays on Colombia's History, Culture, Peoples, and Armed Conflict.* No place of publication: La Nasiona.

The Economist (2015) Gay rights in Latin America: The rainbow tide. 16 October www.economist.com/the-americas/2015/10/16/the-rainbow-tide (Accessed 30 June 2021).

Tremlett, G. (2014) José Mujica: is this the world's most radical president? *The Guardian* 18 September www.theguardian.com/world/2014/sep/18/-sp-is-this-worlds-most-radical-president-uruguay-jose-mujica (Accessed 31 May 2021).

Triquell, A. (2018) Those who are no longer: Photographic image and representation of women's absence(s) in Mayra Martell's *Ensayo de la Identidad* at Ciudad Juarez. *photographies* 11(1): 52–72.

Trouillot, M. R. (1995) *Silencing the Past: Power and the Production of History.* Boston: Beacon Press.

Tuck, E. and Yang, K. W. (2012) Decolonization is not a metaphor. *Decolonization: Indigeneity, Education and Society* 1(1): 1–40.

Tuhiwai Smith, L. (2012) *Decolonizing Methodologies: Research and Indigenous Peoples,* 2nd ed. London: Zed Books.

Tzul Tzul, G. (2015) Sistemas de gobierno comunal indígena: La organización de la reproducción de la vida. *El Apantle: Revista de Estudios Comunitarios* 1: 125–140.

——— (2018) Rebuilding communal life. *NACLA Report on the Americas* 50(4): 404–407.

UDEFEGUA (2020) *Informe de Situación de Personas Defensoras de Derechos Humanos, Guatemala 2019* informe_udefegua_2019_2020_alta.pdf (Accessed 15 August 2020).

UNDP (2018) Paraguay: Sustainable soy and beef. www.greencommodities.org/content/gcp/en/home/countries-and-commodities/paraguay-beef-and-soy.html (Accessed 5 June 2021).

——— (2019) *Human Development Report 2019: Beyond Income, Beyond Averages, Beyond Today – Inequalities in Human Development in the 21st Century.* New York: United Nations.

Uribe-Uran, V.M. (2013) An academic's search for answers to violence against women: An interview with Professor Rita Laura Segato. *Hemisphere* 22: 14–19.

Urkidi, L. and Walter, M. (2011) Dimensions of environmental justice in anti-gold mining movements in Latin America. *Geoforum* 42(6): 683–695.

Valdivia, G. and Lyall, A. (2019) The oil complex In Latin America: Politics, frontiers, and habits of oil rule. In J. Cupples, M. Palomino-Schalsha and M. Prieto (eds) *Routledge Handbook of Latin American Development.* London: Routledge, pp. 458–468.

Valentina González, R. (2019) *Quinceañera Style: Social Belonging and Latinx Consumer Identities.* Austin: University of Texas Press.

Valle, M. M. (2018) The discursive detachment of race from gentrification in Cartagena de Indias, Colombia. *Ethnic and Racial Studies* 41(7): 1235–1254.

Vargas-Chaves, I., Rodríguez, G.A., Cumbe-Figueroa, A. and Mora-Garzón, S.E. (2020) Recognizing the Rights of Nature in Colombia: the Atrato River case. *Revista Jurídicas* 17(1): 13–41.

Vasconcelos, J. (1997[1925]) *The Cosmic Race/La raza cósmica,* translated by D. T. Jaén. Baltimore: Johns Hopkins University Press.

Vegliò, S. (2018) Planetary urbanization and postcolonial geographies: What directions for critical urban theory? In J. Cupples and R. Grosfoguel (eds) *Unsettling Eurocentrism in the Westernized University.* London: Routledge, pp. 116–130.

Vergara-Figueroa, A. (2018) *Afrodescendant Resistance to Deracination in Colombia: Massacre at Bellavista-Bojayá-Chocó*. Cham: Palgrave.

Vergès, F. (2021) *Decolonial Feminism*. London: Pluto Press.

Vidal, J. (2011) Bolivia enshrines natural world's rights with equal status for Mother Earth. *The Guardian* 10 April www.theguardian.com/environment/2011/apr/10/bolivia-enshrines-natural-worlds-rights (Accessed 28 June 2021).

Vílchez, D. (2017) Juliet Hooker: "No pensamos que el nicaragüense puede ser negro". *Confidencial/Niú*, 17 July https://niu.com.ni/juliet-hooker-no-pensamos-que-el-nicaraguense-puede-ser-negro/ (Accessed 30 May 2021).

Viveros-Wacher, P. and Kraus-Weisman, A. (2018) Nuestros terremotos. *Salud Pública México* 60(1): 105–108.

Volk, S. and Schlotterbeck, M. (2007) Gender, order, and femicide: Reading the popular culture of murder in Ciudad Juárez. *Aztlan: A Journal of Chicano Studies* 32 (1): 53–86.

Wade, P. (2006) Afro-Latin studies: Reflections on the field. *Latin American and Caribbean Ethnic Studies* 1(1): 105–124.

———— (2010) *Race and Ethnicity in Latin America*, 2nd ed. London: Pluto Press.

———— (2017) *Degrees of Mixture, Degrees of Freedom: Genomics, Multiculturalism, and Race in Latin America*. Durham, NC: Duke University Press.

Wainwright, J. (2008) *Decolonizing Development: Colonial Power and the Maya*. Malden: Blackwell.

Wainwright, J. and Bryan, J. (2009) Cartography, territory, property: Postcolonial reflections on indigenous counter-mapping in Nicaragua and Belize. *cultural geographies* 16: 153–178.

Walker, L. E. (2009) Economic fault lines and middle-class fears: Tlatelolco, Mexico City, 1985. In J. Buchenau and L. L. Johnson (eds) *Aftershocks: Earthquakes and Popular Politics in Latin America*. 184–221. Albuquerque: University of New Mexico Press, pp. 184–221.

Walsh, C. (2009) *Interculturalidad, estado, sociedad: Luchas (de)coloniales de nuestra época*. Quito: Ediciones Abya Yala.

———— (2010) Development as Buen Vivir: Institutional arrangements and (de)colonial entanglements. *Development* 53(1): 15–21.

———— (2019) (Decolonial) notes to Paulo Freire: Walking and asking. In R. Aman and T. Ireland (eds) *Educational Alternatives in Latin America: New Modes of Counter-Hegemonic Learning*. Cham: Palgrave, pp. 207–230.

Watson, P. L. (2021) Iran's Latin America strategy and the challenges to the balance of power. In G. L. Gardini (ed) *External Powers in Latin America: Geopolitics between Neo-extractivism and South-South Cooperation*. Abingdon: Routledge, pp. 138–152.

Wiarda, H. J. and Kline, H. F (2011) The Latin American tradition and the process of development. In H. J. Wiarda and H. F. Kline (eds) *Latin American Politics and Development*. Boulder: Westview Press, pp. 1–98.

Williams, E. L. (2018) *Mucamas* and *mulatas*: Black Brazilian feminisms, representations, and ethnography. In C. R. Rodriguez, N. Tsikata and A. Adomako Ampofo (eds) *Transatlantic Feminisms: Women and Gender Studies in Africa and the Diaspora*. Lanham: Lexington Books, pp. 122–140.

Williams, S. (2008) Rethinking the nature of disaster: From failed instruments of learning to a post-social understanding. *Social Forces* 87(2): 1115–1138.

Williamson, E. (2009) *The Penguin History of Latin America*. London: Penguin.

Willis, K. (2019) The Sustainable Development Goals. In J. Cupples, M. Palomino-Schalsha and M. Prieto (eds) *Routledge Handbook of Latin American Development*. London: Routledge, pp. 121–131.

Wilson, V. (2021) Afrodescendientes in Paraguay: The 209-year struggle for recognition. *Latin American Perspectives* [blog] Political report 1453 https://laperspectives.blogspot.com/2021/02/political-report-1453.html (Accessed 29 May, 2021).

Wise, C. (2021) China in Latin America: Winning hearts and minds pragmatically. In G. L. Gardini (ed) *External Powers in Latin America: Geopolitics between Neo-extractivism and South-South Cooperation*. Abingdon: Routledge, pp. 44–58.

Wolf, S. (2012) Mara Salvatrucha: The most dangerous street gang in the Americas? *Latin American Politics and Society* 54(1): 65–99.

Woods, N. (2006) *The Globalizers: The IMF, the World Bank, and Their Borrowers*. Ithaca: Cornell University Press.

Wright, M. (2019) Visualizing a country without a future: Posters for Ayotzinapa, Mexico and struggles against state terror. *Geoforum* 102: 235–241.

Wynter, S. (2003) Unsettling the coloniality of being/power/truth/freedom: Towards the human, after man, its overrepresentation – an argument. *The New Centennial Review* 3(3): 257–337.

Wynter, S. and McKittrick, K. (2015) Unparalled catastrophe for our species? Or, to give humanness a different future: Conversations. In K McKittrick (Ed) *Sylvia Wynter: On Being Human as Praxis*. Durham, NC: Duke University Press, pp. 9–89.

Wyss, J. (2021) Haiti is the only country in Western Hemisphere without vaccines. *Bloomberg* 8 June www.bloomberg.com/news/articles/2021-06-08/haiti-is-the-only-country-in-western-hemisphere-without-vaccines (Accessed 11 June 2021).

Yaffe, H. (2020) Leading by example: Cuba in the Covid-19 pandemic. *Counterpunch* 4 June www.counterpunch.org/2020/06/04/leading-by-example-cuba-in-the-covid-19-pandemic/ (Accessed 5 June 2021).

Yashar, D. J. (2005) *Contesting Citizenship in Latin America: The Rise of Indigenous Movements and the Postliberal Challenge*. Cambridge: Cambridge University Press.

Yúdice, G. (2003) *The Expediency of Culture: Uses of Culture in the Global Era*. Durham, NC: Duke University Press.

Zamosc, L. (2004) The Indian movement in Ecuador: From politics of influence to politics of power. In N. G. Postrero and L. Zamosc (eds) *The Struggle for Indigenous Rights in Latin America*. Eastborne: Sussex Academic Press, pp. 131–157.

Zentella, A. C. (2017) "Limpia, fija y da esplendor": Challenging the symbolic violence of the Royal Spanish Academy. *Chiricú Journal: Latina/o Literatures, Arts, and Cultures* 1(2): 21–42.

Zhang, P. (2019) Belt and Road in Latin America: A regional game changer? *Atlantic Council* 8 October www.atlanticcouncil.org/in-depth-research-reports/issue-brief/belt-and-road-in-latin-america-a-regional-game-changer/ (Accessed 30 April, 2021).

Ziai, A. (2019) Post-development. In J. Cupples, M. Palomino-Schalsha and M. Prieto (eds) *Routledge Handbook of Latin American Development*. London: Routledge, pp. 64–74.

Zibechi, R. (2005) Survival and existence in El Alto. *Counterpunch* 14 October www.counterpunch.org/2005/10/14/survival-and-existence-in-el-alto/ (Accessed 13 June 2021).

——— (2010) *Dispersing Power: Social Movements as Anti-State Forces*. Oakland: AK Press.

——— (2011) Ecuador: The construction of a new model of domination. *Upside Down World* 5 August http://upsidedownworld.org/main/ecuador-archives-49/3152-ecuador-the-construction-of-a-new-model-of-domination (Accessed 24 May 2021).

——— (2012) *Territories in Resistance: A Cartography of Latin American Social Movements.* Oakland, CA: AK Press.

——— (2018) People in defence of life and territory: Counter-power and self-defence in Latin America. *TNI* 3 January https://longreads.tni.org/stateofpower/people-defence-life-territory (Accessed 28 June 2021).

——— (2021) Tupak Katari vive y vuelve…carajo. *Desinformemonos* 21 January https://desinformemonos.org/tupak-katari-vive-y-vuelvecarajo/ (Accessed 22 January 2021).

Zurbano, R. (2013) For blacks in Cuba, the revolution hasn't begun. *New York Times,* 23 March www.nytimes.com/2013/03/24/opinion/sunday/for-blacks-in-cuba-the-revolution-hasnt-begun.html (Accessed 2 June 2021).

Index